Drug
and
Alcohol
Use
**ISSUES
AND
FACTORS**

Drug
and
Alcohol
Use ISSUES
AND
FACTORS

Edited by
Stanley Einstein

Middle Eastern Summer Institute on Drug Use (MESIDU)
Jerusalem, Israel

Plenum Press • New York and London

Library of Congress Cataloging in Publication Data

International Congress on Drugs and Alcohol (1st: 1981: Jerusalem)
    Drug and alcohol use.

    "Proceedings of the First International Congress on Drugs and Alcohol, held . . . 1981
in Jerusalem, Israel"—P.
    Includes bibliographical references and index.
    1. Drug abuse—Congresses. 2. Alcoholism—Congresses. 3. Substance abuse—
Congresses. I. Einstein, Stanley. [DNLM: 1. Substance abuse—Congresses. 2.
Alcoholism—Congresses. WM 270 D79365]
RC564.I575   1981   362.2′9   83-19098
ISBN 0-306-41378-7

Proceedings of the First International Congress on Drugs and Alcohol,
held September 13–18, 1981, in Jerusalem, Israel

© 1989 Plenum Press, New York
A Division of Plenum Publishing Corporation
233 Spring Street, New York, N.Y. 10013

Printed in the United States of America

PREFACE

This book is the outcome of the First International Congress on Drugs
and Alcohol to be held in Israel. Both the conference and this volume grew
out of the need to overcome the specious man-made barriers which continue to
separate intervention plans and efforts in the broad field of substance use
and misuse. This need demands that all of us become aware not only of the
differences which may separate our concerns, but also of the similarities in
our goals and endeavors. We are obligated to draw together toward a common
effort for the common good.

The conference was designed to facilitate the exploration of generic
issues. This volume is designed to document a variety of factors which are
basic to the defining, selection, planning, implementation, and evaluation
of substance use and misuse intervention.

This book is not a traditional proceedings volume. Because the needs
of a listening audience are quite different from those of the solitary
reader, and the roles of a workshop or plenary session participant are also
dissimilar from those of the reader, not all of the Congress presentations
are included, and the articles that are included have undergone major revisions. Current intervention needs and options demand a broad spectrum of
clearly defined roles for all who are or should be involved. Hence these
issues, among others, served as guidelines in the preparation of the revised
articles.

An attempt was made to pinpoint some of the more critical issues in the
broad area of substance use and intervention. This was done from a theoretical as well as from an empirical perspective. The joining together of
theses, experiences, data, and programs offers the reader an opportunity to
determine the functional relationships between them and their most effective
interfaces. The resulting volume is intended to be more than just a passive
document. The reader is asked to go beyond the built-in limitations of
paper and ink and to partake of the uncertainties of ideas. A meaningful
reading of any article or of the entire volume should not result in closure.
Each ending should lead to an ongoing beginning. If it does not, the editor
must assume responsibility for adding yet another static object to a world
which continues to be inundated by a flood of paper and paralyzed by conceptual impotence.

The contributors to this volume are to be thanked for their time and
effort in rewriting their articles. What began as a standard Congress
proceedings has become a volume which must be judged on its own merits. The
reader who becomes engaged by the printed materials and their many implications will have to decide how best to use this book. Its most traditional
use will be that of a resource volume. It surely can serve as a catalytic
tool for planning and carrying out needed efforts and programs in the broad
area of substance use intervention. It can also serve as a guide for the
development of much-needed future conferences.

By permitting their articles to be published, the contributors have invited the reader to become part of their professional life space. In so doing, each has invited evaluation of his or her concerns, ideas, positions, and efforts. Presentation to the Congress provided the contributors with an opportunity for feedback. Readers may choose to write to contributors in order to clarify issues, share opinions, ask for advice, or for a myriad of other reasons, thus providing ongoing feedback, which will serve to minimize the stultifying conditions which are all too often associated with substance use intervention.

After finishing this volume, the reader may wish to consider the possibility that his or her views regarding drugs, their use, and their users, as well as the views of the contributors to this volume, are inadequate, wrong, or irrelevant, and may wish to explore the implications of such a conclusion. However, certainty in an uncertain area such as substance use and misuse, in an uncertain and unstable world, is surely an unneeded man-made barrier to much-needed solutions.

Stanley Einstein

Jerusalem, Israel

CONTENTS

INTERVENING IN THE USE AND MISUSE OF DRUGS AND ALCOHOL:

CRITICAL ISSUES -- SPECIOUS ACTIONS

Stanley Einstein

Jerusalem, Israel

The first documented case of drug misuse is well known to most of us. Since it occurred a long time ago, the substance which was misused was an easily available natural product and not a synthetic one. The user had been forewarned not to use it, although the consequences of its use were unknown. Nevertheless the "temptation," the "desire," the "inquisitiveness," the "challenge" -- concepts well known to us -- to satiate man's complex appetites could not be resisted. A "slimy sociopath," devious in his ways, was the apparent drug connection. The first reported drug user was a woman. She spread drug use to man. This is the earliest documented case of "drug contagion." An endless epidemic was initiated. In this same documented case study, signs and symptoms of a drug experience and projected withdrawal are described. Paradise was lost by the ingestion of a natural chemical. Since that awesome event conferences have been held, books and articles have been written, and policies have been made which have resulted in the following rather peculiar scenario.

An increasing number of the world's population has, over time, undertaken to intervene, with questionable knowledge, in the lives of an increasing number of the world's drug using population who use an increasing number of natural as well as synthetic chemical substances in a relatively unknowledgeable way. Apparently the half-life of the tree of knowledge, notwithstanding its potent effect, was quite short.

An open, honest, and in-depth analysis of most past, as well as current, drug use intervention may force us to acknowledge that these efforts are, and always have been, a time-consuming, energy-consuming, fiscal resource, and manpower-misusing ritualized myth.

The challenge facing all of us is whether we wish to continue to maintain the mythology of intervention, and its relatively poor results, or whether we can mobilize ourselves as well as others to plan and carry out realistic intervention, if it is at all needed.

The following overview (Table 1) is designed to achieve two primary goals:

1. To suggest why most drug use intervention efforts have to date been specious in their formulation, have built failure into their content and process, and have succeeded in continuing along in this way irrespective of the state, nature, and meaning of drug use at the specific time.

Table 1. Drug Use: Critical Issues and Factors

Drug use	Characteristics of users	Behavior of users	Semantics of drug use

Type
Frequency
Amount
Pattern
Manner
Meaning
Place of use
Availability
Drug potency
Function(s)
	Age, Sex, Race,
	  Ethnicity, Religion
	Status:
	  Educational
	  Vocational
	  Occupational
	  Marital
	  Socioeconomic
	  Medical
	  Psychological
	  Intelligence
	Social Skills
	  Coping
	  Adaptational
	  Functioning
	  Drug-use related
	  Non-drug use related
		Physical, psychological,
		  social functions, and
		  dysfunctions
		Conventional, deviant,
		  drug-oriented social
		  involvement
			Definitions:
			  Medical
			  Legal
			  Social
			  Religious
			  Scientific
			Street semantics
			Professional semantics
			Concepts
			Issues
			Problems
			Solutions

Attitudes and values	Scientific knowledge	Economics	Public policy

Drug use stereotypes
Drug user stereotypes
Drug stereotypes
Attitudes about drug use
Attitudes about medicine
Attitudes about illness
Attitudes about health
Attitudes about treatment
Attitudes about abstinence
Attitudes about control
 	
Treatment
Why treat?
Site of treatment
Evaluation-screening
Goals
Treatment techniques
Policies and procedures
Treatment roles
Ongoing evaluation
Followup/outcome
Early case finding
Prevention
Criteria: Success/failure
Needed support system
Intervention agent:
  Issues and patterns
Focus: Individual/
  system(s)
 	
Law enforcement
Drug related laws
Medication related laws
Social substance related
  laws
Behavior related laws
Control mechanisms/
  systems
Strategy
Philosophy
	Theories about:
	  Drug use
	  Drug action
	  Treatment
	  Communication
	  Pain
	  Illness-health
	  Deviance
	  Risk taking
	  Decision making
	  Control
	  Research < Prevention
	Classification of:
	  Drug users
	  Drug use
	  Drugs
	  Treatment
	  Prevention
	  Research < control
	    efforts
	  Intervention agents
	
	Genetic aspects
	Lethal/nonlethal
	Drugs
	Medicines
	Social substances
	Environmental chemicals
	Individual/group
	
	Evaluation
	Focus: Individual/
	  system(s)
	Criteria
	Goals
	Techniques
	Strategies
	Utilization
	Policies
	Staff
	Sites
	Costs
	Public availability
	Ethics
		Illegal facets
		Legal facets
		Individual facets
		System factors
		
		Cultural values
		Drug related
		People related
		Activity related
		Life-style related
		
		Social pharmacology
		Substance:
		  Use
		  Meaning
		  Distribution
		  Support/inhibi-
		    tion factors
		  Future personal
		    chemistry
		
		Altered states of
		consciousness
		Drug related
		Nondrug related
		Individual
		Group
		
		Research
		Defining problem/
		  issue
		Data collection
		Data analysis
		Data availability
		Data utilization
		Data ownership
		
		Special populations
		Definitions
		Classifications
		Needs
			Laws
			Policies
			Procedures
			Politics
			
			Education
			Nonrelated alter-
			  natives to living
			Drug related alter-
			  natives to living
			Coping/adaptational
			  skills
			Decision making
			Reality oriented
			  life-style(s)
			"Alternative"
			  life-style(s)
			Respect for people
			Respect for
			  chemicals
			Respect for feeling
			The education agent
			Community roles and
			  responsibilities
			
			Mass media
			People oriented
			  patterns of living
			Object oriented
			  patterns of living
			Classification of
			  goals
			Classification of
			  types
			Temporal Factors
			
			Philosophy and
			religion
			Non-drug-related
			  social rituals
			Drug-related
			  social rituals
			Religious
			  rituals
			Ethics
			
			Drug use interven-
			tion
			Whose need?
			  Individual
			  System(s)
			  Macro-Micro
			  Events
			  Symptoms
			  Behavior
			  Functioning
			  Life-style(s)
			  Source of Support
			  Source of Resis-
			    tance

2.   To suggest a minimal number of types of issues and factors which must be considered if effective drug use intervention is to be planned, carried out, and learned from.

FOUR IRRELEVANT AND MISUSED DRUG USE INTERVENTION FACTORS

The four factors listed above the line in Table 1 have continued to color the focus and structure of most intervention efforts. It appears that the general consensus continues to be that if we were only able to:

1.   define and describe that which concerns us about drugs, their use and their users appropriately -- the semantics of drug use

2.   pinpoint and comprehend the listed factors associated with drug use

3.   characterize the current drug user as well as predict who may or surely will be a future user (or abstainer)

4.   describe, categorize, and perhaps even understand the behavior of drug users

reasonably successful intervention plans could be developed. The unpleasant fact is that generally this has not happened.

The intervention formula of the semantics of drug use + characteristics of drug users + behavior of drug users has become a meaningless repetitive ritual, a magical mantra in a secular world and not the hoped-for predictable foundation or frame of reference for intervention. This is so because these four factors, and their interrelationships represent only part of what must be considered in intervention planning -- perhaps the smallest part. A review of Table 2 makes this quite apparent.

For example, our success in characterizing a chronic poly-drug user (who sniffs opiates, swallows barbiturates, injects amphetamines, smokes hashish, and adds alcohol to his chemical "ecstasy," while enjoying some nicotine), who functions reasonably well in the drug "deviant" world even though we have diagnosed him or her as a "dangerous," "regressed," "sensation-seeking," "escapist," "sociopathic," "addict personality," who is street wise, and who has rejected our world, values, and a nonchemically-based life style, has not been matched by successfully meeting his or her needs, our needs, and the needs of society at large.

One of the key issues still facing us in 1986, after literally centuries of various types of substance-use intervention, is pinpointing, understanding, and then coming to terms with why we want to intervene; with whom; what our viable options are; who should do the intervention; and what the costs may be if we are successful or fail, or if we don't intervene at all?

The why of intervention, or nonintervention, is only tangentially addressed by the semantics of drug use vis-a-vis the concepts, issues, and problems which have come to be associated with the relatively new modern specialized field of drug-use intervention. This is the field whose very existence, as a separate field, is based upon not succeeding in its efforts!

Before continuing on this conceptual trip, which will explore a variety of factors which are critical for drug-use and misuse intervention planning and efforts, as well as for intervention generally, it would be useful to consider the following issues. Let us forget specific drugs for the moment,

Table 2. Types of Factors which Influence Drug Effects

Specific factors	
Drug related	Drug user related
Dose	Age
Purity of the drug	Gender
Dosage form	Presence of disease states
Duration of action	Body area and weight
Route of administration	Genetic factors
Interaction with foods	Tolerance
Interaction with other drugs	Biological rhythms
Frequency of administration	Metabolic rate
Basic pharmacological action	Nutritional status
Nonspecific factors	
Suggestability	Physical and social setting
Personality	Body image and awareness
Information/instructions	Prior mood and body states
Expectations	Sociocultural background
Previous experiences	Attitudes
Communication networks	Symptom sensitivity
Ritual of drug taking	Degree of acquiescence
Motivation for use	Transference
Social learning	Social interaction
Availability of labels	Meaning of drug effects

how they are used and/or why. Let us put the drug user aside (a process which seems to be quite easy to do). One should have expected that this relatively richly endowed "unique" field, under the pressure of legitimate as well as illegitimate concerns and interests of intervention agents and agencies, as well as the public at large, should have made some significant and unique contribution(s) from its many daily experiences. Theoretical contribution(s)! Conceptual contribution(s)! Organizational contribution(s)! Policy contribution(s)! Technique or technological contribution(s)! Ideological contribution(s)!

Has the field done so? Indeed have any of us, whose status, positions, salary, life style, travel, and perhaps even our retirement is owed to the drug problem? We insist upon building and expanding a growing specialized drug-use intervention empire, locally, regionally, nationally, and internationally. Yet what we do, whom we do it to and/or for, and how we do it is taken from a myriad of other fields. Ironically those fields claim daily that they are ill equipped to effectively intervene in our chosen problem(s). As we peruse the literature we are confronted with a sad finding. The concepts and theories which we use and that define the issues and problems which concern us, as well as the parameters of our intervention efforts, are not unique to drug use. They never have been. Many are untestable. Some are unusable. And some are silly if not insulting to man's intelligence. Indeed the words and concepts which are supposedly critical to our work are oftentimes not matched, in their possible meaning(s), by our own actions. For example:

- Physical Tolerance -- should be evaluated against our own physical, social, and behavioral intolerance of drug users.

- Drug Availability -- should be evaluated against our all-too-frequent unavailability to helping those who need our help, concern, and even care and love.

- <u>Pathological Dependency</u> -- should be evaluated against the symbiotic relationship which we may have with the drug intervention field.

- <u>Contagion</u> -- should be evaluated against the content, meaning, and value of what we ourselves are spreading.

Our organizational contributions are realistically best described as simply being

1. more organizations doing variations of a limited number of interventions (treatment, prevention, control, and research),

2. with little meaningful interrelationship or sharing,

3. with surprisingly similar sad results,

4. in which we tend to learn little from our experiences or those of our colleagues

Organizations come and go, leaving behind a word salad of their initials while drug use continues and even increases.

The policies which have set the parameters as well as the tone of drug misuse intervention and which have defined the meaning of success and failure are, as in most other areas of human endeavor, set by transient individuals who are committed to "making it." They are generally poorly trained in policy making or decision making. All too often they choose to be at a distance from the very problem(s) which they have helped to define. They may initiate policies for (or against) that are carried out by others who often have difficulty differentiating between policy, technique, goal, and process. We would be hard pressed to nominate a single drug use intervention policy which to date can be demonstrated as unique for the needs of our own specialized field.

As for the unique techniques which we use and the technological contributions which we have made -- the less said the better. Professionals who work in our field bring with them the traditions, status, roles, and techniques of the disciplines in which they have been trained. Paraprofessionals bring with them:

1. the instability of a job-role which is still being defined,

2. a tradition which is still being built,

3. a second-rate status,

4. a face-saving mythology for a rejecting society which believes that the existential knowledge of drugs and their effects, as well as having coped in the drug world, results in unique skills and abilities which are intrinsic to successful drug use intervention.

As for the unprofessionals -- their techniques are to be admired as we permit them to exploit drug users and nonusers alike as well as the field as a whole.

Lastly, the ideological contributions which have been made in our field are easily subsumed under past and present religious and secular ideologies of various types. Indeed one might say that drug use intervention is a belief system which, while having its roots in a variety of ideologies, has tended to increasingly isolate itself from these generic roots in order to perpetuate its own uniqueness. It feeds on itself while giving needed sustenance to very few.

Has the analysis up to this point been too harsh? Too cynical? Too critical? Obviously I don't think so. What I have included has already passed a process of censorship in order to make it more palatable. The reader can imagine the less than academic tone and terminology of the initial drafts.

The point is that if our field is really unique, and if it really needs and merits continuation as a separate area of endeavor, we had best try to understand why our successes have been so limited. These issues are being raised to serve as stimuli, as catalysts, which hopefully will allow for reasonably planned intervention rather than the continued contribution of reflexive reactions.

Returning to Table 1 we can see that a variety of factors, which are basic to intervention in the drug use field, must be considered -- not only drugs, their use, the user (whoever she or he is or can be and however he or she behaves or however we defined the drug problem). As we review each of these generic factors or areas in Table 1, we should ask ourselves:

1. How has this factor been considered in intervention planning?

2. What is its actual or potential meaning and/or effect upon planning?

3. How can this factor be translated practically into a program?

4. What further knowledge is needed to assure us that this factor will be and can be used in an appropriate way?

5. What has and/or can facilitate or inhibit the inclusion of the selected factor in an intervention program?

6. How is this factor's impact to be evaluated? By whom and when?

No doubt other questions have come to mind which the reader may feel are equally relevant, if not even more meaningful, in assessing the place and task of each type of factor in an intervention effort. Add them at your leisure. Of even greater importance -- use them!

SCIENTIFIC KNOWLEDGE

Since science has seemingly become the newest God by which to judge plans, efforts, and meaning, let us turn to scientific knowledge as a critical factor in drug use intervention planning.

Six types of theories continue to be depended upon for intervention plans and programs. These include theories relating to: drug use, drug action, illness, health, deviance, and control.

Unfortunately, commitment to a theory or type of theory doesn't often result in its functional utility, or even its practical use in our specialized field. Consider the following facts:

1. Notwithstanding the great number of theories which have been offered to explain drug use, we still don't know why a specific person or group turns to drugs, stops or continues such use, or never uses a specific substance or combination of substances. Retrospective analysis passes as

prospective prediction, and description passes for etiology.

2.   Related to this unsettling reality is the fact that few theorists, in any of the intervention areas, have felt the sense of responsibility to suggest how their theories might be tested out. For example, "basic scientists," whatever that means, and whoever they are, theorize about and study biochemical receptors, but don't seem to understand "social" receptors or "community" receptors. And those who pride themselves in being in the front line of drug use intervention where the drugs, their use, the drug users, and the real action are, don't seem to understand what the theories, or even the data, imply, mean, and demand of them as well as their efforts. There is little if any interchange or meaningful interface between basic and applied researchers, "hard" scientists, "soft" intervention agents, and the community which foots the bill.

3.   Our knowledge of drug action has gone through the stages of:

- magic (the use of accidentally discovered edible plants plus a magical sound or movement resulting in a specific observable or describable effect),

- one-sided pharmacological descriptions of the physical effects of drugs -- the notion of the magic bullet,

- effects resulting from the interface of certain chemical structures with certain unique receptors,

- psychosocial and moves toward social pharmacology theories.

Each of these theories or stages demands particular types of intervention. Yet the reality is that we still don't understand -- in a usable way -- either the action or the resulting or associated drug experiences.

4.   As we continue to be committed to abstinence as a goal, as a status, and as a process and life style, we don't understand, nor seem interested in understanding, the action of abstaining or the abstinent experience.

5.   Whereas illness, health, and deviance have been clothed in a scientific aura, the reality is that theoretically and conceptually they are the outgrowth of ever changing value judgments. One hundred years ago drug users were not considered to be "sick." They were considered to be immoral or amoral, and intervention plans and efforts were suited to such a viewpoint or "diagnosis." Modern scientific knowledge, ideology, and technology have helped us to view drug users as being "sick." In the past, unacceptable substances were classified as being "the devil's brew." Now, if used by the right kinds of people -- us and those whom we care about -- they are readily available as recreational substances.

6.   Until recently none of our religious or secular control theories and efforts, whether the focus has been upon the specific substance, its use, the drug experience, the user, the supplier, or the variety of intervention agents and agencies, have been based upon or related to scientific knowledge, tradition, or technology. But since money has been made available to study the effectiveness of control efforts (evaluation or pseudoevaluation) a scientific status has also been projected upon control plans and efforts. But the reality is that controlling the drug user and/or supplier by death, incarceration, fines, or forced treatment, controlling the availability of substances by licenses, taxes, or restricting growth

through crop control and/or destruction, or influencing sale by age, patient status, site, temporal factors, or other restrictions have been insufficiently effective to limit or entirely prevent certain drugs as well as drug use.

7. Aside from the theories in these six areas, which have been basic to intervention planning and programs, a variety of other seemingly relevant theories have had little attention paid to them.

- It is the rare treatment program, mandated to treat the drug user, the alcoholic, the tobacco smoker, or even the overweight or obese, which integrates the demands of a specific theory into its treatment process. In our specialized field we seem to treat "deviant" others in spite of available theories and not with the help of available theories.

- It is the rare prevention program which is aware of the current state of communication theory. The mass media increasingly invade our privacy and life space, helping us to live the so-called good life through conspicuous consumption. The ballot that we select for the political candidate who is supposed to represent our views and our needs is magically attached to printers ink, TV, and radio. We continue to show irrelevant films and slides about drugs and drug users to "populations at risk" -- the living -- lecture to captive audiences who rarely listen, produce books and articles which compete with dust on our shelves, and create a variety of posters to wallpaper halls and walls. Communication maintains a deafening silence in our field.

- Efforts continue to be made to prevent, cure, or at the very least to minimize pain. All types of pain, physical, psychological, social, economic, and political. Specialized congresses with more than 1000 participants meet to discuss new pain-related findings and to share their theories and return home to combat pain. Have theories of pain been integrated into our intervention plans and efforts?

- Risk taking and decision making have increasingly entered the jargon of many disciplines. What facilitates the laboratory rat's path across the unpleasant grid for a specific chemical substance? How are we to utilize the knowledge that a smoker, with a family history of cancer, continues to smoke? Indeed she/he may even purchase cigarettes sold in the gift shop of the hospital where she/he is visiting a relative or friend who is undergoing chemotherapy for cancer. How many intervention agents are aware or care to be aware of or use the current scientific knowledge about risk taking and decision making in their day-to-day intervention roles and efforts?

8. Classification is another crucial factor which should deserve our attention.

This exists for no other reason than it permits us to order what we are facing so that we can set intervention priorities in terms of the resources which we have available to us. We owe it to ourselves to consider the implications, for drug use intervention, of some of the following issues:

1. What types of classification systems currently exist in our unique field in the areas of treatment, prevention, research, policies, and control?

2. Who are the classifiers? Why are they the classifiers? When and under what conditions do they classify?

3. What is the validity of the classification criteria which are being used?

Table 3. Categorizing Chemical Substances by Function(s)

Types of functions	Type of substance	user	Manner of use	Site of use	Support system
Aesthetic					
Aphrodesiac					
Cultural					
Leisure					
Sacramental					
Psychological					
Ego disrupting					
Treatment					
Institutional					
Political					
Ideological					
Economic control					
Social control					
War or other conflicts					
Research					
Nutrition					
Environmental control					

4. What have we done, to date, with classification theories, variables, and systems within our intervention plans and efforts?

An evaluation of the current state of the art of drug use intervention would lead one to conclude that classification is an academic theoretical activity which is related, for the most part, to what we do or even what we want to do.

- In reality the drug user is viewed, as well as responded to, as a unitary type.

- Drug use is a "bad" or negative event or destructive behavior.

- Drugs are a unique, negative, and homogeneous classification of natural or synthetic substances, as distinct from, and generally compared to, the positive status of medicines or desirable social substances.

Table 3 is designed to help the reader develop a classification system of chemical substances in which the substances' function initiates classification. The reader is then asked to consider which drugs fit into each functional category, who is the user, how and where is the specific substance used, and what support system facilitates (or inhibits) its use.

Treatment for the drug user is generally limited to verbal, behavioral, chemotherapy, and environmental techniques, even though an almost endless number of types of available processes and techniques are in use with and for other types of fellow citizens (see Table 4).

9

Table 4. Selected Categories of Treatment Techniques

Type of technique	Examples	Selection criteria	Preferred agent
Verbal (cognitive/ emotional)	Individual, couple or group, psycho drama, marathon, family, scream, role-play, conflict reso- lution		
Chemotherapy	Methadone, antago- nists, antidepress- ants, tranquilizers, etc.		
Behavioral	Token economy, reality therapy, aversive, other conditioning		
Relaxation	Hypnosis, meditation, yoga, biofeedback, breathing exercises, guided imagery		
Nutritional	Vitamins, diet		
Physical	Acupuncture, acupressure, running, sensory awareness, electrical and chemical aversion, muscle relaxation		
Surgical	Plastic surgery (tattoo removal)		
Vocational	Vocational testing- training		
Spiritual	Humanistic therapy, logotherapy, pastoral counseling, prayer		
Expressive	Art, dance, music, crafts, occupational therapy		
Environmental	Therapeutic community, halfway house, commune, drop in center, sheltered workshop, crisis intervention, day hospital, night hospital, survival training		
Legal	Pardon, parole, probation, Court ordered treatment		
Detoxification (Neonatal, self- iatrogenic addiction)	Self or medical (short/long term)		

Prevention has yet to be meaningfully delineated into information-dissemination, education, and training in our field. And confusion reigns as to whether we wish, as part of our prevention focus, to contend with symptoms, behavior, life styles, drugs, people, drug use, the drug experience, culture, subcultures, agencies, policies, etc.

Research is rarely classified in terms of the needs of the intervention system. Most often research is done because money is available for it.

For the sake of clarity one should distinguish between the following treatment foci: detoxification, symptom removal or control, symptom change or control, and life-style change.

The typology of basic and applied research is characterized by tradition, ideology, status, salary, etc., but has had little to do with effective intervention plans or efforts.

Finally, the description of intervention agents as professionals, paraprofessionals, addiction specialists, significant others, and volunteers is no more than a semantic exercise. We still don't know, nor do we appear to be interested in knowing or learning what criteria should be used to permit someone to carry out a specific or general intervention role, task, or program in our field.

In a sense, classification in our field continues to operate as an either-or phenomenon. Those who work in the drug use/misuse intervention field, who focus upon variables which are classified as being fundamental to the field -- drugs, their use, and the user -- are the "drug interventionists." The rest of the world pays us to do this to supposedly protect them.

ATTITUDES AND VALUES

It should be apparent from the focus, manner, and content of this analysis up to this point that attitudes and values are central to drug use intervention. But is being aware of this sufficient? There is an increasing awareness by the professional as well as the public at large about the stereotypes and attitudes which we have permitted to develop and allowed to continue to exist. They are associated with drugs, their use and users, abstinence and the abstainer, medicines as well as social substances, illness, health, treatment, and a variety of other intervention options. Such awareness doesn't mean that we have effectively and creatively utilized attitudes as a critical factor in our plans and efforts. More often than not our attitudes and stereotypes have helped us to be clearer about what we are against. And in this process they have blurred our visions of what we can and should be for. As a result, the field is deluged with ritualistic, repetitive, compulsive programs and intervention foci which are myopic and often elusive. Our current attitudes tend to breed and spread a sense of hopelessness on the part of many people who are active in the field and certainly for the public which foots our bill.

We seem to have become quite comfortable with the concept of burning out but does not seem to be concerned about a most useful notion, burning in. This is the process by which more and more people and agencies have found a niche for themselves in the intervention field and have done, over time, less and less productive and useful work.

And yet the energy underlying these disabling attitudes, and the time consumed by them, could just as effectively be used in selecting any one of a number of satisfying and appropriate intervention options. For example:

Table 5. Selected Attitudes as Intervention Tools which Reinforce
and/or Inhibit Planned Intervention Outcome Regarding Drug Use
and/or the Drug User

Dysfunctional attitudes	Functional attitudes
Insensitive	Sensitive
Indifferent	Curious
Unresponsive	Responsive
Giving up	Persistent
Impatient	Patient
Hopeless	Hopeful
Inflexible	Flexible
Rigid	Firm
Fearful	Anxious
Helpless	Optimistic
Angry	Concerned
Apathetic	Interested
Resentful	Challenged
Cold-Cool	Warm
Rejecting	Accepting
Closure	Open
Hysterical	Relaxed
Suspicious	Exploratory

Why reinforce the self-fulfilling myth (attitude) of the untreatability
of the drug user, or alcoholic, by treating everyone who has at some time
used some drug and who enters the treatment process, for whatever their
manifest or latent reasons, or who is forced into treatment? Why not
employ a very pragmatic treatment attitude? We will seek to engage people in
treatment who need this form of intervention, who want it or can be helped
to want it, who are engageable, and who can effectively utilize treatment --
process as well as technique(s). Our projected attitude, and firm stance,
could readily be that treatment is not the available dumping ground for the
fears and attitudes of society or other institutions.

Why continue to use policies which after much trial and error remain
trial and error? Dysfunctional policies continue to breed maladaptive
attitudes, as well as maladaptive behavior and programs in our field.
Perhaps a more pragmatic viewpoint -- an attitudinal commitment in itself --
would be of help. If a policy facilitates the achievement and the continued
achievement, of a specific goal, or set of goals, we should continue with
it. If the policy has not been successful -- however we define this to be --
we should change it, or get rid of it. The functional and dysfunctional
policy maker should be viewed in the same way.

Should we continue to be the agents of neighborhoods, communities, and
society at large, which have actively and passively helped to create the
drug problems, and which generally refuse to actively take part in the
possible solutions to these problems other than footing the bill in order to
keep the problem away from their doorsteps? The alchemist failed in
creating gold from garbage.

We are bound to fail in our intervention if unrealistic, insensitive,
disabling, and rejecting social community attitudes continue to permeate
what we do. The computer industry paraphrased this quite succinctly:
Garbage in -- Garbage out. Perhaps the time has come for those who choose
to work in the field of drug use intervention to consider whether we are the
collectors of society's "garbage"; trapped alchemists, or creative and
effective innovators and initiators who can carry out our roles only when

Table 6. Economic Factors Associated with the Use/Misuse
of Drugs, Medicines, and Social-Recreational Substances

Legal components	Illegal components
Production/growth	Illegal sale of drugs
Distribution	Cost to the community for
Sale	stolen property
Related taxes	Development of a resale
Insurance (theft, health, etc.)	system for stolen goods
Entertainment (films, songs, etc.)	Drug-related prostitution
Advertising	Organized crime
Research	
Treatment (emergency, ongoing)	
Penal programs (jails, prisons, etc.)	
General law enforcement (courts, police, probation, etc.)	
Specialized drug law enforcement	
Prevention education	
Accidents (highway, industrial, home, etc.)	
Illness	
Absenteeism (work, armed services, etc.)	
Loss of manpower (unemployment, institutionalization, death, etc.)	
Service unavailability (work, armed services, etc.)	
Inadequate education	

the needed attitudinal support system is available and active. To do this,
we, with the help of others, will have to determine what types of attitudes
are conducive to the creation and carrying out of effective needed
intervention. Once this is achieved, and it is achievable, we will have to
learn to make such facilitating attitudes part of the rubric of
intervention.

ECONOMICS

Drug use and its intervention continues to be a costly affair.
What is often overlooked is that the illicit facets of drug use
are actually smaller in number, as well as in potential and actual impact,
than the licit factors (see Table 6).

We also tend to forget that individual economic factors can easily
be overshadowed by system factors. For example, the economic worth of
individual drug-related crimes are likely to be a poor second when measured
and evaluated against dishonest, if not criminal, activities carried out by
parts of the intervention system in the name of the common good. The
economic value of illegal dangerous drugs surely indicates no more profitable
a venture, over time, compared to legal dangerous drugs, produced and sold
for our leisure and comfort, or the medicines which keep us alive longer but
no more fulfilled than many drug users.

A key issue is not that there is a profit to be made in drugs, whatever their source and however they are classified, but rather that the increased commitment to <u>progress through personal chemistry</u> can be and is exploited daily all over the world. Sooner or later the intervention agent, and the system which he represents, or which represents him, must realistically and honestly confront the following issue: What are the economic <u>merits</u> of drug use intervention?

Can we afford to intervene effectively?

Can we afford not to intervene at all?

Will successful intervention be more disabling, monetarily as well as nonmonetarily, for individuals, systems, and society at large, than continuing with the status quo?

PUBLIC POLICY

The latter question is very much related to the factor of public policy. Our policies and laws have determined what our intervention focus and parameters are to be, as well as what they should not or cannot be. For example, our laws define the legal status and availability of drugs irrespective of their structure, action, and consequences. Our policies determine who will be treated and where, who will be educated about drugs and alternatives to drug use, what types of research are to be funded and for how long, who is to be the active focus of law enforcement and who and what is to be overlooked. Our intervention procedures determine who will be aware of intervention programs as well as who will partake of them. Drug use and its intervention has and continues to be heavily politicized even though the fruits of this process have been bitter and barely digestible. Let us review some of the consequence of current drug use intervention public policy:

A "stoned" subculture exists within an overmedicated society.

Admission to treatment is rarely based upon a drug user's current condition, his ability to effectively use available treatment now or even predictions about his future drug- or non-drug-related behavior. More often than not it is based upon the implicit theories held by the "intaker," as well as by his possible mood that day.

Drug and alcohol prevention policies are currently based upon the ephemeral concepts of <u>risk</u> or <u>susceptibility</u>, and the conceptual values of non-drug <u>coping</u> and <u>adaptation</u>. But the reality is that all of us are at risk regarding the use and misuse of an increasing number of potent chemicals in the various physical and social environments which we must somehow learn to adapt to and cope with. But all of us aren't given the option of participating in prevention programs. Whether this is a blessing or a punishment is to be questioned.

Research policies, in, and outside the drug use intervention field, have tended to develop an intriguing and stable tradition. In order to be funded, at least publicly, one must somehow demonstrate that what one plans to do is related somehow, theoretically or in some applied fashion, to what already has been done by someone else. On the surface this appears to be eminently logical as well as a reasonable safeguard against the misuse of public funds, manpower, and time as well as institutions. But one must consider whether the original research, which now serves as a reference if not a yardstick for the currently planned research, was so valid and fruitful that the new project is really needed. One should question whether the tradition of ritualized site visits, the opinions of those who may know little about the specific research area and its factors, complexities, and merits -- the "old boy" system which has easily developed in our field both nationally and internationally -- has helped or hindered in the collection, dissemination, and use of needed data. One should question why much of the research done in our field not only suffer from unoriginality, but is also sterile. Lastly one must question why research policies have to date not resulted in the discovery and availability of usable data for intervention, or as a catalyst for the development of testable theories.

The politicalization of drug use has given us, among other gifts, the following nonrandom list: the Boxer rebellion, Prohibition, CIA supported distribution of opiates from the Golden Triangle, Nixon's war on drugs, the Ayatollah's final solution for drug users, public support for tobacco growers at the same time that public support was given to cancer and cardiovascular disease research, the separation of drug and alcohol intervention efforts which have grown into personal and institutional empires, and the administration of psychotropic drugs to combat soldiers.

None of these policies were really needed by any of us. But we have not dared to challenge these policies for fear that the policies which permit us to intervene in body cells, chemical structures, living and dead organs and organisms, peoples' lives, incarcerated citizens -- or with the printed word, the recorded voice, the cinematic image -- would become unavailable to us. Live and let live has been our intervention motto.

The reality is that a limited public has made major policies which may be irrelevant, counterproductive and at times even dangerous. Many of us have carried out these policies. Many of us still do. Many of us will no doubt continue to do so, even though some of the current policies regarding drug use intervention in our various countries don't facilitate and may even hinder effective drug use intervention.

TREATMENT

Ever since drug use intervention changed its primary conceptualization from morality to criminality to psychopathology, and its focus from the immoral to the criminal-deviant to the "sick," we have come to depend more and more upon treatment as the panacea. There are no panaceas! There are just efforts which are viable or nonviable, evaluable or not evaluable, etc. Much can be said about the treatment process, the treatment techniques currently in use as well as available ones which are not often used, as well as the other factors listed in Table 1. Space limitations do not permit an in-depth analysis. Only one issue will be raised. This is the philosophical issue of why we treat drug users at all and the implications of our answer(s). There are many factors which can and do initiate current treatment efforts. They include among others: health factors and considerations, legal-control factors, social factors, cultural factors, economic factors, political factors, educational factors, and religious factors.

Each factor brings with it its own models with their definitions of who and what the focus of treatment should be, traditions, goals, staff, techniques, temporal commitments, preferred sites for intervention(s), role definitions and delineations for all who are involved in this process, policies, needed support systems, monetary and nonmonetary costs, budgetary sources, criteria for success and/or failure, as well as visibility, status, demands, and desirable and undesirable consequences.

What has happened increasingly over time is that the unique demands, structures, and processes of each of these initiating factors and their models have tended to be overlooked. Treatment, a natural arm of the health

15

model, has been designated by society at large, in many countries, to
intervene in drug use because of legal and control issues, in combination
with secular-cultural and economic concerns in a process which continues to
be highly politicized. This is aside from the economic gain which is
possible. The result has been an increasing number of types of drugs used
by an increasing number of types of people -- young and old -- with much
dissatisfaction being experienced by the drug user and his family, treatment
agents and agencies, policy makers, as well as the community at large. No
doubt there are reasons for this. But one reason seems to be blatant. It
is perhaps best explained in Jewish theological terms -- one simply
shouldn't mix meat and dairy products. Or, continuing with yet another
gastronomic image -- a Hungarian goulash is good when it is planned that way
and not when it is the result of a happening. If our treatment results have
been less than we have planned for, or expected, perhaps it is time that we
stopped blaming the drug user, his drugs, the competing drug experience,
policy makers, budgets or their lack, etc., and put the blame -- if blame is
needed -- where it belongs. We have built failure into our treatment plans
and programs. With the same amount of effort, energy, and time, if not
less, we should be able to build success into what we are doing. The
question is whether we really want to, care to, or dare to!

EDUCATION

     Another favorite area for drug use intervention has been prevention-
education. Two parallel assumptions remain related to this area.
Abstainers -- the drug use virgin(s) -- can be educated not to use drugs,
and the drug user or novice user or novice abstainer can be educated to stop
drug use or not to return to drug use. How is this to be accomplished? By
supplying the right information, at the right time through personal and mass
media technology to a captive audience in a public place (which is society's
agent), and which allows for the dynamic (or dull) effects of cognitive and
emotional integration, peer and nonpeer interaction and pressure, toward an
end goal of satisfying nondrug alternatives to "coping," "adaptation,"
"decision making," and perhaps even "conflict resolution." The reality is
abysmal! In a sense drug prevention-education has become a variety of
activities which treatment, research, and law enforcement have deigned not
to cope with. The dissemination of anything is easily given an educational
stamp and status. Somehow in this process we have forgotten -- if we ever
knew and/or accepted -- that:

  - treatment is an educational process;

  - research -- the discovery, collection, analysis, integration, and
    dissemination of facts, questions, issues, points of view, knowledge --
    is an educational process and ideology;

  - law enforcement, in modern terms, is (or can be) a corrective
    educational experience and process.

     Why is this mentioned? Because the separation of drug use
prevention-education, as it currently is in many parts of the world, from
other concurrent intervention efforts,can only limit its scope and
effectiveness. To date, both of these aspects, scope and effectiveness,
continue to be quite limited.

     The continued primary focus of drug intervention education efforts,
upon drugs, drug use, drug users, the drug experience, and Protestant ethic
non-drug alternatives as goals to be aspired to and activities to be engaged
in by populations at risk, has led to the development of insufficiently
trained and educated intervention agents and policy makers.

The carrying out of drug intervention education efforts by volunteers as well as by paid addiction specialists has permitted neighborhoods, communities, and society at large to abrogate their roles and responsibilities for developing and maintaining individual, group, and communal life styles which foster healthy ethical functioning.

Current drug intervention education efforts have not paid attention to the challenge which increasingly faces modern man in underdeveloped, developing, and developed countries. The psychosocial environment, with the help of increasing technological intrusion, has made it more difficult for individuals and groups to isolate themselves from making a decision as to whether and how they will adapt with and through an increasing number of people in their life space, or with an increasing number of types of available psychoactive substances.

Drug prevention-education is increasingly becoming a shiboleth, a ritual, a slogan, at a time when all of us need it.

SOCIAL PHARMACOLOGY

In this context the concept of social pharmacology may be a most useful one. The reality seems to be that it really isn't drugs, their use, or their actions which really concern most people and institutions. Rather our concern is with the actual or imputed social aspects of drug-taking behaviors and experiences. In an industrial world committed to <u>control</u> and <u>predictability</u>, any possible threat to either one of these unachievable values is viewed with seriousness and is intervened with as quickly as possible. Often this is done with an overkill mentality, technology, and outcome. It is ironic that drug use is imputed to result in the lack of control, unpredictability, and irresponsibility! Social pharmacology is a relatively new concept which has yet to develop the rich tradition (so enjoyed by other professions) or to achieve the status of full membership in the drug use intervention world -- nor we in their world. It is questionable whether they need us or even want us. We certainly need their help. Social pharmacology should offer us insight, from a historical perspective, of the use of natural and synthetic substances, in various cultures, for various reasons, in a variety of ways, with an almost endless

Table 7. The Process of Distribution of Medicines and Drugs:
Factors and Relationships

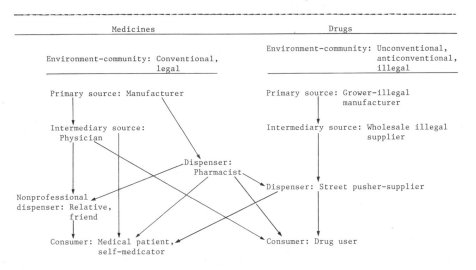

17

number of meanings, social as well as other types, which are imputed to both drug use and the drug experience. It should facilitate our learning about the meaning of drug experience. It should facilitate our learning about the meaning of drug distribution and not just about its technology and structure.

Social pharmacology should help us learn to understand the personal and system support factors, and their interrelationships, which prevent, minimize, or reinforce drug use, misuse as well as abstinence. We should expect, of social pharmacology, the shedding of some light on the future of personal chemistry as well as the future of psychotropics. And yet the concept, as well as the area of social pharmacology, remains relatively unknown to many if not most people working in the field of drug use intervention. As for the public at large, unbeknownst to them, they simply participate in social pharmacology on an almost daily basis. (See Table 7.)

Social pharmacology, its concerns, theories, concepts, technology, documents, and representatives are a blatant example of the nonuse of available tools for effective drug use intervention.

GENETICS

The genetic aspects of drug use intervention present the field with another type of challenge. It has been little more than a generation since research efforts were initiated in order to describe and to comprehend the actual and/or potential relationship(s) and effects of certain types of substances -- drugs -- upon the general makeup of laboratory animals and man. The traditional and broader focus and concern of many studies has been about the mental and physical health of individuals who misuse certain "drugs." During this period of time, much effort has been devoted to questioning or proving the safety and/or danger(s) of certain substances for individuals directly and for society indirectly. The general title covering such work has been genetic aspects of drugs of abuse. Let us take a closer look at what has occurred as well as its consequences for us. Although much has been said as well as "mis-said" about the genetic consequences of drug use, relatively little is known about both the lethal and non-lethal effects of the majority of drugs currently being misused, many of the medications and social substances being used, or many of the chemicals which are part of our work, leisure, treatment, educational, and home environments. One can explain this in terms of still to be developed technology. But that is not the point. Many of our intervention efforts have misused spurious mass media reports about genetic damage. Ironically, given the instant concern for LSD-produced monsters, it is odd that rarely have drug treatment programs utilized genetic counseling as part of their process. Nor have they themselves often been willing and interested to involve their patients in needed genetic research. How does this compare with urine testing -- for guilt by urination -- as a valid and necessary treatment process? It is odd that the making available of 20 cc of blood for genetic research, by a patient who agrees to make this contribution after an explanation is given to him or her, is often looked upon as if the request was for a primitive blood-letting ritual.

This is not to say that geneticists are totally blameless. How many of them, anywhere in the world, are willing to leave the security of working with status conditions and illnesses, or even publish their findings in our specialized journals, for work in the deviant field of drug use intervention? And yet if effective intervention plans are to be made, which include not only current drug users, but their offspring as well, the genetic aspects of drug use must somehow be included and integrated in our efforts.

18

There are a number of other issues which deserve comment as well.

<u>Genetic aspects of drugs of abuse</u> is a false, limited, and limiting combination of concepts and words. It is false in that it seductively permits professionals and the public at large to focus upon only certain substances, <u>dangerous drugs</u>, used in certain ways, <u>illegally-nonmedically</u>, by certain types of people, <u>drug users/addict personalities</u>.

It is limiting in that it prevents individuals and groups from carrying out those efforts which are necessary in order to better understand what are the as yet unanswered questions, and what can/should be done with the knowledge which we currently have (plans, programs, decisions, policies, etc.).

From a geneticist's point of view the significance of any genetic alteration is in its harmful effects upon future generations or upon the organism itself. These alterations are the result of changes in the genes themselves or due to a rearrangement of the chromosomes which have the genetic material. In this context the phrase <u>drugs of abuse</u> is meaningless.

ALTERED STATES OF CONSCIOUSNESS

As we move from the science of genes and chromosomes to the phenomenology of altered states of consciousness we are faced with another dilemma and challenge. Altered states of consciousness (ASC) seem to be as basic to man's existence and adaptation as his genes and chromosomes are to his survival and ability to reproduce as well as his effective functioning. And this seems to be increasingly so as modern life increases in complexity. The nonusual experiencing of oneself and others as well as the various physical and nonphysical worlds of which we are or may be a part, may be absolutely necessary if we are not to be disabled by a technological world which makes increasing demands upon us and intrudes more and more into what we do and who we can or can't be. Unfortunately, ASC continues to be tainted by a cult-like phenomenological status. If we are concerned about people continuing to drop out of society through drug use, we owe it to ourselves to learn what potentially healthy options ASC can offer to them as well as to ourselves. To date we have rarely used this option. And when we have, we have often confused the technique aspect of ASC with the process, the consequent temporary behavior, or the more permanent life style.

Very few treatment and prevention programs in our field have to date shown an interest in ASC. Relatively little research has been funded. And few policies exist to safeguard our concerns about the consequences of ASC. This is so because there are either relatively few available data, or little interest in the available data by which to create needed functional policies.

RESEARCH AND EVALUATION

This brings us to the roles and effects of research and evaluation within and upon drug use intervention. The latter is generally no more than a lip service ritual, and the former has generally been unrelated to the needs of the field.

Let us consider just a selection -- a biased selection -- of the interface of evaluation and research with the field of drug use intervention.

Very few intervention efforts, whatever their content or focus, have been or are currently evaluated in a way which leads to usable data and/or conclusions. This is not because of a lack of money, technology, staff, time, or tradition. We simply don't appear to really want to know the consequences of what we -- or even others -- are doing. Rather than confront this possibility, because it is not a nice one, we spend our efforts in developing and using fine-honed research designs, with supersophisticated statistical techniques, on expensive computers, with specialists who don't really understand what we are doing, or why, and we publish our results in specialized journals which aren't read by policymakers whose decisions will or will not permit us to continue with our intervention efforts.

Does this description seem too harsh? Evaluate it! Investigate the last time a policymaker in our field has kept up with the current available literature, where s/he is aware of the implications of classic literature, and what value(s) s/he imputes to the drug use intervention evaluation process and results. Indeed, does he or she understand what evaluation can and can't offer? Can the pressure which he or she is under, from a variety of legitimate and illegitimate sources, to make a decision, whatever its consequences, await complex and time-consuming evaluation efforts? Probably not. As a result, more often than not, evaluation is perceived as an unwanted stepchild which is needed to prove the wholesomeness and health of a doubtful family of intervention efforts.

In the best of all worlds, research should/could help us to define the issues and factors which make up drug use and its intervention. Research should be able to define its problem nature as well as for whom it is more as well as less of a problem. It should offer hints or guidelines for effective intervention plans and efforts. But we do not live in the best of all worlds. Adam and Eve and the snake saw to that. Instead generally what happens is that vested interest pressures set research priorities. At times the funding is cut off so that insufficient data are collected. At times, when money is available, too many data may be collected because the dynamics of building institutional empires, or meeting the pressures of publish or perish, don't facilitate mutual aid between colleagues. For example, how many more MMPI studies of drug users are really needed? How many more student drug surveys are really needed? Why haven't researchers investigated the effectiveness of intervention agents so that we can have some reasonable criteria for hiring "winners" and not hiring "losers" who "burn out" soon because they may have joined us with depleted batteries? The same may hold for drug use intervention policymaker suitability and effectiveness.

Putting all of this aside, once the data have been collected and assessed, how available are they to those who may or do need them? This includes not only the professional sector but also the public at large. How long were the dysfunctional effects of marijuana known before they became publicly available? Why was the possibility of cancer-producing effects of LAAM not shared with Israelis when Israel was asked to consider using LAAM? Why are fear techniques continued to be used in drug prevention programs when we know that they are either not effective or may even be damaging? How can we explain using the status of scientific research to say that drug use and alcohol use is decreasing when many average citizens as well as professionals could document increased multiple drug use?

Lastly one should briefly consider the concept of research utilization. In simplistic terms this concept asks: once the data are in, what then? The immediate vulgar answer, in our field, has generally been -- not much! The sad reality has been that drug use intervention developed into a specialized field, with its own foci, goals, language, rituals, techniques, staffs, sites of preferred intervention conferences, journals, books, records, slides, films, tapes, prizes, and budgetary sources with very little relationship to research. It has continued, and it has grown in scope but not necessarily in meaningful breadth and depth or effectiveness, without much regard for or use of empirical or other types of useful data. And as a result it is legitimate to say that drug users have suffered, as has the general community, the intervention system, and its agents.

LAW ENFORCEMENT

The lack of reliance upon a research ideology by which to check intervention effectiveness as well as available data has in no way affected drug use law enforcement policies, plans, or programs. In a world committed to having others control drugs, use and the user, the law enforcement system (agents as well as agencies) have willingly filled the control vacuum. After all, someone has to do it. At times they have misinformed the public about the nature of a drug's action and effects, the size and scope and meaning of the drug problem. At times they have even supplied drugs in order to achieve a common good which is a superior goal to drug control -- intelligence material about other criminal activities. At times they have bowed to community pressures and have prevented the development of need programs. At times they have sentenced drug users to treatment when they knew that they themselves wouldn't use such treatment. They have done what we have permitted them to do -- and at times even more -- because they have institutionalized our fears and our stigmas of the dangerous "junky," one of a variety of moral panics. Law enforcement has presented us with a problem which their technology, laws, and policies, budgets, and manpower cannot hope to solve. They wish us to believe that man's appetites, chemical as well as other types, are controllable by external intervention. We know, through the experience of centuries, that generally this has not been so. They wish us to believe that legally mandated treatment of drug users is a viable option. We know through the experience of decades that this has rarely if ever been demonstrated to be so. More specifically, we know, if we choose to, that effective law enforcement may be a threat to the ideology and practice of democracy when its focus is man's appetites. We also know that effective drug use law enforcement intervention cannot operate in a society with a passive, unconcerned, unresponsive citizenry. Law enforcement cannot do for us what we are unwilling to undertake for ourselves. Drug use law enforcement intervention has served a number of unwholesome purposes and roles.

It has permitted professionals and the public at large to believe that since something is being done about drugs, their use and the dreaded user, we can remain uninvolved.

It has permitted us to blame the law enforcement arena when drugs appear to be too available, when drug use has become too visible, and/or when the visible stigmatized drug user has entered our supposedly protected, safe, sanctified personal life space. As long as they remained where they belonged, killing themselves and one another with their drugs, law enforcement was supported and respected. But God help the police officer, or judge, who tampered with our recreational drug using/problem drinking relative or friend.

It has permitted us to believe that the increase in crime is due to drug use, and that effective law enforcement will diminish both. This myth continues in spite of an increase, worldwide, in organized crime, disorganized crime, and terrorism -- which we feel impotent to do anything about. The myth has grown that in an abstinent world there would be little or no crime. This myth is one of the many side effects of a relatively new contagious ailment -- conceptual impotence.

Table 8. A Pragmatic Intervention Schema

		General focus of intervention			
	Drugs	Drug use	Drug experience	Drug user	Other (specify)
Selected inventory factors					
Specification of focus					
Factors initiating intervention					
Factors preventing intervention					
Intervention goals					
Availability of specific factors for intervention					
Policies facilitating efforts					
Policies inhibiting efforts					
General techniques used					
Specialized techniques used					
Availability of generic factors					
Preferred sites for intervention					
Available sites for intervention					
Preferred intervention agents					
Desirable qualities of intervention agents					
Needed training for intervention agents					
Time needed for intervention					
Necessary budget					
Available budgetary source					
Appropriate evaluation process					
Criteria: success/failure Focus of intervention Intervention system Society at large Other (specify)					

Space limitations do not permit me to continue to comment upon the other factors listed in Table 1. At this point there is no doubt in my mind that you the reader will be able to guess or even to predict what I might write about the potential importance and impact of these factors upon drug use intervention. They too have not been appropriately utilized.

My analysis could easily cease at this point. Criticism is a legitimate activity and a time-honored profession. It is not incumbent upon me to match my criticizing analysis with constructive suggestions. But I will try to anyway. After all, I do feel responsible for the state of a field which has given me a living, status, permitted me to travel, has allowed me to intrude into the lives of some as well as to be invited by others, and to edit yet another book. My suggested intervention schema is relatively brief, hopefully to the point, quite usable, and is even testable.

It is schematized in Table 8 in terms of select intervention factors and the general focus of intervention. Your task, as a passive reader, as well as a potentially active and concerned citizen, is to fill out this table in a way that makes it useable for you and for those whom you are concerned about. How does one do this? Certainly not by remaining passive! Not by reverting to instant reflexive responses. Nor by necessarily doing it alone. Nor by joining bandwagons. Somehow each reader, you included, must find--by creating actively and not by passively discovering--and then choose what is generic and not specious.

The challenge in the field of drug use intervention, is to create solutions to real, solvable, problems and not to let our preferred solution(s) to become a new source and/or type of problems. The latter situation appears to be our current situation.

SELF-ESTEEM AND DRUG ABUSE*

R. A. Steffenhagen

Department of Sociology
University of Vermont
Burlington, Vermont 05405

INTRODUCTION

Theories of drug abuse can be categorized according to the disciplines of each individual theorist. While each one of these perspectives sheds a certain amount of light on the process of drug abuse, none of them alone describes the process precisely enough to provide a framework for making predictions about individual behavior.

A fundamental criticism of existing drug research, based on such theories, is that individual studies are too parochial in their focus. Patterns of drug use cannot be explained through an understanding of physiology, psychology, or sociology alone. A predictive model of drug abuse must be able to describe the relationship between physiological effects of drugs, individual motivational and behavioral traits, and characteristics of the individual's social environment.

A second criticism of existing drug research is methodological. Methodological deficiencies in drug research have led to a bias by which researchers simply tend to confirm preexisting stereotypes about drug abuse. First of all, most studies of drug abuse are retrospective, i.e., their measures of personality and/or social environment are gathered after the individual has already developed a pattern of drug abuse. This leads to an inability to disentangle the causes and consequences of drug use. It also allows for the maximum introduction of bias on the part of the researcher since the restrospective data are generally gathered through clinical interviews in which the researcher's professional bias may have a profound effect on how the subject reports his past experiences and on how the researcher interprets these reports. Second, the subjects for these studies' are generally institutionalized patients -- or at least drug users or alcoholics -- who are in serious enough trouble to have drawn themselves to the attention of some legal, penal, health or social service agency. While studying identified addicts is certainly the most efficient way to gather a sample of drug abusers, it introduces a substantial bias because it taps only the portion of the drug-using population that is dealing least effectively with drugs -- the most disturbed segment.

---

*This article is a modification of "An Adlerian approach toward a self-esteem theory of deviance: a drug abuse model" which appeared in Journal of Alcohol & Drug Education, Vol. 24, 1978, pp. 1-13.

Finally, drug researchers have been consistently preoccupied with comparing drug users to nonusers in an attempt to isolate an "abuse-prone" personality type or a typology of "drug abuse personalities," tacitly assuming that all drug users are, in fact, drug abusers. This has led to research designs in which nonusers are compared to users and that obscure important factors which may be related to drug abuse. A predictive model of drug abuse that sorts out psychological and sociological conditions predisposing the individual to drug abuse cannot be established until researchers distinguish between degrees of drug use or abuse. In order for this to be accomplished, comparisons must be made not only between users and nonusers, but also between users and abusers.

Self-esteem theory, originating in Alfred Adler's Individual Psychology (Ansbacher, 1956), provides a framework which is, at once, more comprehensive and more person-specific than traditional physiological psychological, and sociological approaches. Moreover, it dictates a methodology which avoids the biases inherent in much of the research to date. Self-esteem theory provides both a more comprehensive basis for viewing drug use than do many existing theories, and a methodological approach that overcomes many of the limitations of earlier studies.

These theoretical problems result in significant conceptual deficiencies, especially when one attempts to account for the individual who has established a drug use pattern in a social context favorable to drug use. Where drugs are available, and where peer pressure encourages experimentation with drugs (a state of affairs common to the college campus and the ghetto), social forces alone may be sufficient to account for drug use. Degree of drug use, or drug abuse, however, cannot be accounted for by social milieu alone, since individuals in identical social contexts differ in the extent to which they use drugs. If individuals of varying personality types, when exposed to drugs and drug-using subcultures, use drugs, then it is important to search for some underlying process to explain why drug use becomes pathological for some individuals and not for others.

What is this "underlying pathology" in drug abuse? Of central importance to the proposed research is the underlying factor of self-esteem (see William James, 1890, and Ansbacher, 1956). Sociological theory has long been aware of the role that discrepancies between an individual's aspirations and his realistic expectations play in producing deviant behavior (Merton, 1938; Short, 1964), noting that choice of deviant behavior is apt to be influenced by patterns of deviance present in the individual's immediate social environment (Sutherland, 1947; Cloward and Ohlin, 1960). When an individual sees himself as unable to achieve culturally valued goals through socially acceptable means, he is apt either to reject the goals themselves or to attempt to achieve them through unacceptable (deviant) means.

Much of the discrepancy between an individual's aspirations and expectations is structurally determined: the poor youth with an accurate assessment of reality who aspires toward material success would justifiably have lower expectations of achieving his goals than an affluent youth with the same goals. Part of the process, however, is an intrapsychic one, stemming not from perceptions of current reality but from self-perceptions originating in early family socialization experiences. Individuals at all social class levels vary in the extent to which early experiences favor the development of a sense of mastery, or competence, in attaining goals. A psychological sense of adequacy in goal-striving is part of what we mean by high self-esteem.

McAree, Steffenhagen, and Zheutlin (1972) studied illicit drug use on a college campus and used the Minnesota Multiphasic Personality Inventory as well as a sociodemographic schedule in an attempt to discern differences between users and nonusers as well as differences between types of users. We found we could classify three distinct types of users: marijuana only, multiple-drug use, and gross multiple drug use. The marijuana only group showed no personality differences between themselves and nonusers while the multiple users showed significant differences on three of the ten scales of the MMPI (P < .01) and the gross multiple users were significantly different on nine of the ten scales (showing more pathology). These data clearly suggest that an adequate theory must be able to explain nonuse, use, and abuse, as well as the fact that moderate use of alcohol (licit drug use) is different from abuse (alcoholism).

From my years of drug research I have come to the position that it is not the quantity or even the regularity of drug use which distinguishes between use and abuse but rather the intrapsychic dynamic of the "need." I have found regular heavy marijuana users whose use pattern was fundamentally determined by the social milieu (college dorm life) rather than any psychic need; they were easily able to discontinue use with a change in the social milieu. In contrast, I have worked with students who had developed dependency needs for marijuana even if it were only one "joint" per day, and who couldn't give up that one joint. Moreover, often in research we assume quantity to be associated with abuse because it is very difficult to get at the intrapsychic dynamics of drug abuse.

Self-esteem theory, developed within the framework of Alfred Adler's individual psychology, places self-esteem at the apex of personality. This mode assumes that individuals of any personality type vary with respect to self-esteem and that self-esteem itself, as well as distinctive styles of coping with the need for or lack of self-esteem, develops within a social context.

While Adler saw a feeling of inferiority as a common denominator of individuals, he placed a great emphasis upon each individual's unique style of dealing with this feeling. Styles of coping with feelings of inferiority are developed according to each individual's choices concerning his/her own goals, or life plan, and cannot be understood apart from the social context in which the individual makes these choices. Furthermore, individuals differ in the degree to which they feel inferior. The variability of feelings of inferiority, or the variability of feelings of self-esteem, affords an excellent opportunity for exploring the relationship between inferiority and individual behavior as well as offering an explanation of deviant behavior.

Inferiority feelings on the part of the individual reflect the extent to which he sees himself as unable to attain his goals. Both psychological (the extent to which the individual, through early experiences, comes to see himself as able to achieve his goals) and sociological (the extent to which the society in which the individual lives offers viable routes for the individual to attain his goals) processes are relevant. Furthermore, the fashion in which an individual chooses to deal with inferiority or with an inability to attain goals is also socially as well as psychologically determined.

Self-esteem develops along with the rest of the personality through the socialization process. Its lack can be the by-product of a pampered life-style or the polar opposite, a neglected life-style. In the pampered life-style the individual is given everything -- he develops no feeling of self-worth through his own accomplishments (see Anabacher, 1956).

In juxtaposition to the pampered life-style is the neglected life-style. Here the personality develops in a situation of deprivation and neglect where the child gets no support from within the family. He finds the world hostile and reacts in a hostile manner, displacing the very hostility he perceives in others. In neither case do the individual's own efforts lead to a sense of accomplishment. In the case of the pampered individual, his efforts are irrelevant because he is given everything he needs whether he strives for it or not. In the case of the neglected individual, his needs remain unmet regardless of how much he strives.

It is our contention that self-esteem becomes the basic psychodynamic mechanism underlying behavior -- social or antisocial -- and thus deviance. A person with a low self-esteem may well resort to sociopathic acting-out techniques for coping with his poor self-image. Behavior problem children in school are merely acting-out their aggression, their guilt, their inner inadequacies. Dea (1970) concluded that delinquents have a poorer self-image than nondelinquents and feels that society sees them less positively.

It is suggested here that deviant behavior can be explained by the interaction of self-concept (the psychodynamic component) and specific situations (the social component). In the instance of drug abuse, this concept can be illustrated as follows:

Self-esteem	Social situation	Drug use pattern
Low	No pressure to use drugs*	nonuse
	Pressure to use drugs	abuse
High	No pressure to use drugs	nonuse
	Pressure to use drugs	drug use but not abuse

Any theory of drug abuse which fails to account for the social situation is destined to fail, but to emphasize the social situation to the exclusion of psychological factors is equally useless. As shown in the example above, it is rather the interaction of psychological and social variables which leads to a particular pattern of drug use.

Self-esteem and Goal Orientation

Self-esteem has been defined by William James as

$$\frac{Success}{Pretensions} = self\text{-}esteem$$

This would fit within the Adlerian framework if pretensions are viewed as what Adler called goal orientations. An individual who has "god-like" goal expectations will not be able to achieve them and, thus, his self-esteem will be low. It is important to realize that what may appear as success to an outside observer may not be success for the actor in that his goal orientation may be so far beyond reality that the achievement for him is not success but failure. Goal orientations become a very important concept within the framework of self-esteem theory because if one is to explain drug abuse by this approach, it is necessary to understand the actor's goals and not look at success or failure in terms of the researcher's perspective.

---

*Drugs, as used here, is not restricted to illicit drugs but includes the socially accepted drugs, e.g., alcohol, caffeine, and prescription medications of the mood altering group.

A good example is an individual setting his goals so high as to make them unattainable. He may get excellent grades in school but his goal becomes to "know" everything there is to know in the course, so when he receives an "A" it does not give a feeling of satisfaction--he rationalizes it as "The course was easy and I really didn't do the amount of work necessary to really understand the subject in depth."

Over-life-sized goal orientation and low self-esteem are, as Adler says, two approximately fixed points in the psychological scheme of personality development, and in the attempt to deal with these the person may resort to drug use as a coping mechanism.

Self-Esteem and Life Style

The term "life style" was borrowed by Adler from Max Weber. Originally, Adler used the term "Lebensplan" or life span and later called it "Lebenstil" (style of life), and finally life style. He defined life style as "the wholeness of his individuality" (Adler, 1933:189) (the need to have a guiding image).

In a sociological (group) context Weber (1947:191) uses "life style" to refer to the role of social status in providing the normative structure.

In essence they are both saying the same thing, except one is writing in terms of a guiding principle for individual behavior and the other in terms of a guiding principle for group behavior.

Charles H. Cooley clearly understood Descartes' dictim, "Cognito Ergo Sum," as the starting point for man's cognitive development, since man can only start with a societal given: the language and "life style" of the groups are superimposed upon him through the socialization process. The teleological basis of man's behavior from his social class -- with the family as mediator -- becomes his "life style," around which his personality revolves. The group is logically prior to individual; the "I" is the result of a relatively advanced stage of consciousness subservient to a "we" consciousness (society -- the group). As Cooley (1956) expressed it, "it is absurd to think of the 'I' apart from society." By confusing the order, life style has become style of life, and has become equated with other terms, such as ego, self, even "man's personality" (Adler, 1933). Every individual represents a unity of personality and creation of that unity. The personality of the child is a prototype of the personality of the immediate group (the Geminschaft) which is a prototype of the personality of the larger group (the Gesellschaft). However, while the individual is like everyone else, the individual is also different -- just as each snowflake is different, each individual is different both physically and psychologically.

The existentialists see man as a summation of his "choices" or, as Tillich says, "Man is his choices."

What are these choices? This is just another way of viewing man as a summation of life style, because it is the life style which provides and permits the choices. Dealing with behavior, social scientists deal with the milieu interior and the milieu exterior. The choice within each is derived from the life style of the child developed through the socialization process.

The uniqueness and unity of the individual are the yin-yang of the universe. Sociologists have also looked at individual adaptations. Merton (1938), in the development of his theory of anomie, emphasized the role of the congruence or disjunction between the cultural goals and the institutional means: the role of the social order. Adler, on the other hand, emphasized the psychological process, although he never underestimated the importance of the social order, which he called "social interest." All behavior has to be seen in light of the individual's efforts to achieve success, to overcome minus situations and achieve plus situations. Adler sees each individual as striving for superiority and sees life as a process of goal striving. In psychology we speak of motivation as goal-directed behavior. There is no behavior which is not ultimately goal-directed. While goals are treated under goal-orientation in the previous sections, the idea of striving for superiority, or goal striving is particularly important in light of the individual's uniqueness as derived from his life style.

The role of "apperception" is most important in understanding the life style: the world is organized by the individual -- life is seen in terms of the individual's "Weltansicht," which is created from meanings layered upon the perceptions of the individual's past experiences. A stimulus is interpreted not merely in light of its sense experience but in terms of this apperceptive mass, these past perceptions. Ideology may be more important than reality.

Generally speaking, culture helps make for constancy of behavior, so that an individual's life style has much in common with everyone else's but, because of subjectivity, each individual is also unique. Thus while social deprivation in the ghettos may provide the cultural basis for heroin addiction, not all ghetto youth become heroin addicts. In this environment many of the youth, but not all, because of curiosity and peer pressure, try and use heroin; while some become "hooked," others are able to "kick the habit" without help (medical intervention). Consistency and uniqueness exist simultaneously in the same individual.

In the Adlerian typology, such deprivation is called a neglected life-style, and it is certainly important in explaining drug abuse. On the opposite end of the continuum is the pampered life-style, which helps account for such drug abuse as alcohol or polydrug abuse, which is frequent in the middle-class and college environments.

The understanding of the individual's life style and its relation to the social class life style is important in understanding drug abuse. While the individual's life style becomes a guiding image for him, much of his guiding image is also part of the guiding principle of the group. A middle-class youth with low self-esteem (often manifesting a pampered life-style) who has developed unattainable goals will probably move toward counseling or polydrug abuse, and/or occult membership, which are consistent with the middle-class college subculture. On the other hand, a youth from the slums with low self-esteem (generally manifesting a neglected life-style) probably will move in the direction of delinquency and/or drugs as a coping mechanism. In attempting to predict the direction behavior will take, it is essential to deal with both individual and group life styles, and the life style concept becomes crucial in developing a theory of drug abuse.

Self-Esteem and the Social Milieu: Differential Association and the Role of Peer Group Pressure

Within the social milieu we include the role of differential association. Sutherland (1947) established nine propositions concerning behavior which would be important in establishing norms for the individual which would then serve as guideposts for behavior. This theory is based upon learning theory -- socialization. The individual is socialized into being a conformist or deviant. Sutherland's original theory is concerned with crime and attempts to explain crime in modern society. In the present content, it is useful to substitute the word deviance for crime; thus Sutherland's first proposition would read "Deviant behavior is learned."

The nine propositions are:

1. Criminal behavior is learned.

2. Criminal behavior is learned in interaction with other persons in a process of communication.

3. The principal part of the learning of criminal behavior occurs with intimate personal groups.

4. When criminal behavior is learned, the learning includes: (a) techniques of committing the crime, which are sometimes very complicated, sometimes very simple, and (b) the specific direction of motives, drives, rationalizations, and attitudes.

5. The specific direction of motives and drives is learned from definitions of legal codes either favorable or unfavorable.

6. A person becomes delinquent because of an excess of definitions favorable to violation of law over definitions unfavorable to violation of law (this is the principle of differential association that is the crux of the theory).

7. Differential associations may vary in frequency duration, priority, and intensity.

8. The process of learning criminal behavior by association with criminal and anticriminal patterns involves all of the mechanisms that are involved in any other learning.

9. While criminal behavior is an expression of general needs and values, it is not explained by those general needs and values, since non-criminal behavior is an expression of the same needs and values.

There is no question as to how important the role of socialization is in molding the personality and consequently in determining the behavior of the recipient. Becker and Strauss (1956) refer to the role socialization plays in the process of "turning on" to drugs. The alcohol user is socialized into the use of alcohol -- he is taught what to expect and how to consume. The marijuana user is likewise taught how to "smoke" and what a marijuana high is. In reference to "pot" smoking differential association is particularly important because the individual must associate with users if he or she is to be able to obtain the marijuana. (This theme has been developed in a paper by McCann, Steffenhagen, and Merriam, 1976.)

It is evident that peer group association is an important factor in the socialization process of the individual. The adolescent peer group, as a subculture, may be either an agent of social change or an agent supporting the status quo. In this capacity it plays a vital role in putting pressure upon youth to conform to the values "it" has established for itself, be they delinquent or nondelinquent (drug use or non-drug use).

Related to the theory of excess of definitions favoring deviance or conformity is the role of the psychological support system, which is essential in helping the individual to develop and maintain his self-esteem. Psychological support systems may come from the family, peer group, intimate friend, or counselor, and can help one cope with stress and at times "buck the system" or reject pressure from one of the groups. Also relevant is David Riesman's (1953) typology of social character and his emphasis upon modern youth as being other-directed so that the peer group plays a much more important role in socialization than it had in the tradition- or inner-directed person. Here the peer group may provide the excess of definitions, not merely by virtue of time spent with the group (frequency or duration) but because of the importance of the group's opinions to the individual (intensity). One's contemporaries provide the major source of direction for the individual, and modern youth are particularly sensitive to the opinions and values of others.

Self-Esteem and Deviance

Self-esteem alone cannot predict drug use or abuse; it must be coupled with an understanding of the total social situation and the life style of the individual. However, it is postulated that, in contemporary society, an individual with low self-esteem is particularly vulnerable to drug abuse, because of the widespread prevalence and availability of drugs and because the mass media have so thoroughly dispersed drug information through the population. As the following illustration suggests, drug abuse is only one form of deviant behavior to which such an individual is prone.

Condition	Effect
Low self-esteem	Psychosis
	Delinquency
	Crime
	Drug abuse
	Occult
	Neurosis
	etc.

The particular behavior accompanying low self-esteem can be explained by the individual's social situation and life style. Drug abuse and delinquency may go together in the ghetto environment, whereas drug abuse may exist independently or with a fascination with the occult among the college population (see Steffenhagen, 1974).

There is a close relationship between drugs and cult membership, many cult members having had previous drug abuse histories, with many cult members dropping out and becoming drug abusers. The relationship is more clearly understood in light of Adler's reference to the role of the exaggerated goal or self-enhancement. Adler indicates that there are two relatively fixed points in this scheme: over-life-sized goals which may reach proportions of god-likeness and the low self-esteem of the person who feels inferior. The striving for a goal set beyond all possibility of attainment is in a sense a striving for power. The "normal" person with social interest seeks success in accordance with the dictates of reality, whereas the neurotic with the over-life-sized goals may seek success in terms of immediate gratification and little energy output on his part: "I will pray to the devil for immediate power."

A similar relationship may exist between drugs and delinquency, both having a possible basis in inferiority feelings and neurosis. Not merely drug abuse, but deviance in general may be the earmark of the person with low self-esteem who uses these various techniques (deviance) as coping mechanisms for dealing with feelings of inferiority. Just as the drug abuser is distinguished from the normal drug user by deep-seated inferiority feelings, the "neurotic" delinquent must be distinguished from the "normal" delinquent whose delinquency is a result of an adverse environment or an excess of definitions for delinquency. As previously stated, it is not uncommon to see individuals engage in more than one type of deviant behavior simultaneously, e.g., drugs-occult, drugs-delinquency, drugs-psychosis, or even drugs-occult-delinquency.

Becker and Strauss (1956) have discussed the socialization process involved in drug use. Every drug researcher who has studied marijuana use is familiar with the idea that most users enjoy "turning on" another person and the social camaraderie which emanates from "doing" drugs together. Thus, a normal person can easily become involved in drug use without having any prior emotional instability. McArgee, Steffenhagen, and Zheutlin (1972) concluded that the gross-multiple user showed more emotional pathology than the non-user, as measured by the MMPI. However, on an individual basis, there were also gross-multiple drug users who had normal profiles with no scales above 70, further supporting the idea that there is no clinical profile of drug abuse.

McCann, Steffenhagen, and Merriam (1976) demonstrated that the marijuana user is put into a situation where he is forced to become a part of the deviant subculture in order to pursue a behavior pattern which he finds pleasurable and satisfying. This study provides strong support for the notion that frequency of drug use is fostered by socialization into the subculture. By way of a path analysis, McCann et al. arrived at the following model:

Low age at turn on	Higher frequency of use	Other drugs	Selling marijuana	Selling other drugs

The key point, with respect to the present discussion, is that the frequency of marijuana use had a strong positive effect upon the use of other drugs -- our gross-multiple use.

It is possible for a heavy drug use pattern to emerge from the conditioning process itself without pathology. However, this does not negate the contention that drug abuse is related to low self-esteem because "heavy" drug use is not necessarily abuse. As was stated earlier, heavy use may result from the pressures of the social milieu and not from any inter-psychic need. When the peer pressure subsides the use diminishes, thus heavy use may be socially conditioned, whereas abuse is due to psychic dependency. As with physical addiction, abuse may be reflected in the difficulty incurred in abstinence. If someone can become a polydrug user through this normal entrance into the drug behavior pattern, then how can one distinguish between the person with high and low self-esteem? It is hypothesized that, given heavy drug use, the high self-esteem individual will be able to give up his heavy drug use behavior in a therapeutic setting; or, as in the case within delinquent gangs, when the group decides a particular drug is no longer "cool," the person with high self-esteem will give it up whereas the person with low self-esteem will not be able to.

Self-esteem is the basic psychodynamic mechanism underlying deviance (drug abuse). It is, however, not sufficient to predict who will engage in deviant behavior -- an understanding of the individual's life style, goal orientations, and social milieu is also needed. Given the psychodynamic mechanisms we also need to understand the input of the social structure. This, then, leads to the following paradigm:

self-esteem + life style + personality traits + goal orientation + primary group + social milieu = behavior where

> self-esteem -- high or low
> life style -- self-centered or contributive
> personality traits -- normal or neurotic
> goal orientation -- socially useful or useless
> primary group -- supportive or unsupportive
> social milieu -- friendly or hostile

TREATMENT

If drug theories are to be of value, they should provide a paradigm for treatment whereby success in treatment strengthens the theory. Generally, psychoanalysis has not proven successful in dealing with drug use addiction (including alcoholism) which is a pragmatic indicator of the weakness of the psychoanalytic theory of drug use/addiction.

Self-esteem theory posits the idea that low self-esteem is the underlying psychodynamic mechanism of deviance; therefore, the modus operandi of therapy is the goal of raising self-esteem. The therapist need not deal with the problem (drug abuse) but focuses entirely on self-esteem. The goal is to raise self-esteem and help develop social interest. When the person feels good about himself, s/he can then deal with problems from a perspective of strength. Self-esteem therapy will not necessarily eliminate drug use, as in the case of marijuana on a college campus because of peer group pressure and a social milieu conducive to smoking, but the psychic dependency will be eliminated; use will become a more rational decision. The author has had success with alcoholics, polydrug abusers as well as with phobias, anorexia nervosa, homosexuality, intermittent explosive disorders, etc.

PREVENTION

Within self-esteem theory, prevention lies within the educational system. The recognition of low self-esteem in children (within the educational institution) and the providing of child mental health clinics, as a prototype of Adler's work in Vienna, would do the most to prevent abuse.

CONCLUSION

Our theory postulates that the psychodynamic mechanism underlying deviance is low self-esteem. Self-esteem develops in the individual through repeated experiences of mastery, i.e., repeated experiences in which efforts to achieve a goal are met with success. As such, an individual's self-esteem is a reflection both of his level of aspiration and of his self-confidence. A person may be quite competent, but set his goals too high; or a person may have realistic goals, but feel unsure of his ability to achieve them. If we view all behavior as goal striving and note that the individual evaluates himself in terms of his ability to achieve the goals he sees as important, we begin to understand why self-esteem is the psychodynamic mechanism underlying deviance as well as normal behavior. If the individual feels inadequate, he feels insecure and needs to protect his poor self-image. This results in feelings of inferiority which are frequently compensated for in behavior. These compensatory behaviors generally create further problems in interpersonal relations and increase inferiority.

Even when the achievement of culturally prescribed goals is precluded by an individual's economic disadvantage, low self-esteem is not inevitable. Family structure and early socialization may play an important role in the development of an individual's self-esteem. An individual whose early family experience provided him with opportunities to set realistic goals and to achieve them through his own efforts may carry that perspective with him in later years. On the other hand, an individual who was unable to cope successfully in early life, either due to pampering or to neglect, may continue to see himself as unable to live up to his goals even when he does possess the means for achieving them.

Within the framework of the self-esteem theory we can explain nonuse, social use, and abuse of drugs as well as why therapeutic models are and are not successful.

REFERENCES

Adler, A., Social Interest (1933). Trans. by J. Linton and R. Vaughan, New York: Capricorn Books, 1964.

Ansbacher, Heinz L. and Rowena R. Ansbacher. The Individual Psychology of Alfred Adler. New York: Harper Torchbooks, 1956.

Becker, H. S. and Anselm Strauss. "Carrers, personality, and adult socialization." American Journal of Sociology 62:253-263, November, 1956.

Cloward, Richard A. and Lloye E. Ohlin. Delinquency and Opportunity: A Theory of Delinquent Gangs. New York: The Free Press, 1960.

Cooley, C. H. Social Organization. Illinois: The Free Press, 1956.

Dea, K. K. "Concept of self in interpersonal relationships as perceived by delinquent and non-delinquent youth." Dissertation Abstracts International, Vol. 31 (9-A), 4893, 1970.

James, William. Principles of Psychology, Vol. 1, New York: Henry Holt and Co., 1890.

McAree, C. P., R. A. Steffenhagen, and L. A. Zheutlin. "Personality factors and patterns of drug use in college students." American Journal of Psychiatry 128:890-892, 1972.

McCann, H. G., R. A. Steffenhagen, and George Merriam. "Drug use as a model for a deviant subculture." Prepublication manuscript.

Merton, Robert K. "Social structure and anomie." American Sociological Review 3: 672-682, October, 1938.

Riesman, David. The Lonely Crowd. New York: Doubleday & Co., Inc. Anchor Books, 1953.

Short, James F. "Gang delinquency and anomie." in: Anomie and Deviant Behavior, ed. by Marshal B. Clinard. New York: The Free Press, 1964.

Steffenhagen, Ronald A. "Drug abuse and related phenomena: an Adlerian approach." Journal of Individual Psychology 30:238-250 (November, 1974).

Sutherland, Edwin H. Principles of Criminology. Philadelphia: J. B. Lippincott, 1947.

Weber, Max. The Theory of Social and Economic Organization. Trans. by A. M. Henderson and T. Parsons. New York: Oxford University Press, 1947.

THE MULTIPLE RISK FACTORS HYPOTHESIS:

AN INTEGRATING CONCEPT OF THE ETIOLOGY OF DRUG ABUSE

Brenna H. Bry

Graduate School of Applied and Professional Psychology
Rutgers -- The State University
Box 819
Piscataway, New Jersey 08854

INTRODUCTION

Recently, researchers have been studying the etiology of drug abuse by sampling large numbers of adolescents, assessing a wide range of psychosocial characteristics, and applying multivariate statistical methods to determine which combination of the characteristics best explains initiation or extent of drug use (Jessor & Jessor, 1978; Kandel, Treiman, Faust, & Single, 1976; Segal, Huba, & Singer, 1980b; Smith & Fogg, 1978; Pandina & Schuele, 1983). While some intriguing convergences have occurred among the findings of these studies, some puzzling discrepancies have also been noted. In the past, such discrepancies could be explained by problems in research design, but current studies are well designed and executed. Consequently, the field is now faced with the challenge of integrating disparate findings.

Smith and Fogg (1978), for instance, assessed 30 psychosocial variables in 651 junior high students and found that a combination of five of these variables best differentiated students who would become early marijuana users from those who would start late or not at all. They summarized their findings by stating that low personal competence and low sense of social responsibility seem to predict early drug use.

In contrast, in another large study, Kandel, Treiman, Faust, and Single (1976) found peer use to be the best predictor of marijuana use initiation. They offered an "adolescent subculture" explanation for their results that suggest a subculture exists which revolves mainly around drug use.

A third view of etiology is suggested by the work of Pandina and Schuele (1983). In a cross-sectional study involving 44 characteristics of 1,951 seventh through twelfth graders, they found results that suggest an etiology involving personal dissatisfaction and alienation.

In a longitudinal study of more than 20 characteristics of 432 junior high students, Jessor and Jessor (1978) also found personality variables to be important, but their salient personality variables were different from Pandina and Schuele's. They found personal characteristics reflecting unconventionality and a tolerance for deviance to be predictive.

The latest relevant study also found personality variables to be useful in predicting drug abuse, but the set is different from all of those above. Segal, Huba, and Singer (1980b) tested 48 general personality and life-style variables in 1,095 college freshmen and concluded that drug use variance is best explained by differences in life-style -- the extent to which a young person seeks varied and unusual experiences.

Kandel (1978) has offered a "stage theory" to explain these conflicting findings. She states that different sets of etiological variables are associated with different stages of drug use development. Her theory is supported for what it claims (Kandel, Kessler, & Margulies, 1978), but it does not account for the fact that different sets of predictors have been associated with the same stage of drug behavior development. Other researchers have introduced moderator variables to reconcile the above discrepancies (Braucht, Kirby, & Barry, 1978; Kaestner, Rosen, & Appel, 1977; Segal, Huba, & Singer, 1980a). While some significant isolated findings have emerged from these studies, the use of moderator variables will probably not generate broad principles explaining drug abuse. Instead, the population is likely to become divided into smaller and smaller subgroups until the researchers conclude, as have Nathan and Harris (1980) and Dunnette (1975), that there are probably as many combinations of salient etiological variables as there are drug abusers.

Given the above findings, an integrating conceptualization which needed to be tested was that extent of drug use, from abstinence to very heavy use, is a function of the number of diverse etiological variables instead of any particular set of them. Bry, McKeon, and Pandina (1982), accomplished this by analyzing the responses of 1960 secondary school students to a survey assessing drug use (including alcohol) and six psychosocial predictors. They also assessed to what extent knowledge about number of risk factors might improve ability to predict problematic drug use, or very heavy drug use. A more detailed description of the study appears in the above paper (Bry, McKeon, and Pandina, 1982), but the method and results will be summarized here as the basis for a discussion of their implications for understanding drug abuse and its prevention.

Subjects

Subjects were 1960 high school students from a North Atlantic, working class county, who were surveyed in their schools in the Spring of 1975 as part of another study (Pandina, 1978). Sixty-two percent of the subjects were attending an urban public high school; 16% were attending an urban parochial high school, and 22% were suburban secondary school students. Forty-four percent were male; 56% were female. Sixty-nine percent were white, and the remainder consisted of Puerto Ricans, blacks, other Latinos, or Orientals. The subjects were fairly evenly divided among the four grades of high school, with a few more (28%) in ninth grade and a few less (23%) in twelfth grade.

Survey Instrument

Data were collected via a 26-page self-administered, anonymous questionnaire which took about 1 1/2 hours during the school day to complete. The survey included sociodemographic questions, the John Hopkins Symptom Checklist (SCL-90) (Derogatis, Rickels, & Rock, 1976), the Piers-Harris Test of Self Concept (Piers, 1968), the Streit-Schaefer Family Perception Inventory (Streit & Oliver, 1972; Streit, Halstead, & Pascale, 1974), and a drug and alcohol usage inventory developed by Pandina, White, and Yorke (1981).

Extent of Drug Use

A single composite drug use score (SUI) was calculated for each subject based upon the reported nature of use (extent of lifetime use, frequency of current use, and recency of last use) for ten substances (beer, wine, hard liquor, marijuana, amphetamines, barbiturates, inhalants, hallucinogens, cocaine, and opiates) relative to others in the sample. SUIs are distributed as standard z scores with an overall mean of 50.11 and a standard deviation of 9.78. Some calculations in this study required that subjects be grouped according to the extent of drug use. Subjects who reported that they had never used any substance were assigned to the Abstainers Substance Involvement Group (SIG); those reporting that they had used one or more substances in the past but were not users at the time of the survey were assigned to the Stoppers SIG. The current users were assigned to four involvement groups -- Low Users, Moderate Users, Heavy Users, and Very Heavy Users, based upon the distance of their SUI (expressed in standard deviations) from the overall SUI mean (50.11). These calculations are described fully in Pandina, White, and Yorke (1981) and are based upon the work of Lu (1974).

Number of Risk Factors

The number of risk factors exhibited by each subject was determined by counting how many of the following risk factor criteria appeared in the subject's data:

1. a grade point average of D or F

2. a self-report of "no religion"

3. the independent (outside of the family) use of alcohol before 13 years old

4. a Global Symptom Index on the SCL-90 above 1.24

5. a Behavioral scale score on the Piers Harris below 12

6. an Overall Parental Love scale score on the Streit-Schaeffer below -.54

These criteria were based on pretesting (Bry, McKeon, & Pandina, 1982) and previous studies (Gossett, Lewis, & Phillips, 1972; Pandina & Schuele, 1983; Tennant & Detels, 1976; Tennant, Detals, & Clark, 1975).

RESULTS

It was found that the range of risk factors among the subjects was zero to four with the majority of the students (n = 535) exhibiting none. A one-way analysis of variance revealed a highly significant relationship between number of risk factors and extent of drug use, as represented by the SUI, $F(4,968) = 27.04$, $p < .00001$. Figure 1 shows that the mean SUI, which is in parentheses, for each succeeding risk factor level (0-4) is greater than the one before, representing a linear trend, $F(1,968) = 101.30$, $p < .00001$.

A Duncan's Multiple Comparison's Test indicated that the SUIs of subjects with one risk factor were significantly higher than those with no risk factor, $t(482) = 6.85$, $p < .0001$. Further, the SUIs of those with four risk factors were significantly higher than those with three, $t(8) = 2.01$, p

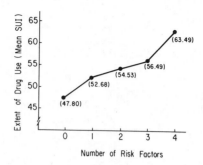

Figure 1. Extent of drug use associated with
each quantity of risk factors in the hypothesis
testing sample (SUI = substance use index). From
Bry et al. (1982), p. 277.

< .04. There was no evidence, however, that extent of drug use among those
with three risk factors was significantly greater than among those with two
risk factors or that drug use among those with two risk factors was
significantly greater than among those with one.

In order to determine if extent of drug use is related to one or two
predominant combinations of risk factors or to many diverse combinations,
frequency analyses of the risk factor combinations were conducted.
Thirty-one combinations of risk factors were found in the sample out of the
51 possible ways that six risk factors can combine into groups of one, two,
three, and four. None of the combinations accounted for the majority of the
cases, and 17 of the 31 appeared more than five times.

The final analysis was conducted to investigate to what extent
knowledge of this relationship between number of risk factors and the whole
range of drug involvement might contribute to our understanding of the most

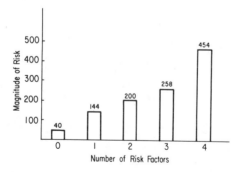

Figure 2. Magnitude of risk for very heavy
drug use associated with each quantity of
risk factors in the hypothesis testing sample
(magnitude of risk = ratio of observed to
expected very heavy drug use). From Bry et
al. (1982), p. 278.

extreme end of the range, very heavy drug use. Figure 2 shows the magnitude of risk for very heavy use that is associated with each quantity of risk factors. Magnitude of risk is an index used in epidemiological research (Dawber, Kannel, & Lyell, 1963).

A magnitude of risk of 100 represents the base rate of very heavy drug use, which was 11.1% in this sample. The presence of two risk factors is associated with a 100% increase above that base rate, or up to a 22% occurrence rate. There was a 50% chance that a subject in our sample who exhibited four risk factors also would report very heavy drug use.

DISCUSSION

The concept that extent of drug use is a function of the number of risk factors received support in the study. These results must be replicated, of course, and expanded to include other etiological variables. Further research must also explore the possibility that a number of risk factors may operate differentially according to age, subculture, the meaning of abstinence, and different situations such as work, war (army), and leisure. It is noteworthy that the findings link drug abuse research to other research areas. Multiple risk factor models, similar to the one tested here, are being used to study maladaptation of children (Sandler & Block, 1979), depression (Brown & Harris, 1978), heart disease (Meyer & Henderson, 1974), and mental illness (Dohrenwend, 1973). This linkage implies that drug abuse, often viewed as a unique phenomenon, obeys laws similar to those observed for other problem behaviors. It also suggests that drug abuse research can contribute to and benefit from other areas of inquiry.

The findings could also provide a framework for integrating past drug abuse research. Rather than contradicting previous findings, this study supports most other predictive models. The current findings suggest that there are many different pathways, many different combinations of etiological variables which lead to drug abuse. The integrating generalization which links most recent research is that the relationship between number of etiological variables and probability of drug abuse is additive. The etiological variables could be as disparate as high sensation seeking or high tolerance for deviance or psychological distress; but the more etiological factors young people exhibit, the higher is their probability of abusing drugs.

This multiple risk factors concept has implications for prevention programs too. It suggests that no one prevention program will be adequate to do the job. Since there are many different constellations of etiological factors and since research has not yet shown one intervention to be effective in reducing them all, multiple interventions must be applied. Specific programs must be applied to reduce specific etiological factors.

Depending upon the needs, family counseling programs that improve family relations (Klein, Alexander, & Parsons, 1977; Streit, 1977) could be combined with lunchtime school rap groups that improve self-esteem (Penfield & Whiteley, 1977), and/or the Early Secondary Intervention Program, which improves school grades and attendance (Bry & George, 1979, 1980). Outward Bound (Kelly & Baer, 1969) and summer youth job programs, which provide constructive avenues for sensation-seeking, could be combined with substance abuse education programs in the schools which decrease early independent heavy use (Blum, Blum, & Garfield, 1976; Kleber, Berberian, Gould, & Kasl, 1975).

In summary, the current findings suggest that prevention efforts begin by assessing which of the many possible etiological factors are present in a given population and then providing specific prevention programs to reduce each one.

REFERENCES

Blum, R. H., Blum, E., & Garfield, E., Drug Education: Results and Recommendations. Lexington, MA: D. C. Heath & Co., 1976.

Braucht, G. N., Kirby, M.W., & Berry, G.J."Psychosocial correlates of empirical types of multiple drug abusers," J. Consult. Clin. Psych. 46:1463-1475, 1978.

Brown, G. W., & Harris, T. Some Origins of Depression: A Study of Psychiatric Disorder in Women. New York: Free Press, 1978.

Bry, B. H., & George, F. E., " The preventive effects of early intervention upon the attendance and grades of urban adoloescents," Prof. Psych., 11:252-260, 1980.

Bry, B. H., & George, F. E., "Evaluating and improving prevention programs: A strategy from drug abuse." Evaluation and Program Planning, 2:127-136, 1979.

Bry, B. H., McKeon, P., & Pandina, R. J., "Extent of drug use as a function of number of risk factors," J. Abn. Psych., 91:273-279, 1982.

Dawber, T.R., Kannel, W. B., & Lyell, L. P., "An approach to longitudinal studies in a community: The Framingham Study," Annals of the New York Academy of Sciences, 107:539-556, 1963.

Derogatis, L. R., Rickels, K., & Rock, A. F., "The SCL-90 and the MMPI: A step in the validation of a new self-report scale," Brit. J. Psychiat., 128:280-289, 1976.

Dunnette, M. D., "Individualized prediction as a strategy for discovering demographic and interpersonal/psychological correlates of drug resistance and abuse," In D. J. Lettieri (Ed.), Predicting Adolescent Drug Abuse: A Review of Issues, Methods and Correlates. Research Issues 11. Rockville, MD: National Institute on Drug Abuse, 1975.

Dohrenwend, B. S., "Life events as stressors: A methodological inquiry," J. Health Social Behav., 14: 167-175, 1973.

Gossett, J. T., Lewis, J. M., & Phillips, V. A., "Psychological characteristics of adolescent drug users and abstainers: Some implications for preventive education," Bull. Menninger Clinic, 36:525-535, 1972.

Jessor, R., & Jessor, S. L., Theory testing in longitudinal research on marijuana use. In D. B. Kandel (Ed.), Longitudinal Research on Drug Use: Empirical Findings and Methodological Issues. Washington, DC: Hemisphere, 1978.

Kaestner, E., Rosen, L., & Appel, P., "Patterns of drug abuse: Relationships with ethnicity, sensation seeking, and anxiety," J. Consult. Clin. Psych., 45:462-468, 1977.

Kandel, D. B., Convergences in prospective longitudinal surveys of drug use in normal populations. In D. B. Kandel (Ed.), Longitudinal Research on Drug Use: Empirical Findings and Methodological Issues. Washington, DC: Hemisphere, 1978.

Kandel, D. B., Kessler, R. C., & Margulies, R. Z., "Antecedents of adolescent initiation into stages of drug use: A developmental analysis," In D. B. Kandel (Ed.), Longitudinal Research on Drug Use: Empirical Findings and Methodological Issues. Washington, DC: Hemisphere, 1978.

Kandel, D.B., Treiman, D., Faust,R., & Single, E., "Adolescent involvement in legal and illegal drug use: A multiple classification analysis," Social Forces, 55:438-458, 1976.

Kelly, F. J., & Baer, D. J., "Jessness Inventory and self-concept measures for delinquents before and after participation in Outward Bound," Psych. Rep., 25:719-724, 1969.

Kleber, H. D., Berberian, R. M., Gould, L. C., & Kasl, S. V., Evaluation of an adolescent drug education program (Final Report of NIDA Grant No. DA-00055). Rockville, MD: National Institute on Drug Abuse, 1975.

Klein, N. C., Alexander, J. F., & Parsons, B.V., "Impact of family systems intervention on recidivism and sibling delinquency: A model of primary prevention and program evaluation," J. Consult. Clin. Psych. 45:469-474, 1977.

Lu, K. H., "The indexing and analysis of drug indulgence," Int. J. Addict., 9:785-804, 1974.

Meyer, A. J., & Henderson, J. B., "Multiple risk factor reduction in the prevention of cardiovascular disease," Prev. Med., 3:225-236, 1974.

Nathan, P. E., & Harris, S. L., Psychopathology and Society, 2nd edn., New York: McGraw-Hill, 1980.

Pandina, R. J. (Ed.). Coping with Adolescent Substance Use. Final Research Report. New Brunswick, NJ: Rutgers Center of Alcohol Studies, 1978.

Pandina, R. J., & Schuele, J. A., "Psychosocial correlates of alcohol and drug use of adolescent students and adolescents in treatment," J. Stud. Alcohol, 44:950-973, 1983.

Pandina, R. J., White, H. R., & Yorke, J., "Estimation of substance use involvement: Theoretical considerations and empirical findings," Int. J. Addic., 16:1-24, 1981.

Penfield, D. A., & Whiteley, R. Final evaluation of an experimental three-year program in drug abuse prevention involving the school system, the community, and the student population in the Piscataway, New Jersey schools (Final Report, NIDA Grant No. R25-DA00832-02). Piscataway, NJ: Piscataway Public Schools, August 1977.

Piers, H. V., Manual for the Piers-Harris Children's Self-concept Scale. (The Way I Feel About Myself). Nashville, TN. Counselor Recordings and Tests, 1968.

Sandler, I. N., & Block, M., "Life stress and maladaptation of children," Am. J. Commun. Psych., 7:425-440, 1979.

Segal, B., Huba, G. J., & Singer, J. L., "Reasons for drug and alcohol use by college students," Int. J. Addict., 15:489-498, 1980(a).

Segal, B., Huba, G. J., & Singer, J. L., "Prediction of college drug use from personality and inner experience,"Int. J. Addict., 15:849-867, 1980(b).

Smith, G. M., & Fogg, C. P., Psychological predictors of early use, late use, and nonuse of marijuana among teenage students. In D. B. Kandel (Ed.). Longitudinal Research on Drug Use: Empirical Findings and Methodological Issues. Washington, DC: Hemisphere, 1978.

Streit, F. Evaluation of Open Door (Final Report on NIDA Prevention Grant to Open Door). Highland Park, NJ: Streit Associates, June 1977.

Streit, F., Halstead, D. L., & Pascale, P. J., "Differences among youthful users and nonusers of drugs based on their perceptions of parental behavior," Int. J. Addict., 9:749-755, 1974.

Streit, R., & Oliver, H. G., " The child's perception of his family and its relationship to drug use," Drug Forum, 3:283-289, 1972.

Tennant, F. S., Jr., & Detels, R., "Relationship of alcohol, cigarette, and drug abuse in adulthood with alcohol, cigarette, and coffee consumption in childhood," Prev. Med., 5:70-77, 1976.

Tennant, F. S., Jr., Detels, R., & Clark, V., "Some childhood antecedents of drug and alcohol abuse," Am. J. Epidem., 102:377-384, 1975.

DEVELOPMENTAL ASPECTS OF MOTIVATION TO DRINK:

A CROSS-SECTIONAL INVESTIGATION

L. Gliksman

Addiction Research Foundation
Community Programs Evaluation Centre
University of Western Ontario
London, Ontario N6A 5B9, Canada

Numerous studies in recent years have investigated the reasons individuals give for using alcohol and drugs. Theories about these motives have run the gamut from a unidimensional motive to as many as twelve dimensions. The investigations themselves have varied widely and have ranged from those that merely list the reasons for drinking to those that attempt to understand the relationship between these various motivations and alcohol use.

Descriptions of a small number of these studies may provide some insight into the nature of these investigations (see Table 1 for summary). Jessor, Carman, and Grossman (1968) defined four categories of drinking functions: positive social functions, conforming social functions, psychophysiological functions, and personality effects functions. However, the authors did not attempt to directly relate these findings to alcohol consumption. Hanson (1973) suggested that drinking occurred for two main categories of reasons: social effects drinking, which is directed by social norms, and personal effects drinking, which involves idiosyncratic decisions about alcohol consumption. He speculated that heavier consumption was associated with the personal effects motivations, but failed to test this hypothesis. Shearn and Fitzgibbons (1973) investigated the reasons youthful psychiatric in-patients gave for their initial use of drugs, continuing use of drugs, discontinuing the use of drugs, and never using drugs. Although some analyses were done in terms of hospitalization, no evidence was provided to substantiate the link between these reasons and the quantity of drug use. Kohn and Annis (1977) speculated that five reasons for drug use existed. They found that only one, internal sensation seeking, which was defined as a liking for unusual dreams, fantasies, or internally generated feelings, was related to the use of drugs, specifically tobacco, alcohol, and cannabis. Barnes and Olson (1977) list seven motives for drug use: anxiety reduction, depression reduction, hostility reduction, adventure seeking, camaraderie, pleasure seeking, and physical pain alleviation. Their concern, however, was with the non-drug alternatives to these motivational states; thus they did not relate each to drug use. Bowker (1977) suggested that eight motives exist for the use of various drugs. He found that, with respect to alcohol, more adults endorsed the social/interpersonal motive than any other motive, while for school age children, the most endorsed motive was adventure curiosity. Scott (1978) looked at the motives of nonusers, users, and exusers of drugs. However,

Table 1. Motives for Substance Use and Summary of Selected Studies

Author(s) and Year	Type of subjects number and sex	Type of substance	Technique used for study	Movtives for use	Conclusions or study limitations
1. Jessor, Carman, & Grossman, 1968	38 male and 50 female intro-ductory psycho-logy students	alcohol	questionnaire	positive social, conforming social, psychophysiologi-cal and personality effects	did not relate motives to use
2. Hanson, 1973	students	alcohol	archival	social effects, personal effects	archival and did test hypotheses
3. Shearn & Fitzgibbons, 1973	167 youthful psychiatric inpatients (about half female)	numerous drugs	survey	multiple reasons for initial use, continuing use, discontinuing use	no evidence pro-vided linking uses or showing relationship with use or non-use of different drugs
4. Kohn & Annis, 1977	232 female and 198 male grade 12 students	tobacco, alcohol, cannabis, painkillers	questionnaire	internal sensa-tion seeking	limited conclu-sions
5. Barnes & Olson 1977	50 males and 50 females at each age level: 13-14, 15-16, 17-18.	general drug use	questionnaire	anxiety, depres-sion, hostility, adventure, camaraderie, pleasure, physical pain	not related to usage
6. Bowker, 1977	students in grades 7-12, college and an adult sample	numerous drugs	questionnaire	adventure/ curiosity, emotional, social	questions asked which drugs satisfy which motive; did not relate use to motives
7. Scott, 1978	120 male and 60 female high school students	general drug use	open-ended questions of users and nonusers	escape, positive experience	no comparisons-- all subjective feelings
8. Winstein, 1978	--	marijuana	archival	knowledgableness self-fulfillment	relies on exist-ing data and uses post hoc interpretations
9. Schilling & Carman, 1978	106 male and 90 female high school drinkers	alcohol	questionnaire	social effects, personal effects	suggest develop-mental sequence with change from social effects to personal effects

no analyses were performed and the relationships reported were based on the subjective feelings of the experimenter. Weinstein (1978), in an archival study, looked at the motives previously interpreted as being involved in initial use, continuing use, discontinued use, and non-use of marijuana. He concluded that initial use was most attributable to a <u>desire for knowledge,</u> while continued use was maintained by <u>self-fulfillment</u> and <u>appeals to psychological drives</u>. Schilling and Carman (1978) looked at the relationships between two general motives for drinking, <u>personal effects</u> and <u>social effects</u>, and <u>internal-external control</u> relating to alcohol use among high school students. They found that "externals" reported more personal effects motivations for their drinking and greater alcohol consumption than did the "internals." However, the direct relationship between the two motives and alcohol consumption was not investigated. Cutter and Fisher (1980) looked at family experiences and motives for drinking. Although they found some relationships between parental attitudes and behaviors and some motives, they did not relate these motives to alcohol consumption. Jessor and Jessor (1977) proposed a comprehensive theory to explain problem behaviors among youth. Subsumed within the domain of problem behaviors was drinking. Each of the problem behaviors was viewed as a symptom of a general, deviant life-style. The theory itself included distal and proximal components of the environment, personality, socialization, and demographic variables. Subsumed under these broader categories were individual motives for behavior.

These studies are fairly representative of the research in the area of motivation and alcohol/drug use. Although they seem to indicate a belief in the utility of the concept, we have seen little systematic effort to define the number of motives and to see their relationships to actual consumption. In addition, although these studies have used elementary, secondary, and university students, in addition to adults, no attempt has been made to determine whether these motives are dynamic, and whether the same motives are equally important determinants of behavior with different aged populations.

The present study will describe the results of a study that addressed these issues. As a starting point, I took Schilling and Carman's (1978) two-factor motivation model and attempted to determine whether these two dimensions adequately described the data. Students from five different age groups completed the questionnaire. This approach allowed me to determine whether differences existed across the various age groups and which motive accounted for the most variance in consumption and behaviors associated with alcohol consumption.

METHOD

Subjects

A total of 998 students were used in the analyses for this study. One hundred and twelve were enrolled in grade 7, 150 in grade 8, 253 in grade 9, 337 in grade 10, and 146 in a community college. The students in grades 7 to 10 (inclusive) were part of a larger study and were selected for this study because they responded affirmatively to a written question: "Do you drink alcoholic beverages?" Only those students who answered yes to this question were allowed to answer the section dealing with motivation and behavior. The community college students were tested solely for the purposes described in this paper. The data were collected in 1979 from the elementary and secondary school students and in 1980 from the students in the community college.

Materials

1. <u>Motivation</u> -- 19 items, each expressing a reason for drinking, were presented to the students. The students' responses were made on the basis of how important they felt each reason was. The 19 items, which are described elsewhere (Gliksman, Smythe, Gorman, & Rush, 1980) were constructed in order to assess the two motives described by Schilling and Carman (1978).

2. <u>Consumption</u> -- This was based on the students' self-reports of their consumption for every day of the week prior to the test session. For purposes of this study, the total week's score served as the measure of consumption.

3. <u>AAIS</u> -- The Adolescent Alcohol Involvement Scale was developed by Mayer and Filstead (1979), and is used as an index of alcohol abuse. The community college students did not complete this measure.

RESULTS AND DISCUSSION

The preliminary analyses consisted of five principle axis factor analyses, one for each sample. The intent of this procedure was to determine whether the different age samples produced similar dimensions of motivation and what these dimensions were. The factor analyses showed great similarity across all samples. Three factors emerged, composed of 17 of the 19 variables used which accounted for approximately 50% of the variance in each sample. Factor 1 is composed of nine items and was labeled <u>Anxiety Reduction Motivation</u>. This factor describes a motivation to drink in order to relax, remove depression, to get to sleep, to forget one's worries, to become less tense, to relieve boredom, for energy, because of a habit, and to prepare one to face things.

Factor 2 is composed of four items, all of which describe drinking because of social circumstances and has been labeled <u>Social Pressure Motivation</u>. These items specify that these people drink because it is difficult to refuse, because most of their friends drink, because other people are watching, and because they feel it is sophisticated to drink.

Factor 3 is also composed of four items and has been labeled <u>Personal Enjoyment Motivation</u>. The items composing this dimension involve drinking because of taste, because of thirst, to feel good, and for enhancement of meals.

The internal consistency reliabilities for these dimensions are good, with mean Cronbach Alpha Coefficients of .86, .77, and .53 for Anxiety Reduction, Social Pressure, and Personal Enjoyment, respectively.

Having established the reliability of the scales, we proceeded to determine the role each motive played on the two criteria, consumption and AAIS. The results of the regression analyses are presented in Table 2.

It is apparent from Table 2 that certain patterns are emerging. Discussion will center on the results of the individual consumption and AAIS regressions and then on the similarities between the two sets of data.

1. Consumption

In grades 7 and 8, the most important determinant of consumption is <u>Anxiety Reduction</u>. Although <u>Social Pressure</u> is marginally significant in

48

Table 2. Regression Analyses of Motivation Dimensions on
Consumption and AAIS

Grade	Consumption				AAIS			
	Order	r	P Value of Beta	R	Order	r	P Value of Beta	R
7.	Anxiety	.39	.05	.39	Anxiety	.58	.01	.58
	Social	.35	.09	.42	Social	.52	.01	.62
	Personal	.34	NS	.43	Personal	.40	NS	.62
8.	Anxiety	.35	.01	.35	Anxiety	.58	.01	.58
	Social	.18	NS	.35	Social	.35	.06	.59
	Personal	.18	NS	.35	Personal	.34	NS	.59
9.	Personal	.42	.01	.42	Anxiety	.44	.01	.44
	Anxiety	.35	.02	.44	Personal	.38	.01	.47
	Social	.08	NS	.45	Social	.13	NS	.48
10.	Personal	.24	.01	.24	Anxiety	.34	.01	.34
	Anxiety	.23	.01	.28	Personal	.30	.01	.38
	Social	.01	.04	.30	Social	.20	NS	.39
C.C.	Anxiety	.39	.01	.39				
	Personal	.26	.07	.41				
	Social	.24	NS	.42				

grade 7 it is nonsignificant in grade 8. In addition, Personal Enjoyment is non-significant for both these grade levels. In grades 9 and 10 both Personal Enjoyment and Anxiety Reduction are significant predictors of alcohol consumption with Personal Enjoyment being the more important contributor. Although Social Pressure adds significant additional variance in grade 10, it does not do so in grade 9. In the community college situation, Anxiety Reduction plays a significant role, while Personal Enjoyment plays a less important role.

It is apparent that great similarities exist between students in grades 7 and 8 and between students in grades 9, 10, and the community college. The Personal Enjoyment motive does not begin to contribute to the prediction of alcohol consumption until grade 9, at which point it contributes significantly for what appears to be many years. It is interesting to note that grade 9 marks the transition from elementary to secondary school, and also to the importance of doing things for personal enjoyment.

It is also interesting to speculate on the reason why Anxiety Reduction plays a significant role in the prediction of consumption with young students in grades 7 and 8 who, one would think, have fewer anxieties and less experience with alcohol than older individuals. This result may reflect the fact that these students do not know why they are drinking. Their responses may reflect the reasons given by older people like their parents or young adults (as evidenced by the results of the community college students).

2. AAIS

The results of the analyses with the measure of adolescent involvement with alcohol show that Anxiety Reduction motivation is most predictive of alcohol involvement across all four grade levels. However, Social Pressure contributes significant variance in grades 7 and 8, but not in grades 9 and 10. Conversely, Personal Enjoyment motivation adds to the prediction of the AAIS in grades 9 and 10, but not in grades 7 and 8. The AAIS measures past involvement with alcohol. The results suggest that the behaviors measured by the AAIS result from alcohol consumption which is initially influenced by the motives measured in this study (primarily <u>Anxiety Reduction</u>).

Over all, the results strongly support the notion that people's beliefs about the tranquilizing effects of alcohol are related to consumption and problems arising from alcohol use. However, they also suggest that with individuals approximately aged 15 and older, <u>Social Pressure</u> does not relate to either variable. <u>Social Pressure</u> is related in grades 7 and 8. These same individuals also drink because of the personal enjoyment they derive from the consumption of alcohol.

CONCLUSION

The results of this preliminary investigation cannot be considered conclusive, but merely suggestive of several things. First, they suggest that the consumption of alcohol is not affected by an individual motive, but by the interaction of several motives, whose importance varies as the individual matures. Furthermore, they suggest that the role of social pressure may exert its greatest influence on younger students (e.g., students in elementary school). By the time these students reach secondary school they seem to underscore the importance of social pressure for their drinking. At this stage of their development the <u>Personal Enjoyment</u> motive is the best predictor of alcohol consumption.

Second, the results have ramifications for alcohol education programming in the school systems. The results suggest that all programs, regardless of the age of the students being addressed, should emphasize that alternatives to alcohol exist which can reduce anxiety. After all, it is the belief that alcohol reduces anxiety that appears to play an important role in the alcohol consumption of individuals of all ages. In grades 7 and 8 emphasis in the programs should also be placed on resistance to social pressure. These students seem to be susceptible to pressure from their peers in influencing their drinking. In the upper grade levels, however, emphasis should shift to alternatives to drinking for enjoyment in certain situations because these students seem to place a great deal of importance on their personal enjoyment. Classroom teachers can capitalize on this knowledge and emphasize alternatives to drinking within the content of an alcohol education program.

Finally, the present results indicate an area where further research in this domain is required. More extensive research on the motivational domain should be conducted in order to produce a more exhaustive list of motives and examine their relationships to each other and to alcohol consumption and to abstinence (although different motives must be assessed for abstinence). In addition, the role of attitudes on motivation and consumption should be more fully investigated. These future studies should be longitudinal in nature to allow us more confidence in discussing developmental aspects of motivation and consumption.

REFERENCES

Barnes, C. P., & Olson, J. N., "Usage patterns of nondrug alternatives in adolescence," <u>J. Drug Educ.</u>, 7:359-368, 1977.
Bowker, L. H., "Motives for drug use: An application of Cohen's typology," <u>Int. J. Addict.</u>, 12:983-991, 1977.
Cutter, H. S. G., & Fisher, J. C., "Family experience and the motives for drinking," <u>Int. J. Addict.</u>, 15:339-358, 1980.
Gliksman, L., Smythe, P. C., Gorman, J., & Rush, B., "The adolescent alcohol questionnaire: Its development and psychometric evaluation," <u>J. Drug Educ.</u>, 10:209-227, 1980.

Hanson, D. J., "Social norms and drinking behavior: Implications for alcohol and drug education," J. Alcohol and Drug Educ., 18:18-24, 1973.

Jessor, R., Carman, R. S., & Grossman, P. H., "Expectations of need satisfaction and drinking patterns of college students," Quart. J. Studies on Alcohol, 29:101-116, 1968.

Jessor, R., & Jessor, S. L., Problem Behavior and Psychosocial Development: A Longitudinal Study of Youth. New York: Academic Press, 1977.

Kohn, P. M., & Annis, H. M., "Drug use and four kinds of novelty-seeking," Brit. J. Addic., 72:135-141, 1977.

Mayer, J. E., & Filstead, W. J., "The Adolescent Alcohol Involvement Scale: An instrument for measuring adolescents' use and misuse of alcohol," J. Studies on Alcohol, 40:291-300, 1979.

Schilling, M. E., & Carman, R. S., "Internal-external control and motivations for alcohol use among high school students," Psychological Reports, 42:1088-1099, 1978.

Scott, E. M., "Young drug abusers and non-abusers: A comparison," Int. J. Offender Ther. & Comp. Criminol., 22:105-114, 1978.

Shearn, C. R., & Fitzgibbons, D. J., "Survey of reasons for illicit drug use in a population of youthful psychiatric in-patients," Int. J. Addic., 8:623-633, 1973.

Weinstein, R. M., "The avowal of motives for marijuana behavior," Int. J. Addic., 13:877-810, 1978.

SENSATION SEEKING, ANXIETY, AND RISK TAKING IN THE ISRAELI CONTEXT

Steven E. Hobfoll,[1] T. Rom,[2]
and Bernard Segal,[2]

[1]Ben Gurion University of the Negev
Beersheva, Israel
[2]University of Alaska
Anchorage, Alaska

Sensation seeking may be viewed as the need or tendency for individuals to seek novel or risky stimulating experiences. Research on sensation seeking suggests that, as a general drive or desire, if other exciting avenues were available to youth, these might also be pursued. For example, a number of studies have shown that participation in risky occupations or sports is related to high sensation seeking (Bacon, 1974; Hymbaugh & Garret, 1974). Sensation seeking may even be related to positive traits of curiosity and self-exploration (Segal, Cromer, Hobfoll, & Wasserman, in 1982 a,b; Segal, Huba, & Singer, 1980).

Among some samples in a number of studies drug use has been shown to be a product of a desire or a drive for increased stimulation (Segal et al., in press a,b; Hobfoll & Segal, in press) and, among other samples, drug use has been shown as a chemical solution that decreases uncomfortable anxious emotionality (Conger, 1951; Masserman, Jacques, & Nicholson, 1945). Most of the more recent research has been carried out on samples of young people within the United States, where drugs have been readily available for high sensation-seeking or troubled youth.

Although anxiety may be related to the use of sedative type drugs [approach behavior] (Hobfoll & Segal, 1983), Zuckerman (1976) has hypothesized that it is related to the avoidance of the type of behaviors to which high sensation seekers are drawn. So, whereas sensation seeking is related to the approach toward dangerous and exciting behavior, anxiety may be related to the avoidance of these behaviors (with the exception of use of tension reducing drugs).

The comparatively low rates of drug use, including hashish and alcohol, among Israeli youth (Berman, 1972; Yavetz & Shuval, 1980) may be, in part, a by-product of both the prevailing taboo on drugs and the availability of other experiences which offer exciting alternatives. With respect to the latter, volunteering for combat units within the required military service period was seen as a possible outlet for Israeli youths, especially those who are high in sensation seeking. Additionally, those who do use drugs may be more high trait-anxious youths who are seeking a chemical solution to their discomforting emotionality (Spielberger, 1980, p. 90).

Applying these assumptions to Israeli youth culture, a number of hypotheses were formulated:

1. It was predicted that sensation seeking would be predictive of participation in high risk or stimulating behavior.

2. More specifically, it was predicted that for an Israeli university sample, sensation seeking would be related to participation in special combat units of the armed service and to participation in risky and stimulating sports.

3. It was predicted that those students who indicated that they used drugs and alcohol would be higher on trait anxiety than nonusers, but that high trait-anxious persons would avoid dangerous sports and combat military participation. Trait anxiety is defined as the tendency of an individual to experience states of uncomfortable anxious emotionality over time.

METHOD

Sample

Forty male undergraduates from a variety of faculties of study were tested in the winter of 1980/1981 during an English class required of all students. One subject requested not to participate in the study and two subjects did not come to the second class, in which the Anxiety Scale was administered. Failure to answer more than three questions on the anxiety scale resulted in the nonuse of one of these tests. The average age of this sample of students was 21.38. This is somewhat older than comparative American university or military recruit samples due to the minimum mandatory three years of military duty for Israeli males following high school, and the increasingly common pattern of one year taken to travel or work following demobilization. No additional demographic data were collected for this group in order to be more convincing as to the guarantee of anonymity due to the sensitive area of inquiry (both drugs and military). It can be assumed from experience with similar samples at this university that many of these students were married, and that almost all were working in addition to studying.

Instruments and Procedure

Each student completed a brief assessment battery consisting of the Interest and Preference Inventory (IPI) (Segal, 1979), Hebrew version (Hobfoll & Rom, 1980), the trait measure of the Spielberger, Gorsuch, & Lushene State-Trait Anxiety Inventory (STAI), Hebrew version (Teichman & Melanik, 1978), and a behavioral inventory which was designed for the present study to assess participation in stimulating and high risk behaviors. Participation in this study was voluntary and anonymous.

The IPI is an empirically developed scale developed to assess one's interest in pursuing varied and unusual experiences. The measure is analogous to Zuckerman's (1979) Sensation Seeking Scale in that it represents the extent to which one strives for new experiences or stimulation. The scale (Form I) is a 45-item true/false questionnaire which yields a score representing a tendency toward high or low experience seeking. The Hebrew translation had to substantially change some questions due to cultural differences and the use of idioms. Correlation between the two scales for a separate sample of 50 subjects was found to be only moderate ($r = .51$, $p < .001$). The fact that the reliability sample expressed considerable difficulty with the idioms in the English version is seen as the primary factor in limiting this correlation (e.g., I like my friends to be "straight").

The STAI is a widely used measure of state (situational) and trait anxiety which has been shown to be a valid instrument in numerous studies in the U. S. and more recently in Israel (Hobfoll, Anson, & Bernstein, 1983; Margalit, Teichman, & Levitt, 1980). The 20-item measure assesses the general trait of anxiety by having subjects indicate the degree (never, sometimes, usually, almost, always) to which a series of statements is descriptive of them. Trait anxiety is a predictor of the frequency andmagnitude for which an individual will have elevations in state anxiety reactions across situations (Spielberger, 1966).

The 18-item behavioral scale was constructed for this study to assess former participation in risk taking and stimulating behavior in the areas of military, sports, and substance use, and is presented in Table 3. Weightings were assigned a priori in order to give greater weight to what as subjectively assessed by the authors as being riskier or more taboo behaviors. These weightings are listed in parentheses in Table 3. More dangerous ("hard") drugs were given greater weight, as were more potentially dangerous sports and combat-oriented military activity. The drug items were weighted to give emphasis to continued drug use rather than occasional or one-time experimentation, although experimentation does receive a somewhat higher score than non-use. Drugs of use were almost exclusively hashish with only three subjects having once used valium or opiates. While this study used this scale as a criterion measure, it was nevertheless assumed that the study's results might also lead to revisions in the scale itself for future research.

RESULTS

Means and standard deviations for all measures are presented in Table 1. In order to analyze the data, scores on the behavioral scale and its three subcategories (military, sports,and substance use) and the IPI and STAI trait scale were compared using Pearson product-moment correlations. The results of these analyses for all pairs of scores are presented in Table 2. The .05 level of significance was assigned as acceptable, but less convincing results are also presented for the reader's interest with their corresponding significance levels.

The behavioral subscale scores for military and sports were highly correlated with the behavioral total score, but negatively correlated with the substance use score, as shown in Table 2. The military and sports subscales were also strongly correlated to each other.

Sensation seeking as measured by the IPI was significantly correlated with the behavioral total score; its correlation with the military subscale,

Table 1. Means and Standard Deviations for Scores on Trait Anxiety, Sensation Seeking (IPI), and Risk Taking·Behavior

	Mean	SD
Trait anxiety (STAI)	43.40	3.21
Sensation seeking (IPI)	33.33	9.31
Behavioral total	20.05	6.70
Military	8.11	3.31
Sport	6.66	3.10
Substance	6.36	1.02

Table 2. Pearson Correlations for Behavioral and Personality Measures

	Behavioral total	Military	Sport	Substance	Trait anxiety
Military	.81 P < .001				
Sport	.80 P < .001	.47 P < .002			
Substance	.04 P < .41	.12 P < .24	.39 P < .01		
Trait anxiety	.00 P < .49	−.04 P < .40	−.06 P < .37	.24 P < .09	
Sensation seeking	.29 P < .04	.24 P < .08	.06 P < .37	−.05 P < .39	.04 P < .41

Table 3. Pearson Correlations for Item Behavioral Scores
with Sensation Seeking and Trait Anxiety

	Items (weightings)[+]	Sensation seeking	Trait anxiety
1.	During military, did you volunteer for dangerous duty? (3)	.16	−.12
2.	Were you in a performing unit? (3)	−.07	.02
3.	If yes, did you volunteer to be in this unit? (3)	.36**	.13
4.	Were you in special combat forces? (3)	.23*	.05
5.	Did you volunteer for this unit? (3)	.06	.07
6.	During a dangerous assignment, did you feel excited (with positive connotation)?	.14	−.21
7.	Do you fly gliders? (3)	.14	−.21
8.	Do you play soccer? (1)	−.10	.22
9.	Do you scuba dive? (3)	.28**	−.04
10.	Do you play basektball? (1)	.14	.04
11.	Do you mountain climb? (3)	.13	.08
12.	Do you race cars? (3)	.05	−.25
13.	Have you ever used drugs? (1)	−.02	.01
14.	Which drugs? (hashish-1, pills-2, opiates-3, highest scored)	.09	.26*
15.	How often? (once or twice-1, a few times-2, once a month-3, once a week-4)	−.15	.28*
16.	Do you drink? (1)	−.23*	−.12
17.	How often? (as item 15)	.02	.15
18.	Have you ever been "drunk?"	−.03	.03

+   Items keyed such that positive correlation reflects
    behavioral participation.
*   P < .10.
** P < .05.

while approaching significance, was not significant. The IPI has zero-order relationships to the sports and substance use categories.

The STAI trait scale was not found to be correlated with the overall behavioral score. Trait anxiety had a negligible negative correlation with the military subscale score and was not found to be related to sports participation. The positive correlation of trait anxiety with substance use, while approaching significance, was also not significant.

Item level correlations for behaviors with sensation seeking and trait anxiety are presented in Table 3. Alpha was defined as .10 for these analyses due to the limitations of the yes/no format on correlations. Volunteering to be in a performance (combat) unit, participation in combat forces, scuba diving (the only risky sport commonly practiced in Israel), and not having drunk alcohol were significantly positively correlated to sensation seeking. Avoidance of car racing, use of "harder" drugs, and frequency of drug use were significantly correlated to trait anxiety.

DISCUSSION

The behavioral measure appears to follow the hypothesized direction. Military and sports participation were found to be positively related while substance use was negatively related to military and sports participation. In a sense, this supports the authors' assumption that drug usage in Israel was seen as taboo for this sample. In this university student sample those who chose to participate in socially accepted behavior such as combat units in the military or risk taking or stimulating sports tended not to be the same young men who use drugs.

Correlations between the two personality measures and the behavioral measure were modest. While the overall correlation between the behavior score and sensation seeking was significant, the relationship with trait anxiety was not. As predicted, sensation seeking was found to be related to the general score on the risk behavior scale. Further inspection suggests that this relationship is based, for the most part, on the positive relationship between military risk-taking behavior and sensation seeking. Sensation seeking was not found to be related to overall substance use, but was related to avoidance of alcohol. The lack of significant findings for sports in general were precluded due to the low number of subjects participating in the riskier sports contained in the scale. It may be noted, however, that scuba diving, the only risky sport commonly practiced in Israel, was related to sensation seeking.

Sensation seeking may not have been related to substance use because of the types of drugs used by this sample. As found in previous research, sensation seeking is not necessarily predictive of the use of alcohol and sedative drugs. In the current sample a very limited use of any high sensation drugs, such as stimulants of hallucinogens, was reported. Thus, the findings for sensation seeking and drug use may be a simple confirmation of past research. Had hallucinogens and amphetamines been more readily available in Israel, perhaps a relationship between sensation seeking and drug use would have been found.

As predicted, trait anxiety was found to be related to substance use, but was not significantly correlated to overall avoidance of military or sports risk taking. Those who scored higher on trait anxiety were more likely to score higher on the substance use subcategory, but the level of correlations was not significant. Item analysis of this subscale found trait anxiety to be significantly correlated to the use of "harder" drugs and the more frequent use of drugs, but not significantly correlated to use versus nonuse. It is interesting to further note that trait-anxious individuals did not feel excited when on dangerous missions, and avoided car racing and glider flying.

The findings of the current study tend to support Zuckerman's (1976) two-factor (sensation seeking, anxiety) theory of approach or avoidance from novel or risky situations with regard to sensation seeking, but not necessarily with regard to anxiety. As predicted by the two-factor theory, sensation seeking did relate to participation in risky behavior especially in the military context. Trait anxiety was related to avoidance of some risky behaviors, as predicted by the two-factor theory, but was related to approach to drug use, perhaps due to the drugs' tension-reducing effect.

These findings have implications for future cross-cultural research. While sensation seeking tendencies may be channeled into military activities in a developing country at war, they may not necessarily be directed into other nondrug risk taking in a developed country at peace. Also, the pace of Israeli life in general may have been a factor affecting sensation seeking. Future research between the U. S. and Israel and other countries might offer interesting insights into these questions.

REFERENCES

Bacon, J. Sensation levels for members of high-risk volunteer organiza-
    tions. Unpublished manuscript, 1974. (cited in Zuckerman, 1976).
Berman, Y. Drug Abuse in Israel. Ministry of Social Welfare, Jerusalem,
    Israel. No. 5432, 1972.
Conger, J. J. The effects of alcohol on conflict behavior in the albino
    rat. Quarterly Journal of Studies on Alcohol, 12:1-29, 1951.
Hobfoll, S. E., and Rom, T. Interest and Preference Inventory: Hebrew
    version. Unpublished manuscript, Ben-Gurion University of the Negev,
    Beersheva, Israel.
Hobfoll, S. E., and Segal, B. A factor analytic study of the relationship
    between sensation seeking, trait anxiety, and drug use among detained
    and adjudicated adolescents. International Journal of the Addictions
    18:539-549, 1983.
Hobfoll, S. E., Anson O., & Bernstein, J. The effect of consecutive ego-
    threats on high versus low trait anxious individuals. In R. Schwarzer
    and C. D. Spielberger (Eds.), Advances in Test Anxiety Research, Vol.
    2. Hillsdale, New Jersey: Erlbaum Associates, 1983, pp. 81-86.
Hymbaugh, K., and Garret, J. Sensation seeking among skin divers.
    Perceptual and Motor Skills, 38:118, 1974.
Margalit, C., Teichman, Y., and Levitt, R. Emotional reaction to physical
    threat: Reexamination with female subjects. Journal of Consulting and
    Clinical Psychology, 3:403-404, 1980.
Masserman, J. H., Jacques, M. G., and Nicholson, M. R. Alcohol as a preventive
    of experimental neuroses. Quarterly Journal of Studies on Alcohol,
    6:281-299, 1945.
Segal, B. Interest and Preference Inventory. Unpublished manuscript,
    University of Alaska, Anchorage, 1979.
Segal, B., Cromer, F., Hobfoll, S. E., and Wasserman, P. Z. Reasons for
    alcohol use by detained and adjudicated juveniles. Journal of Drug and
    Alcohol Education, 28:53-58, 1982(a).
Segal, B., Cromer, F., Hobfoll, S. E., and Wasserman, P. Z. Patterns of
    reasons for drug use among detained and adjudicated juveniles.
    International Journal of Addictions, 17:1117-1130, 1982(b).
Segal, B., Huba, G. J., and Singer, T. L. Drugs, Daydreaming, and
    Personality: A Study of College Youth. Hillsdale, New Jersey:
    Erlbaum Associates, 1980.
Spielberger, C. D. Theory and research on anxiety. In C. D. Spielberger
    (Ed.), Anxiety and Behavior. New York: Academic Press, 1966.

Spielberger, C. D. Understanding Stress and Anxiety. New York: Harper and Row, 1980.

Spielberger, C. D., Gorsuch, B. L., and Lushene, R. E. Manuel for the State-Trait Anxiety Inventory. Palo Alto, California: Consulting Psychological Press, 1970.

Teichman, Y., and Melanik, C. Spielberger State-Trait Anxiety Inventory: Hebrew version. Unpublished manuscript, University of Tel-Aviv, Israel, 1978.

Yavetz, R., and Shuval, Y. Use of Drugs Among Adolescents in Israel, Hebrew University, School of Medicine, Hadassah Hospital, Jerusalem, 1980.

Zuckerman, M. Sensation seeking and anxiety traits and states, as determinants of behavior in novel situations. In L. G. Sarason and C. D. Spielberger (Eds.), Stress and Anxiety, Vol. 3. Washington, D.C.: Hemisphere, 1976, pp. 141-170.

Zuckerman, M. Sensation Seeking. Hillsdale, New Jersey: Erlbaum Associates, 1979.

RELATIONSHIP BETWEEN HEAVY DRUG AND ALCOHOL USE

AND PROBLEM USE AMONG ADOLESCENTS

Helene Raskin White

Center of Alcohol Studies
Rutgers University
New Brunswick, NJ

INTRODUCTION

Adolescent drinking and drug use[1] has attracted a lot of attention from policy makers, researchers, and the public at large. There are concerns about use because of the unknown effects of alcohol and drugs on maturation in terms of psychological and physiological development and the acquisition of skills such as coping. Of greater importance, however, is the issue of problem use called abuse, misuse, or heavy use.

Some researchers claim that any alcohol use by a teenager is misuse because it is illegal behavior (Marden & Kolodner, 1977). These researchers would probably make the same claim regarding drug use given that it is also illegal. Other investigators define heavy use as abuse. It is our contention that simple use does not equal abuse and that even "heavy" use may not be problematic for every teenager. Therefore, what is needed is a measure of problem use applicable to teenagers. An operational definition is necessary in order to examine the relationship between problem use and other variables for the purpose of elucidating the etiology of alcohol and drug problems and identifying risk factors. Only through such empirical investigations can we hope to improve our prevention and intervention programs.

Much of the research to date has relied upon "problems associated with" as a standard nonclinical measure of abuse of a substance (Akers et al., 1979). A handful of consequences are asked about. But these questions are formulated arbitrarily; there is no rationale for their inclusion. Each study uses a different set of questions and there is no commonality in the final definitions they arrive at. Therefore, there is no generalizability across studies (Mayer & Filstead, 1980). In a sense, these measures of consequences are measures of single events rather than conditions, so they measure drug problems but not necessarily problem drug users.

Also, these types of measures use arbitrary cutoff points as indicants of problem use. Some studies stipulate only one occurrence of a consequence (e.g., Park, 1958), while others specify a number of times in a number of areas. For example, Donovan and Jessor (1980) require an adolescent to experience problems two or more times in three or more areas to be defined as a problem drinker. One study of alcohol problems in college students counted some problems if they were experienced at least "sometimes" while drinking and other problems only if they were experienced "frequently" (Wechsler & Rohman, 1981). Likewise, the time frame chosen is often arbitrary ranging from ever, (O'Donnell et al., 1976) to within the last four weeks (Smart et al., 1978).

Another frequent measure of problem use focuses on use patterns, that is, the frequency and quantity of use. Some arbitrary quantity (e.g., 2 or more, 4 or more, 5 or more drinks) at least once a week or more than once a week is chosen as indicating "heavy" or "problematic" use (Straus & Bacon, 1953; Wechsler & Rohman, 1981; Rachal et al., 1980, respectively). Again, the amount and the time frame chosen are capricious and there is a lack of commonality and generalizability across studies.

In addition to quantity and frequency, many studies define problem use by an arbitrary number of times high, drunk, or stoned over a specific time frame. In a national youth study, problem drinkers were defined as those adolescents who were drunk six or more times in the last year (Rachal et al., 1980). Interestingly, these same researchers, in a national study conducted four years earlier, defined problem drinkers as those adolescents who were drunk four or more times in the last year (Rachal et al., 1975).[2]

Unlike alcohol studies, drug researchers rarely use a quantity/frequency measure or specific number of times high measure as an indicator of problem drug use. This absence may reflect a more accepted feeling that any illicit drug use is problematic. It may also result from an emphasis on examining adolescent drug users in general regardless of their level of problems.

Adolescent and adult substance use cannot be judged by the same set of standards. Quantity and frequency can have a different impact upon individuals at various developmental levels. The concepts of drug use experience and tolerance must also be considered. In addition, teenagers often experience problems because of their underage status. For example, they could get in trouble with police for drinking in a car, which might actually reflect normative behavior as judged by teenage standards. Also, adolescents may experience problems with their parents which adults would not encounter. Therefore, measures of adult problem drinking and drug use may not reflect the types of problems experienced by adolescents. Thus, while good typologies of problem drinking exist for (male) adults (e.g., Calahan, 1970), they are not applicable for teenagers.

The purpose of this paper is to develop a measure of problem use relevant for teenagers and then to determine the relationship between problems and use patterns.

METHOD

Design

The Rutgers Health and Human Development Project is a multiple cohort prospective longitudinal study which examines the acquisition, maintenance, and control of alcohol-and-drug-using behaviors. Each year, subjects within three selected birth cohorts are randomly selected by telephone survey from the State of New Jersey; these represent nominal 12-, 15-, and 18-year-old individuals. After the initial survey, subjects and their parent(s) are interviewed in the home by field staff. Following this contact, subjects come to the test site for a full day of testing, including: physical examination, blood tests, physiological and perceptual-behavior tests, psychological inventories and the completion of questionnaires eliciting information about social networks, parent perceptions, alcohol and drug use, criminal and delinquent activities, and many other aspects of the adolescent's lifestyle, behavior, and attitudes. Although participants are volunteers, self-selection may not threaten the samples' representativeness in terms of the variables of interest. Comparison of demographic characteristics and drinking behaviors of eligible households who agree to participate to those who refuse indicate high comparability (see Lester et al., 1984, for greater detail on subject selection, research design, and measures).

Sample

The sample consists of 472 New Jersey adolescents grouped into three birth cohorts: 1961 (17-18 year olds) n = 155 (M = 78, F = 77); 1964 (14-15 year olds) n = 163 (M = 82, F = 81); 1967 (11-12 year olds) n = 154 (M = 81, F = 73). The sample is predominantly white (90%). About half are Catholic (49%), the others are represented as follows: Protestant (31%), Jewish (9%), and another or no religion (11%). The median income of the sample ($22,940) is comparable to that of the State of New Jersey ($24,510). However, fewer numbers are represented in the lowest income levels, while the majority cluster in middle levels.

The majority of subjects have tried beer (81%), wine (78%), and distilled spirits (65%). Slightly less than half have tried marijuana (44%), while fewer than one-tenth have tried the rest of the substances, ranging from 9% for cocaine to less than 1% for heroin.

The sample for the present analysis consists of all subjects who reported at least minimal initial experience with the substances in each analysis. We have limited the analysis to the 1961 and 1964 cohorts because very few of the 12-year-olds (1967 cohort) had any drug experience, including alcohol outside of the family. The Ns for each analysis will be indicated in the tables.

Data Collection and Measurement

Data are obtained using self-report questionnaires administered on site. Efforts are made to establish rapport with subjects before questionnaire administration so as to assure maximum honesty. All questionnaires contain an identification number for longitudinal analysis. Subjects are reassured as to confidentiality of all data.

Other authors have addressed the issue of bias in self-reports in alcohol and drug surveys. The findings of Whitehead and Smart (1972) lend considerable credibility to self-reported drug use. Single et al. (1975) found that measures of adolescent use of illicit drugs at one point in time were reliable and valid, although these measures are less reliable over time. The authors report, however, that there is a greater probability of underreporting than overreporting. It has also been suggested that self-report measures of alcohol and drug use are slightly higher than those reported in personal interviews (Rachal et al., 1980). The data on alcohol and drug use in our sample are comparable to national surveys using other methods of data collection (e.g., Fishburne et al., 1980; Johnston et al., 1979).

Subjects who reported at least minimal experience with a substance (alcohol, marijuana, and other drugs besides marijuana) were asked to respond to two sets of items regarding negative consequences and concomitants of use. Items included were derived from over 50 questionnaires used in a variety of studies focusing upon adolescent and adult substance (primarily alcohol) use. Items were modified when necessary to reflect the adolescent lifestyle. The following are examples of the types of items:

Not able to do homework or study for a test.

Friends or neighbors avoided you.

Had withdrawal symptoms, that is, felt sick because you stopped or cut down on drinking (smoking, using drugs)[3]

For events projected to occur with relative low frequency, subjects reported the extent (i.e., number of times of occurrence) to which events had ever been experienced. The questions read: How many times did the following things happen to you while you were drinking alcohol or because of your drinking alcohol? A five-point scale comprised the response mode: Never, 1-2 times, 3-5 times, 6-10 times, more than 10 times. For events projected to occur with higher frequency, respondents were asked to report the frequency of each occurrence: How often do the following things happen to you while you are drinking alcohol or because of your drinking alcohol? Responses included: Never, Rarely, Sometimes, Often, and Almost always/always.

In addition, respondents completed a questionnaire detailing their use of 15 substances. Questions were asked about frequency, quantity, number of times high, etc. Substance use indices were used as composite scores of substance use involvement for the present analyses. The Substance Use Index (SUI) combines extent, frequency, recency, and quantity of use of alcohol, marijuana, inhalants, heroin, PCP, cocaine, psychedelics, and nonmedical use of analgesics, stimulants, sedatives, and tranquilizers. The SUI represents a composite score which reflects overall substance use involvement relative to the rest of the subjects in the sample (Pandina et al., 1981). In addition to the SUI, an alcohol use (combining beer, wine, and distilled spirits) and marijuana use index were similarly constructed.

DEVELOPMENT OF THE PROBLEM INDICES

Scores were derived by multiplying each of the potential 54 problems by the ordinal value (0 to 4) indicating the extent of frequency of occurrence reported by the subject, summing the total for all 54 problems, and finally dividing the summed total by 54. (The division step accounts for missing values.) The resultant problem index scores ranged from 0 to 4. Three problem indices were developed: (1) Alcohol Problem Index (API), (2) Marijuana Problem Index (MPI), and (3) Drug (other than alcohol and marijuana) Problem Index (DPI).

Separate analyses of the number of problems experienced by participants regardless of frequency were also conducted. The zero-order correlation coefficients for the relationship between the problem index scores and the number of problems experienced are shown in Table 1. Note that the lowest correlation coefficient is .87 and the correlations are extremely high. Given that the indices were developed based upon the number of problems and their frequency of occurrence, this finding is not at all surprising. It does suggest that the less "refined" measure, i.e., the number of problems, could have been used instead of the indices in the following analyses and would have provided almost identical results (see Table 1).

Table 1. Zero-Order Correlations of Problem Indices
with the Number of Problems

	1964 Male	1964 Female	1961 Male	1961 Female	Total
API with alcohol problems	.98*** (N=48)	.97*** (N=46)	.94*** (N=76)	.89*** (N=68)	.94*** (N=238)
MPI with marijuana problems	.91*** (N=25)	.92*** (N=35)	.87*** (N=57)	.95*** (N=48)	.89*** (N=165)
DPI with other drug problems	+	+	.97*** (N=20)	.92*** (N=19)	.94*** (N=39)

+ DPIs were not calculated for the 1964 cohort because the N of drug users was to small to permit meaningful analysis.

    * p ≤ .05
   ** p ≤ .01
  *** p ≤ .001

Table 2. Zero-Order Correlations between Problem Indices

	1964 Male	1964 Female	1961 Male	1961 Female	Total
API with MPI	.61** (N=24)	.45** (N=33)	.28* (N=58)	.52*** (N=48)	.42*** (N=163)
API with DPI	+	+	.27 (N=19)	.37 (N=18)	.20 (N=37)
MPI with DPI	+	+	.63** (N=19)	.63** (N=18)	.56*** (N=37)

+ DPIs were not calculated for the 1964 cohort because the N of drug users was too small to permit meaningful analysis.

    * p ≤ .05
   ** p ≤ .01
  *** p ≤ .001

RESULTS

Intercorrelations between Problem Index Scores

The intercorrelations between the three problem indices are presented in Table 2. The data indicate that the degree of alcohol problems is related somewhat to the intensity of marijuana problems, but not to other drug problems. Subjects who experience marijuana problems are also likely to experience problems with other drugs. In general, the correlations are higher for the older females than males (1961 cohort).

Sex and Birth Cohort Differences in Problem Indices

    Figure 1 shows the sex and birth cohort differences in the mean number
of problems experienced by subjects. Two-way analyses of variance were
performed on these mean differences. The 1961 males experienced far more
alcohol problems per person as compared to an average of 10 problems for the
other three groups. The sex-by-cohort interaction (F = 10.13, df = 2, p ≤
.001) as well as the sex difference (F =10.63, df = 1, p ≤ .001) and cohort
difference (F = 9.29, df = 1, p ≤ .001) are statistically significant (see
Figure 1).

    Note that the 1964 females are similar to the 1964 males, while for the
1961 cohort the sex difference is great. Also, the difference between
cohorts is apparent for the males, but not for the females. While there are
statistically significant sex differences for the number of marijuana
problems (F = 4.98, df = 1, p ≤ .05), cohort differences are not

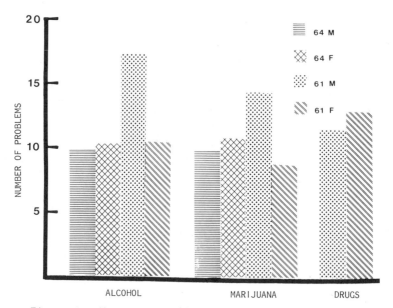

Figure 1. Mean number of problems by sex and cohort.

significant. The sex difference is again magnified in the 1961 cohort in
which the males averaged 14 marijuana problems as compared to 9 for the
females. For the 1964 cohort males ($\bar{x}$ = 10) and females ($\bar{x}$ = 11) were
nearly identical. Note that the older females experienced, on the average,
even fewer problems than both males and females in the younger cohort. The
number of drug problems and the Drug Problem Index were calculated only for
the 1961 cohort. The number of other drug users in the 1964 cohort was too
small to permit meaningful analysis. The difference between males ($\bar{x}$ = 12)
and females ($\bar{x}$ = 13) in the mean number of other drug problems is not
statistically significant.

    The sex and cohort differences in the problem index scores are
presented in Figure 2. Note that the histograms are almost identical to
those in Figure 1. The analyses of variance yielded similar results to
those discussed above (see Figure 2).

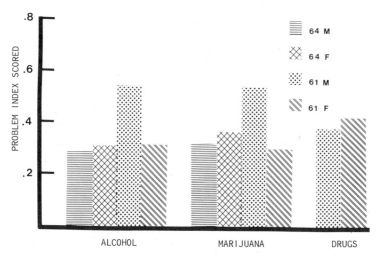

Figure 2. Mean problem index scores by sex and cohort.

Table 3. Zero-Order Correlations between Alcohol Problems
and Use Patterns

	1964 Male	1964 Female	1961 Male	1961 Female	Total
API with:	(N=48)	(N=46)	(N=76)	(N=68)	(N=238)
Beer					
How often drunk	.60***	.30*	.42***	.13	.38***
Number of times drunk past year	.44***	.48***	.51***	.41***	.52***
Extent (number of times ever)	.34**	.43**	.44***	.28**	.42***
Frequency	.39**	.45***	.31**	.33**	.42***
Quantity	.39**	.39**	.34***	.05	.33***
Total involvement	.44***	.45***	.35***	.20	.40***
Age first use outside family	-.27*	-.14	-.45***	-.32**	-.27***
Distilled spirits					
How often drunk	.55***	.27*	.34***	.14	.32***
Number of times drunk past year	.42***	.56***	.44***	.17	.42***
Extent (number of times ever)	.30*	.42**	.34***	-.11	.26***
Frequency	.27*	.53***	.16	-.02	.24***
Quantity	.47***	.37**	.11	-.05	.19**
Total involvement	.39**	.48***	.22*	-.08	.25***
Age first use outside family	-.32*	-.07	-.27**	-.11	-.11*
Alcohol use index	.45***	.41**	.22*	.09	.32***
Marijuana use index	.16	.36*	.50***	.27	.41***
Substance use index	.30*	.75***	.53***	.11	.44***

   * p ≤ .05
  ** p ≤ .01
 *** p ≤ .001

Table 4. Zero-Order Correlations between Marijuana Problems
and Use Patterns

	1964 Male	1964 Female	1961 Male	1961 Female	Total
MPI with:	(N=25)	(N=35)	(N=57)	(N=48)	(N=165)
How often stoned	.47*	.11	-.04	.25	.13
Number of times stoned past year	.08	.05	.25	.48***	.25***
Extent (number of times ever)	.35	.43**	.46***	.55***	.46***
Frequency	.26	.45**	.33**	.34*	.35***
Quantity	.14	-.05	.20	.29*	.19*
Age first use	-.34	-.13	-.40**	-.36**	-.25***
Marijuana use index	.21	.17	.30*	.39**	.30***
Alcohol use index	.30	-.04	.28*	.18	.22**
Substance use index	.30	.24	.30*	.46***	.35***

   * p ≤ .05
  ** p ≤ .01
 *** p ≤ .001

Relationship between Problem Indices and Use Patterns

The zero-order correlation coefficients and their level of significance for the relationship between API and various measures of alcohol use are presented in Table 3. In general, the API correlates more strongly with the measures of beer use than those for distilled spirits. This difference may be accounted for by the greater variability in beer use as compared to distilled spirits use. Most of the relationships between the API and wine use variables were so weak that they were not included in the table. It may be that variation in frequency, quantity, etc., of wine use is smaller than for beer and distilled spirits, thus accounting for lower correlations. The strongest relationships are observed for the API with number of times drunk on beer in the past year, frequency of beer use, number of times beer was ever used, and total beer use involvement (a weighted value combining frequency, recency, extent, and quantity of beer use relative to the rest of the subjects in the sample) (see Table 3).

All correlations are positive except for the age of first use, which indicates that the greater the involvement, the greater the number and frequency of problems, and that the younger the person first drank alcohol outside of the family, the more alcohol problems they have.

There are some differences in the strengths of the relationships by sex and cohort. The relationship between alcohol problems and use patterns is weakest for the 1961 females. The relationships with distilled spirits are stronger for the 1964 than the 1961 cohort.

Correlations between the API and the substance use indices were also calculated and are reported at the bottom of Table 3. The API correlates more strongly with total marijuana use involvement and overall substance use involvement than with total alcohol involvement. This finding is probably an artifact of the higher API scores as well as higher MUI and SUI scores for 1961 males. In fact, when the correlations are examined as a function of sex and birth cohort, the API/alcohol use correlations are higher than the API/marijuana use correlations for the 1964 cohort, while the reverse is evident for the 1961 cohort. However, the highest correlations are observed for the API with SUI for the 1964 females and the 1961 males.

The correlation coefficients between the Marijuana Problem Index and various measures of marijuana use are presented in Table 4. In general, these correlations are lower than those between alcohol problems and beer use. The highest correlations are for the number of times ever used and the frequency of use. All other correlations are relatively low, although all except one are statistically significant to at least the .05 level of significance (see Table 4).

The correlations are higher for the 1961 cohort than for the 1964 cohort and this difference may reflect greater variations in marijuana-using behavior among the older subjects. Contrary to the findings for alcohol, among the 1961 cohort, females display slightly stronger relationships between problems and use than males (except for age of first use). The correlations between the marijuana problem index and total marijuana use involvement are statistically significant but low for the 1961 cohort and are not significant for the 1964 cohort. The relationship between the MPI and total alcohol use is relatively weak while the reverse was noted in Table 3. A statistically significant relationship between the MPI and overall substance use is observed for the 1961 cohort. For all groups the MPI/SUI correlation is higher than the relationship between the MPI and the other two overall use measures.

The Drug Problem Index was correlated with measures of overall alcohol, marijuana, and substance use. None of these correlations are statistically significant (and thus are not presented in a table).

SUMMARY AND CONCLUSIONS

Statistically significant relationships were observed between alcohol problems and beer use, while fewer significant relationships existed between alcohol problems and distilled spirit use. There were even fewer significant relationships between the MPI and measures of marijuana use and none between the DPI and substance use. The highest correlation coefficients for any sex-cohort groups was .60, several were .50, but the majority was less than .40. Thus, at the very most, some use variable can account for 36% of the variance in problem scores and most use variables account for less than 16% of the variance. These findings indicate that although there is a definite relationship between problems and use patterns, other factors are important in distinguishing between heavy and problem use.

Our task is to identify those factors so that we might be better able to predict teenage problem users and design prevention approaches. Although it is important to stress moderate use patterns among adolescents in order to prevent problems, we must also concentrate on other factors which are yet to be uncovered.

Perhaps certain types of problems such as withdrawal symptoms, loss of weight, etc., may be directly related to use patterns, while other problems such as school problems or problems with parents, etc., may be related to other variables such as personality characteristics, family environment, or luck. In order to examine these possible differences, we have begun to develop a typology of teenage alcohol problems (to be followed by a typology of marijuana and other drug problems) similar to that developed for adults by Calahan (1970) and his associates. We believe that by classifying consequences of use into qualitatively meaningful categories, we will better be able to look at their relationship to use patterns as well as to other subject characteristics. Throughout our longitudinal study we hope to develop measures of problem use which are relevant to each stage of the life cycle.

NOTES

1.   Drug use refers to the use of illicit substances such as marijuana, cocaine, psychedelics, etc., and the nonmedical use of licit substances such as analgesics, amphetamines, barbiturates, etc. Substance use refers to drug and alcohol use.

2.   It has been argued that frequent drunkenness itself may represent problem or problem-prone behavior for the adolescent or for those with whom he or she interacts (Rachal et al., 1980). Yet, being drunk a couple of times a year may not be an indication of a serious behavior disorder. In fact, it may be normative for teenagers to get drunk -- is getting drunk once every other month a real problem? There appears to be no immediate or long-term harmful consequences for many adolescents who get drunk. In fact, intoxication may have certain benefits for teens, such as providing them with time out, peer solidarity, shared rituals, and enjoyable sensations (Finn, 1979). Blane and Hewitt (1977) state that intoxication is not a problem in and of itself, but it bears an association to problems consequent to the acute effects of alcohol and that frequent heavy drinking is the basic cause of adolescent and young adult drinking problems. The literature points to an association between alcohol problems and use patterns although this relationship is not as strong as one might predict. In the national study of teenage drinking, the correlation between frequency of drunkenness and frequency of alcohol-related negative events was low enough to suggest that they were not measuring the same thing (male $r = .53$; female $r = .44$). In fact, 22% of the males and 31% of the females had no problems, yet were defined as alcohol misusers because they got drunk six or more times in the previous year (Rachal et al., 1980).

3.   The list of consequences and concomitants is available from the author.

ACKNOWLEDGMENTS

Preparation of this manuscript was supported, in part, by a grant from the National Institute on Alcohol Abuse and Alcoholism (No. AA 03509-03). The author wishes to acknowledge Dr. Robert J. Pandina for his ideas and comments on earlier drafts of this paper, and Robert N. Krupnick for the computer analyses.

REFERENCES

Akers, R. L., Krohn, M. D., Lanza-Kaduce, L., and Radosevich, M. Social learning and deviant behavior: A specific test of a general theory. American Sociological Review, 44 (4):636-655, 1979.

Blane, H. T. and Hewitt, L. E. Alcohol and Youth: An Analysis of the Literature, 1960-1975. Washington, DC: National Technical Information Service, 1977.

Calahan, Don. Problem Drinkers. San Francisco: Jossey-Bass, 1970.

Donovan, J. E., and Jessor, R. Adolescent problem drinking; Psychosocial correlates in a national sample study. In Filstead, W. J. and Mayer, J. E. (Eds.), Adolescence and Alcohol. Cambridge, MA: Ballinger Publishing Co., pp. 9-20, 1980.

Finn, P. Teenage drunkenness: Warning signal, transient boisterousness or symptom of social change? Adolescence, 14:819-834, 1979.

Fishburne, P. M., Abelson, H. I., and Cisin, I. National Survey on Drug Abuse: Main Findings. Rockville, MD: National Institute on Drug Abuse, 1980.

Johnston, L. D., Bachman, J. G., and O'Malley, P. M. Drugs and the Class of
    '78: Behaviors, Attitudes, and Recent Trends. Rockville, MD:
    National Institute on Drug Abuse, 1979.
Lester, D., Pandina, R. J., White, H. R., and Labouvie, E. W. The Rutgers
    Health and Human Development Project: A longitudinal study in alcohol
    and drug use. In Mednick, S. and Harway, M. (Eds.), Handbook of Longi-
    tudinal Research. New York: Praeger Press, 1984.
Marden, P. G., and Kolodner, K. Alcohol Use and Abuse Among Adolescents,
    NCAI Report No. NCAI026533, Washington, DC: National Institute on
    Alcohol Abuse and Alcoholism, 1977.
Mayer, J. E., and Filstead, W. J. Empirical procedures for defining adole-
    scent alcohol misuse. In Filstead, W. J. and Mayer, J. E. (Eds.),
    Adolescence and Alcohol. Cambridge, MA: Ballinger Publishing Co.,
    pp. 51-68, 1980.
O'Donnell, J. A., Voss, H. L., Clayton, R. R., Slatin, G. T., and Room, R.
    G. Young Men and Drugs -- A Nationwide Survey (NIDA Research Monograph
    5). Rockville, MD: National Institute on Drug Abuse, 1976.
Pandina, R. J., White, H. R., and Yorke, J. Estimation of substance use
    involvement: Theoretical considerations and empirical findings. Int.
    J. Addic., 16(1):1-24, 1981.
Park, P. Problem Drinking and Social Orientation: A Sociological Study of
    Pre-Alcoholic Drinking. Unpublished dissertation. Yale University,
    1958.
Rachal, J. V., Williams, J. R., Brehm, M. L., Cavanaugh, B., Moore, R. P.,
    and Eckerman, W. C. A National Study of Adolescent Drinking Behavior,
    Attitudes and Correlates. Washington, DC: National Institute on
    Alcohol Abuse and Alcoholism, 1975.
Rachal, J. V., Guess, L. G., Hubbard, R. L., Maisto, S. A., Cavanaugh, E.
    R., Waddell, R., and Benrud, C. H. Adolescent Drinking Behaviors,
    Vol.1. The Extent and Nature of Adolescent Alcohol and Drug Use: The
    1974 and 1978 National Sample Studies. Washington, DC: National
    Institute on Alcohol Abuse and Alcoholism, 1980.
Single, E., Kandel, D., and Johnson, B. The reliability and validity of
    drug use responses in a large-scale longitudinal survey. J. Drug
    Issues, 5(4):426-433, 1975.
Smart, R. G., Gray, G., and Bennett, C. Predictors of drinking and signs of
    heavy drinking among high school students. Int. J. Addic.,
    13(7):1079-1094, 1978.
Straus, R. and Bacon, S. D. Drinking in College. New Haven: Yale
    University Press, 1953.
Wechsler, H., and Rohman, M. Extensive users of alcohol among college
    students. J. Studies on Alcohol, 42:149-155, 1981.
Whitehead, P. C., and Smart, R. G. Validity and reliability of
    self-reported drug use. Can. J. Criminology and Corrections,
    14(January):1-8, 1972.

THE SOCIAL NETWORKS OF DRUG ABUSERS

BEFORE AND AFTER TREATMENT

J. David Hawkins   Mark W. Fraser

Center for Social Welfare   Social Research Institute
  Research   School of Social Work
School of Social Work   University of Utah
University of Washington   Salt Lake City, Utah 84112
Seattle, Washington 98195

INTRODUCTION

The term "drug abuse treatment" is, in a sense, a misnomer. Drug
treatment programs do not seek merely to stop the illicit use of drugs,
but rather to assist people who no longer are functioning effectively
personally, socially, or economically in legitimate society and to alter
their patterns of living (Drug Abuse Council, 1980: 14). In this regard, it
has been widely recognized that social rehabilitation services are essential
to successful treatment (Bloom and Sudderth, 1971:172; Lewis and Sessler,
1980:120). To be effective, social rehabilitation services in drug
treatment should address the social factors that are related to the
initiation, maintenance, and return to drug abuse. Unfortunately, there is
not yet agreement about how social factors interact in the etiology of drug
use and abuse, and how they should be addressed in drug treatment. As noted
by the Drug Abuse Council 1980:5:

> The underlying social dynamics and problems that lead to drug
> misuse are so exceedingly complex so as yet to elude totally
> satisfactory solutions.

This chapter extends understanding of the relationships between social
interaction and drug abuse by using social network analysis methods in
research on street drug abusers. The descriptive study reported here seeks
to clarify how interactional and structural aspects of drug abusers' social
lives can be better considered and appropriately addressed in drug abuse
treatment. Following definitions of the population of interest and the
social network analysis approach used in the study, the design and
results of the study are presented, and the implications of the results
for drug abuse treatment are briefly explored.

The research reported in this paper was supported by Grant No. H81DA02071
from the Services Research Branch, National Institute.

POPULATION

The misuse of drugs is becoming recognized as a widespread phenomenon that transcends heroin addiction and the abuse of other "street drugs" (Drug Abuse Council, 1980:4). However, this paper is concerned primarily with the misuse of psychoactive drugs other than alcohol which are not legitimately prescribed by a physician. The focus includes the abuse of opiates, including heroin, which is a type of drug abuse which has received a preponderance of the attention of policymakers concerned with drug abuse in the United States (Lewis and Sessler, 1980). Sixty percent of the client slots nationally are designated for heroin abusers (National Institute on Drug Abuse, 1978). Those who use heroin and other opiates are the sole objects of the rehabilitation efforts of methadone maintenance and detoxification clinics, as well as pilot treatment projects using L-alpha acetyl methadol (LAAM). Opiate abusers represent a major portion of clients who enter residential drug-free therapeutic communities as well. Because the treatment of opiate abusers remains a major thrust of drug treatment efforts and expenditures, and because this form of drug misuse continues to be viewed as a serious problem in this society, this paper specifically investigates the social aspects of opiate abuse in certain sections.

Social Network Analysis

In the last decade, social network analysis has emerged as a tool for measuring the structural and interactional aspects of people's social environments. As defined by Mitchell (1969), a social network is a

...specific set of linkages among a defined set of persons, with the additional property that the characteristics of these linkages as a whole may be used to interpret the social behavior of the persons involved.

Social networks are distinct from social support. Networks are the connections between people (Hammer, 1981:47). Social network connections may support conforming, conventional patterns of behavior (Phillips, 1981:121), they may support patterns of deviance (Cohen, 1955), or they may provide little or no support at all. While network research is in its infancy (Leinhardt, 1977:4), two approaches have begun to emerge. The first defines a focal person or "ego" and identifies persons or "alters" who are network members, usually friends, family, peers, co-workers, etc. The second defines a distinct group of persons, and maps their patterns of interaction within prescribed boundaries. In drug abuse research there are few published network studies of either type. Killworth and Bernard (1974) applied network analysis to offenders, but their study was limited to a small prison living unit. Pattison et al. (1979) studied addict, alcoholic, normal, and a variety of psychiatric populations, but reported no sampling procedures and little empirical evidence to support their conclusions. The present study focuses on the personal or "ego-centered" networks of drug abusers.

THE STUDY DESIGN

The Sample

The subjects of this investigation were clients in four residential drug treatment centers (therapeutic communities) between November 1979 and June 1980. To guard against local geographic effects, residents in two treatment programs in a northwest city and two in a western urban area were purposively selected. Eligibility for the sample was defined as residence in treatment for three or more months. All eligible clients at two small treatment programs were sampled. A random sample of eligible clients

Table 1. Drug Use and Crime Participation Characteristics
of Sample (n = 106)

Drug use, ever, regular use or addiction	
heroin	69.8%
morphine, codeine, darvon	47.2%
amphetamines	66.2%
Age at first regular use	
heroin	14.3
amphetamines	11.7
Crime, ever	
drug-related	95.3%
burglary, robbery, theft	29.2%
none	4.7%
Arrests	
ever	96.2%
average number	11.0
average age at first arrest	15.9 years
Currently on probation	67.0%
Committed illegal act in	
30 days prior to treatment	74.5%

Table 2. Comparative Demographic Data of Two Samples
of Treated Drug Abusers (Percent)

Variable	Study sample (n = 106)	DARP study (n = 20,630)
Male	76	77
White	65	54
Black	20**	43
Chicano	10	10
Never married	56	48
Currently married	14*	28
Age, older than 25	64**	43
Parents marriage intact until subject aged 12	51	64
Previous drug abuse treatment	62	51
Age at first arrest, younger than 18	38	41
Cost of drug habit, more than $50/day	44	34

*χ^2, 1df, $p < .05$. **χ^2, 1df, $p < .001$.

stratified by time in treatment at two larger programs was also constituted. All sampled residents elected to participate in the research, producing a total sample of 106 respondents (Program One = 21; Program Two = 16; Program Three = 20; Program Four = 40).

As shown in Table 1, respondents have long histories of lifestyles characteristic of drug use "on the streets," including the sale and use of heroin, other opiates, and amphetamines, as well as barbiturates, cocaine, and hallucinogens. They are not incidental drug users. Many have also engaged in criminal behaviors more serious than the sale of illegal drugs. These include crimes against persons (murder, assault, rape), crimes of profit (robbery, burglary, shoplifting), and crimes against property (vandalism). Only a small percentage have never committed a crime beyond illegal drug use.

As shown in Table 2, subjects range in age from 16 to 48 years ($\bar{X}$ = 28.3), and 76% are male. Ten percent are Chicano, 20% are Black, 5% are Native American, Asian, or other, and the remaining 65% are white. Table 2 compares the subjects of this study with the Drug Abuse Report Program (DARP) national sample of 20,630 opiate users referred to drug treatment from 1969 to 1974 (Bale et al., 1980; Curtis and Simpson, 1977). The DARP study oversampled methadone maintenance programs which were not sampled in this study. Nonetheless, the two samples are similar in sex distribution and various measures of deviance. The present sample has a significantly lower proportion of Black respondents, and is slightly older. Fewer residents are currently married. Respondents appear to have experienced more family disruption during childhood and more previous involvement with drug treatment programs.

Overall, the current data, when compared with DARP data, are not inconsistent. The observed differences may be due to time of test or cohort effects. For example, given the price increases of drugs on the streets, the current respondents cannot be said to use more drugs than the DARP sample on the basis of a cost-of-habit variable. Similarly, differences in marital status may be due to changes in the social institution of marriage and may not reflect greater deviance within the current sample. The distribution within the race/ethnicity variable appears to corroborate others' observations that therapeutic communities tend to serve more whites, white methadone maintenance programs serve more minority clients (Bale et al., 1980). Finally, the fact that the DARP population is younger suggests a possible cohort effect. It is possible that fewer young people are becoming involved in street-level addict life styles, and that the population of addicts on the streets is getting older. Such speculation is beyond the domain of the current investigation. In summary, no great departures from the expected characteristics of drug abusers in treatment in 1980 are noted. Where departures from DARP data are observed, plausible alternative explanations exist. In general, sampling bias does not appear to pose a serious threat to external validity.

The prospective study design called for interviewing the first ten clients who left treatment from each of the four programs after collection of baseline date. The follow up sample from each program was: Program One = 12; Program Two = 12; Program Three = 11; Program Four = 14. Two residents from Program Two who were incarcerated within the first week after leaving treatment were excluded from the follow up sample. As shown in Table 3, 41 (83.7%) of the sampled clients were successfully interviewed at one month following treatment, and 43 (87.7%) of the sampled clients were interviewed at three months following treatment. These include approximately equal numbers of clients who graduated and those who left treatment before graduation. Thirty-eight (77.6%) of the 49 clients sampled for follow up were interviewed at both one month and three months following treatment.

The use of time as a basis for sampling from the initial respondent pool may introduce bias in the results of prospective follow up studies. The risk of oversampling those who left before completing treatment is minimized by the fact that respondents were initially interviewed at different points in their treatment careers. To ascertain the extent to which the sample of 38 respondents with complete follow up date can be viewed as an unbiased sample of the population of 106 clients, z-scores were calculated on 42 sociodemographic, drug, crime, and social network variable. The means of the sample (38) were compared with those of the population of 106 respondents. None of these were significant at the $\alpha = .05$ level using a 2-tailed test of significance. However, the means varied to a degree, though nonsignificantly, on three variables, as reported in Table 4. While their own drug use did not differ from that of the population before treatment, the sample followed may have had somewhat more deviant networks

Table 3. Follow-up Sample and Completed Interviews
(n = 49)

	Number sampled		Number completed one month follow-up interview		Number completed three month follow-up interview	
	Splits	Graduates	Splits	Graduates	Splits	Graduates
Program 1	5	7	3	7	3	7
Program 2	8	4	4	4	5	4
Program 3	6	5	4	5	5	5
Program 4	8	6	8	6	8	6
Totals	27	22	19	22	21	22
Grand totals	49		41		43	

Table 4. Differences in Means of the Most Divergent Pretreatment Characteristics of the Study Population and the Sample Followed (percent)

	Mean of population n = 106	Mean of follow-up sample n = 38	
Network density	50.4	58.7	z = 1.73
Proportion of daily and frequent drug users in network	42.5	50.8	z = 1.62
Proportion of network with negative attitudes toward drugs	42.1	35.8	z = 1.396

before treatment, as shown in Table 4. Their networks contained more daily and frequent drug users and fewer members with negative attitudes toward drugs.

Data Collection Method

Structured interviews were conducted at three points in time with each respondent. Data traditionally collected in studies of drug treatment program clients were collected at each interview. These include self-reports of the average frequency of use of ten illicit drugs. These data are used in the results reported below to compare the social networks of those who used any illegal opiates (heroin, illegal methadone, morphine, codeine, or other opiates) in the first three months following treatment with those who did not. A client history interview focusing on clients' experiences prior to treatment entry was administered while respondents were in treatment. Recall is a potential source of bias in these data, since respondents had been in treatment for varying lengths of time at the point of administration of this instrument. To minimize this bias the following procedures were used in collecting social network date in this interview. Respondents were asked to list on cards the first names of people with whom they had contact in the last four weeks before incarceration, hospitalization, or treatment. Because the reliability and validity of self-report of social contacts continues to be controversial (Bernard and Killworth, 1977; Killworth and Bernard, 1976, 1979), interviewers were trained to use frequent prompts to enhance the recall of participants. Respondents were asked about specific events and activities related to residence, employment, schooling, participation in group, clubs, or organizations, receipt of services, illegal activities, and drug use during the month specified. After each set of questions, respondents were asked to add personal network members with whom they interacted in that activity area. Sixteen of the respondents added a total of 18 names to the originally constructed lists as a result of the use of this procedure. By using these probes and by focusing on major actors in the networks, it is hoped that biases due to recall are minimized (Hammer et al., 1969; Hammer, 1980a, 1980b, 1980c). Respondents then placed the network members' names on a map according to domain (household, family work or school, organizations, informal business, formal business, neighbor, other friend), and connected those members who knew one another to provide an estimate of network density. They then sorted network members' names into specified categories to provide data on duration of acquaintance, frequency of contact, intensity, attachment, members' drug attitudes and use, and reciprocity. Much previous research on social interaction and drug use has focused on "best friend" and family relationships, generally ignoring the effects of significant to others who are not immediate family members or intimate friends (cf. Hirschi, 1969; Kandel, 1973; Akers et al., 1979). The social network data collection method of this study permits the description of size, density, composition, and quality of respondents' personal networks as these may be linked to patterns of drug usage.

A one-month followup interview was scheduled for the fourth to sixth week following termination from the program for the first 49 respondents to leave treatment after client history interviews were completed. Seven sampled subjects could not be located for one month interviews, and one refused an interview at this point. The remaining 41 respondents were interviewed. In the social network portion of this interview, respondents were asked to list on cards the first names of people important to them and with whom they had contact in the first four weeks following treatment. Again, the names were mapped and sorted.

All subjects sampled for followup were followed up again at 12 to 17 weeks after they left treatment. A total of 43 of the 49 sampled subjects were interviewed at this point. Five could not be located, and one refused to be interviewed. The social network portion of this interview focused on network interaction in the four weeks prior to the three-month followup interview.

RESULTS

Three sets of findings are presented below. First, the pretreatment networks of the original sample of 106 respondents are examined. Next, the changes in network characteristics from pretreatment to one-month follow up to three-month follow up are examined for the 38 respondents who were interviewed at all three points. Finally, the networks of nine respondents who reported using opiates during the first three months following treatment are compared with the networks of 29 respondents who did not report any opiate use during this period.

Pretreatment Networks of Drug Abusers: Structural Characteristics

On the average, respondents named 14.4 people in their personal networks. This corresponds with Pattison et al.'s (1979) report that the networks of heroin addicts contain 14.6 persons. On the average, the networks are smaller than the averages of 22.2 and 22.6 for "normal" respondents reported by Pattison et al. (1979) and Sokolovsky et al. (1978), and somewhat larger than the average of 10.2 members reported by Sokolovsky et al. (1978) for schizophrenics with active symptoms. These abusers maintain interaction with others prior to treatment, though their networks appear somewhat smaller than those of less deviant populations.

On the average, approximately two-fifths (42.5%) of all respondents' network members use hard drugs two or more times a week. In contrast, an average of 32.4% of respondents' network members prior to treatment use only alcohol or no psychoactive drugs. Similarly, an average 46.1% of the network members have positive attitudes toward drugs, while 43.1% are reported by respondents as having negative attitudes toward drugs. While these pretreatment networks have a substantial proportion of drug users, they are not exclusively peopled with those who use or are favorable to use. Nor do network members consistently influence respondents to use drugs. An average of 22.6% influence respondents to use drugs, while 33.8% of the network members are reported to influence respondents not to use drugs, during the month prior to treatment. Apparently, drug abusers whose lives are seriously enough disrupted to enter drug treatment maintain pretreatment networks in which sizable proportions of members influence against drug use.

To assess the issue of network support for opiate use in particular, the sample was divided into those who used opiates several times a week or more in the month prior to treatment and those who abstained from opiate use during this period. The drug use of this latter group included use of nonopiates (amphetamines and barbiturates, cocaine, or marijuana) during this pretreatment period. Partial correlations, using treatment programs as covariates,[1] were computed between network members' use and attitudes toward drugs and respondents' opiate use . As shown in Table 5, opiate users' networks provide stronger support for drug use and contain significantly fewer members who use only marijuana or alcohol than do the networks of opiate abstainers. There appear to be particularly strong pressures toward drug use in the networks of those who use opiates prior to treatment, while there is more pressure against use in the networks of abusers of other substances.

Table 5. Partial Correlations between Respondents' Opiate Use and Network Members' Attitudes and Drug Use Behaviors, Controlling for Program Differences (n = 106)

	Partial correlation(r)	F-test	p-value
Use of drugs including opiates	.398	13.717	.001
Use of marijuana or alcohol only	-.393	13.239	.001
Use of any drug	.297	7.075	.010
Use of alcohol only	-.239	4.234	.039
Supportive attitudes toward use of drugs	.401	13.978	.001

Table 6. Cross Tabulations of the Proportion of Network Members Involved in Informal Business with Subjects Participation in Non-violent Crimes of Profit and the Sale of Illicit Drugs (n = 106)

		Type of crime Nonviolent, for profit*	
		None	Any
Proportion of network members with whom subject is involved in informal business	Low	26	6
	Medium	22	21
	High	10	21
		Illicit drug sale**	
		None	Any
Proportion of network members with whom subject is involved in informal business	Low	24	8
	Medium	14	29
	High	5	26

$*\chi^2 = 15.622$; $p < .001$. $**\chi^2 = 24.559$; $p < .001$.
$Tau_c = .410$; $p < .001$. $Tau_c = .494$; $p < .001$.

Few members of respondents' pretreatment networks are acquaintances from work or school (7.4%), or from organizations, social groups, or clubs (3.3%). Instead, an average of 21.7% of the members are identified as informal business associates with whom respondents have financial dealings outside the course of a job or legitimate business. These include drug dealers, fences, and crime partners. As shown in Table 6, the proportion of informal business acquaintances in the network is positively related to respondents' own involvement in nonviolent crimes of profit and in illicit drug sales, reinforcing the picture of a significant proportion of

Table 7. Partial Correlations between Respondents' Opiate Use and the Proportion of Network Members in Various Domains, Controlling for Program Differences (n = 106)

Domain	Partial correlation(r)	F-test	p-value
Work/school	-.251	4.910	.030
Organizations	-.279	6.139	.016
Friends	-.102	.767	.384
Informal business	.402	14.050	.001
Formal business	-.062	.280	.598
Neighbors	-.142	1.510	.223
Abode	.187	2.657	.107
Family	-.166	2.066	.155

respondents pretreatment networks as criminogenic. Nonetheless, respondents' networks are not restricted to interactions with drug or crime contacts; 21.3% of the network members are relatives, 24.9% are friends (some of whom may be drug users), and 2.1% are neighbors. While these data again suggest the presence of a substantial minority of conventional others in users' networks, it is not evident from the data how these network members should be viewed from a rehabilitative perspective. Arling (1976) has suggested that a large proportion of family members, as compared with friends and neighbors, in networks is predictive of low morale. Stanton and Todd (1978) have argued that the family relationships of opiate users are themselves pathogenic.

Partial correlations were conducted to assess the relationship between the proportion of network members from various domains and abuse or abstention from opiate use in particular. As shown in Table 7, opiate abusers have significantly fewer contacts from work or school, or organizations, and significantly more informal business contacts than do respondents who used other drugs prior to treatment. The social networks of opiate users before treatment appear particularly deviant in their composition. It appears that, prior to treatment, opiate addicts are likely to be reinforced by network relationship in which drug use and criminal behavior are valued positively by peers.

Pretreatment Networks of Drug Abusers: Interactional Characteristics

The interactional data suggest that prior to treatment, drug abusers are involved in a relatively stable and positively valued set of relationships. Respondents report seeing 66.3% of their network members daily or several times per week. They view 63.1% of their members as close friends and enjoy seeing 68.1% of them. Surprisingly, they report remembering birthdays or sharing holidays with 41.2% of their pretreatment network members, an indication of considerable personal reciprocity in these networks. Fifty percent of the relationships are characterized by several different kinds of activities, and 35.2% involve reciprocal lending of household items, clothes, tools, or money. The networks are stable, with 55.7% of the members known over three years, and only 8.1% known less than four months. Finally, they are hierarchical, with respondents viewing themselves as higher in status than 21.7% of their members, of lower status than 38.5% of their members, and of equal status with 39.6% of their network members. In sum, abusers' interactions with network members appear to be stable, hierarchical, enjoyable, friendly, and reciprocal exchanges focused

Table 8. Comparison of Social Network Characteristics of Those Who Did and Those Who Did Not Use Opiates in the Three Months after Treatment at Three Points in Time (Mean Proportions of Networks)

	Nonusers n = 29			Opiate users n = 9		
	Pre-treatment $\bar{X}$	1 month post-treatment $\bar{X}$	3 months post-treatment $\bar{X}$	Pre-treatment $\bar{X}$	1 month post-treatment $\bar{X}$	3 months post-treatment $\bar{X}$
	n = 16.2	n = 17.9	n = 17.3	n = 14.7	n = 15.4	n = 12.2
Mean proportion From T.C.		23.5	21.4		16.9	26.2
Density	59.6	44.0a	47.7b	55.8	44.8	54.0
Known over 1 year	68.4	48.6a	54.4b	80.2	49.9a	35.4b
Known over 6 months	17.3	33.1a	28.4b	15.6	40.0a	41.8b
Frequent drug users	49.4	3.4a	5.8b	55.1	20.2a	11.2b
Influenced $\bar{S}$ to use	26.0	2.6a	3.0b	28.8	20.4	16.2
Positive attitude to drugs	47.8	7.8a	d	58.9	18.9a	d
Negative attitude to drugs	38.3	77.4a	d	27.7	69.5a	d
Household members	18.9	12.4	12.9	18.3	19.7	31.4
Family members	15.6	16.7	18.8	16.5	15.8	12.2
Work/school members	6.6	21.3a	25.1	10.7	13.1	18.0
Organization members	3.2	11.8	6.7	0	9.6	10.5
Informal business members	25.8	0.5a	1.5b	28.1	8.7	1.2b
Service agency members	6.4	7.0	3.9	5.2	6.5	5.4
Other friends	21.4	29.9a	29.1	21.1	24.2	19.8
Multistranded relations	29.6	17.1a	16.7	25.4	29.9	27.3
Instrumental reciprocal	40.9	70.1a	20.abc	32.0	44.5	18.9c
Personal reciprocal	39.3	38.2	d	39.4	27.9	d
Of lower status than $\bar{S}$	19.0	15.4	d	24.4	17.4	d
Of equal status to $\bar{S}$	35.3	54.2a	d	50.0	41.3	d
Of higher status than $\bar{S}$	45.6	30.5	d	25.7	41.3	d
Very close friends	36.3	40.8	35.6	29.6	28.8	28.4
Like seeing	62.2	70.2	d	68.9	68.1	d
Share thoughts often	30.3	43.9a	d	28.4	36.7	d
Never share thoughts	24.5	8.5a	d	34.5	27.5	d
Trust	57.1	51.4	d	52.4	45.4	d
Don't trust	31.9	26.4	d	33.7	37.1	d
Want to be like	23.5	25.4	25.4	30.1	19.9	13.5
Don't want to be like	44.1	28.7	34.0	42.4	41.9	43.0

a = t test for paired dependent observations from pretreatment to 1 month follow-up significant at .05.
b = t test for paired dependent observations from pretreatment to 3 months follow-up significant at .05.
c = t test for paired dependent observations from 1 month to 3 months follow-up significant at .05.
d = data not available from 3 month follow-up.

on a range of activities which include, but are not limited to, drug use. To this point, network size and the apparent support of certain network members for drug use are the major characteristics distinguishing the networks from those of less deviant populations.

However, one finding stands in apparent opposition to those reported above. In response to the question "How much did you want to the kind of person this was?" respondents report that they wanted to be like only 15.2% of their pretreatment network members. In contrast, they report that they did not want to be like 43.1% of the members of their personal networks.

Apparently, though they liked many of their network members before treatment, respondents did not hold many of them in high esteem and did not, according to their self-reports, view them as role models to be imitated. Pearson correlations indicate that the desire to imitate pretreatment network members is negatively related to the proportion of network members from the most deviant domain, informal business associates (r = .405, p = .001). In contrast, respondents who report wanting to be more like network members have networks with more members from work or school (r = .228, p = .001), formal business (r = .202, p = .019), and family (r = .219, p = .012) domains. These results suggest that even when active in street life, drug abusers are not necessarily committed to deviant social definitions. They do not want to be like their network members who engage in the most deviant activities. Rather, they appear to emulate network members when their networks include those who subscribe to more conventional values and who pursue more legitimate lines of action.

Pre- to-Posttreatment Changes in Social Networks

The 38 respondents for whom complete data from client history, one month, and three months follow-up interviews are available, are the subjects of this section. As noted earlier, this sample is partitioned here into two subgroups: those who report one or more instance of opiate use in the three months following treatment (n = 9) and those who report no opiate use during this period (n = 29). As shown in Table 8, participation in a residential treatment program for more than three months is associated with changes in drug abusers' networks when they return to the larger community. Networks remain approximately the same size, but they are populated by new people. Both those who use opiates in the first three months following treatment, and those who do not use opiates during this period report significant decreases in the proportion of network members known over one year (from 68.4% pretreatment to 48.6% in the first posttreatment month, to 54.4% in the third posttreatment month for those who do not use opiates, and from 80.2% to 45.9% to 35.4% for the opiate users). There are corresponding increases in the proportion of network members known less than six months (from 17.3% pretreatment to 33.1% in the first month, to 28.4% in the third month following treatment for the nonusers and from 15.6% to 40.0% to 41.1% for the opiate users). The return to the community appears to require the reconstitution of a social network for all residential drug program clients. Neither those who use opiates immediatly following treatment nor those who do not return entirely to their old pretreatment social networks in the first three months after treatment.

Furthermore, during the first month following treatment, both groups of returning clients have significantly less members in their social networks who favor drug use than they did before treatment (a decrease from 47.8% pretreatment to 7.8% in the first month following treatment for the nonusers, and a decrease from 58.9% to 18.9% for the posttreatment opiate users). Correspondingly, after treatment both groups have significantly more network members who are opposed to drug use, and both groups have significantly fewer informal business associates (street drug and crime contacts) during the third month following treatment than they had prior to treatment. The new networks which clients develop following treatment include new individuals. Moreover, they include different types of people than they included before residential treatment. Residential treatment is a dislocation which is associated with short-term changes in social network composition. These changes do not include the establishment of networks predominated by fellow clients from the treatment program or by program staff members. The average number of network members from the treatment program in the third month following treatment is 3.6 people or an average of only 22% of the posttreatment networks. Following treatment, clients do not remain well connected to the social support provided by residential programs. Rather, they establish new networks which include only a few members from treatment.

A final difference in interaction patterns over time is notable. For both opiate users and nonusers, reciprocal exchanges and lending of goods and money increase during the first month following treatment. This is a time when former clients are reestablishing themselves in the community. This increase is significant for the nonusers. However, by the third month following treatment, the proportions of network members whith whom respondents exchange goods and money decrease significantly in both groups to below pretreatment levels. The proportion of such instrumental exchanges is significantly below the pretreatment proportion for the nonusers. Fararo and Sunshine (1964) have reported greater receiprocity in delinquent dyads than in nondelinquent dyads. It is possible that after former clients have participated in the instrumental exchanges necessary to establish themselves in the community in the first month following treatment, they increasingly refrain from becoming entangled in exchanges of goods and money characteristic of the street drug user's life style (cf. Johnson, 1980; Stephens and Smith, 1976). In this respect, a decrease in instrumental reciprocity may be a desirable part of establishing a more conventional life style following treatment.

The posttreatment networks of both nonusers of opiates and those who use opiates in the first three months after treatment change from before treatment. Exploration of the differences between the posttreatment networks of these two groups provides a beginning understanding of the relationships between network characteristics and a return to drug use. The two groups were not significantly different in sex or previous education prior to treatment. However, those who used opiates following treatment were slightly older than the nonusers (a mean of 29.3 years as compared to 27.0 years). Further, all of those reporting opiate use in the first three months following treatment where white. Eight of the 29 clients who did not report opiate use following treatment were of ethnic minority backgrounds. Thus, the sample of posttreatment opiate users underrepresents minority opiate users, a bias which limits the generalizability of the findings reported below.

Interestingly, prior to treatment, the two groups' social networks differed significantly on only one of 54 social network variables assessed using separate t-tests. (The small samples precluded use of multivariate analysis of variance procedures in this analysis.) Using a $\alpha = .05$ significance level, three of these t-tests could have been expected to achieve significance by chance alone. Yet the only statistically significant pretreatment social network difference between the two groups was the proportion of network members whom the respondent did not like seeing at all (t = 2.03, DF = 36, p = .05). Nonusers did not like seeing 10.1% of their pretreatment network members, and those who used opiates after treatment claimed to dislike seeing only 3.2% of their pretreatment networks. Prior to treatment, the networks of the two groups did not differ significantly on any other structural, affective, or drug attitude and use measures. Thus, it would appear that posttreatment differences in networks between those who use opiates in this period and those who do not cannot be attributed to pretreatment network differences.

While their networks were similar prior to treatment, in the first month following treatment the networks of the two groups are significantly dissimilar in several respects. Those who do not use opiates in the first three months report significantly fewer regular users of drugs (3.4%) in their networks than do those who use opiates (20.2%) (t = 2.54, DF = 28 [separate variance estimate], p = .032). Similarly, while the proportion of network members who use no "hard drugs" increases for both groups from pretreatment to one-month follow up, this increase is significant only for the nonusers of opiates who have significantly more nonusers in their networks (70.1%) than do those who use opiates (44.5%) (t = 3.12, DF = 35, p = .004). Similarly, the proportion of network members influencing the nonusers toward drug use decreases significantly from 26.0% before treatment to 2.6% during the first month following treatment. The proportion of network members influencing the opiate users to use drugs shows only a modest nonsignificant decline, with one-fifth (20.4%) of the network still influencing toward use, posttreatment. Again, the networks of the two groups differ significantly at one-month follow up in this respect (t = 3.43, DF = 8.38 [separate variance estimate], p = .009). In sum, while the social networks of both groups change in membership from before treatment to one-month follow up, those who use opiates after treatment have posttreatment networks which provide significantly greater support for drug use than do those who do not use opiates.

There are few significant differences at the α = .05 level in attachment variables for the two sets of networks at one-month follow-up. However, trends which approach significance (i.e., p < .10) emerge. The opiate users have fewer network members with whom they share and maintain personal reciprocity, i.e., share birthdays and holidays (2.9% vs 38.2%), fewer very close friends (28.8% vs 40.8%), more informal business contacts (8.7% vs. 0.5%), more members with whom they never share thoughts and feelings (27.5% vs. 8.5%), and more members whom they do not trust (37.1% vs. 26.4%). While these differences are not significant at α = .05, they suggest that there may be somewhat less affective support and attachment in the networks of those who return to opiate use in the first three months following treatment.

Surprisingly, by three months following treatment, the networks of the posttreatment opiate users are reported as including more conventional members than they did at one-month followup. The regular users of drugs have decreased to 11.2% of the networks, and those who use no illegal drugs now make up 61.2% of the networks. Yet these networks continue to include a greater proportion of menbers who influence the respondents to use drugs (16.2%) than do the networks of nonusers of opiates (3.0%). Interestingly, the posttreatment networks of the opiate users appear to be constricting (to a mean of 12.2 members), while those of the nonusers remain virtually unchanged in size from the one-month followup (with a mean at three months of 17.3 members). This difference in network size between the two groups approaches significance (t = 1.98, DF = 36, p = .055).

As the opiate users' network constrict, they become more restricted in membership and more dense. Household members make up nearly a third (31.4%) of the users' networks at three months, and the networks are as dense as they were prior to treatment. Further, the proportion of members whom the users emulate ("want to be the kind of person this person is") decreases to 13.5%.

While the small and racially biased sample and short followup period limit the generalizability of results regarding changes in networks over time, the results are suggestive. Major changes in social network composition appear to follow residential treatment. Returning clients appear to establish more conventional networks of interaction during their first months back in the community. However, the use of opiates following treatment is accompanied by the establishment of a network which includes significant social influences toward use. There is some evidence that opiate use is more likely if returning clients do not establish networks which provide consistent affective support and if their networks do not include role models deemed worthy of emulation. In these circumstances, the networks may constrict to include less representation from domains beyond the immediate household. This may be detrimental in two respects. More pressure may be felt by household members to meet the former addict's affective and attachment needs. At the same time, the network is likely to have a diminished capacity to provide alternative sources of support when household interactions themselves become stressful (Gove and Geerken, 1977).

IMPLICATIONS FOR SUPPORTIVE NETWORK DEVELOPMENT

The results reported here have important implications for drug treatment programs. Opiate abusers' pretreatment social networks appear especially supportive of illicit drug use. As long as abusers remain embedded in them, they are likely to experience strong influences to return to opiate use. This study indicates that such social influences are related to a relatively rapid posttreatment return to opiate use among clients of therapeutic communities. Clients in this study who use opiates in the first three months after treatment report significantly more people in their networks who influence them to use drugs. In light of these findings, it appears that a desirable goal for community-based treatment programs which serve opiate abusers is to help clients alter the composition of their social networks in order to eliminate the members who are supportive of drug use.

The drug abusers studied here have supportive, stable relationships with the members of their pretreatment networks. They like these people, they like spending time with them, and they have established enduring relationships which include a broad range of shared activities. Therefore, it is not enough to provide opiate abusers with someone to replace a presumed void in their prior lives. If relationships with more conventional people are to be established, they should offer at least an equal measure of social support to that provided by pretreatment networks. An intentional effort to cut the ties that influence addicts to use illegal opiates is likely to be necessary (Waldorf and Biernacki, 1981). Drug abusers are likely to curtail interactions with other users and informal business associates only as they become connected with people involved in more conventional activities (Waldorf and Biernacki, 1981). Yet, opiate abusers in particular, appear to be isolated from people in the domains of work, school, organizations, groups, and clubs. They are simply not connected with many people with whom they could develop attachments to replace their associations with drug-abusing peers.

Nevertheless, the followup results from this study of therapeutic community subjects suggests that a replacement of network members is possible. The composition of respondents' networks changed from pre to posttreatment to include fewer drug users and fewer longtime acquaintances. Moreover, both opiate users and nonusers in the three-month followup period developed some degree of attachment to their posttreatment networks. Both groups report networks in which approximately three-fifths of the members are close or very close friends. Both groups report sharing their thoughts and feelings with over a third of their network members, and both groups would trust nearly half of their networks at one-month followup with personally damaging information (see Table 8). It would appear that new social ties can be developed to replace old bonds to drug users.

However, the fact that new relationships are possible does not guarantee that they will be supportive of more conventional lives. The data suggest that a key to ensuring that the new relationships are more supportive of rehabilitated lives is to assist clients to connect with nondrug users whom they admire and respect, people whom they can view as role models (see also Waldorf, 1973). Only this esteem element of the affective bond appears to be consistently lacking in the realtionships of drug abusers with their pretreatment networks, and it is precisely this element of esteem which appears to be absent when networks are predominated by drug users and informal business associates. New networks supportive of rehabilitation are likely to take the place of pretreatment attachments when clients become linked with more conventional others whom they respect as role models and to whom they feel close, with whom they share thoughts, whom they like seeing, and with whom they share a degree of personal reciprocity.

In summary, three conditions appear to be important if the social factors contributing to drug abuse are to be adequately addressed in programs:

1.   Support in the social network for drug abuse should be minimized or eliminated.

2.   Clients should be linked to new network members who are engaged in conventional activities, and who can provide the attachment and support formerly provided by a network more strongly supportive of drug use (Stanton, 1979; Waldorf and Biernacki, 1981; Wolf and Kerr, 1979).

3.   Special care should be taken to ensure that the new network contains viable role models whom the former abuse holds in high esteem.

These criteria simultaneously imply broadened roles for drug treatment program staffs in the rehabilitation process and suggest very real limits of what treatment personnel can expect to achieve through traditional treatment services. Counseling or other client-focused services are not likely to affect the primary relationships between clients and their social networks, nor are they likely to provide, on a daily basis, the range of attachment and support necessary for the replacement of a network supportive of drug abuse (Hawkins, 1979). For these reasons, treatment programs should consider models of service delivery which expand traditional treatment. This will require a basic shift in orientation away from the accustomed technology in which the client comes to the treater for resolution of problems in the treatment setting (Brown, 1979:19). To accomplish transformations in community networks, treatment personnel may need to become consultants and facilitators to social network members and community volunteers who themselves become the bonding nuclei of new networks supportive of rehabilitation. The tasks of treatment staffs will shift away from direct services to clients toward recruitment, orientation, training, and consultation to community volunteers and natural network members (see Hawkins, 1979, 3-45). While the consultant role is likely to be unfamiliar and difficult to treatment personnel, there is evidence that it is effective and cost-efficient in the community reintegration of former mental patients (Weinman and Kleiner, 1978).

This approach also implies an expanded role for the larger community in drug abuse treatment. If involvement in school, work, and other organizational settings is to produce supportive links with members of these groups, group members will need to become actively involved with former drug abusers. Yet such involvement is in direct opposition to the dominant trend in the last century:

increased urbanization and industralization in society have been accomplished by increased professional concentration in formal agencies charged with responsibilities for solving social problems....Furthermore, these agencies and their programs have tended to become "disengaged" from the community, or at best are peripheral to the mainstream of community life. They have not functioned as a central part of the life of the community. (Pink and White, 1973:29)

Rather than viewing drug abusers as people who must somehow be effectively treated, resocialized, and reintegrated, community members have generally viewed them as "outsiders" (King, 1969;218-219). This attitude has been reflected in the community resistance to the very presence of drug treatment programs in some neighborhoods (Lowinson and Langrod, 1975; Ruiz et al., 1975). Nevertheless, there is evidence from the mental health field of the importance of community acceptance and support in determining client outcome. These factors have been shown to be the best predictors of the social functioning of former mental patients in board-and-care facilities (Segal and Aviram, 1978). Active community participation is an element generally missing from rehabilitation programs for drug abusers. How can fraternal, civic, labor, and business groups, as well as community individuals be engaged in the task of rehabilitating street drug abusers? Only if they somehow come to own the problem as their responsibility.

If this is to happen, professionals and paraprofessionals engaged in the treatment of drug abusers must take the initiative. They must become entrepreneurs for a community involvement and responsibility, approaching community groups and convincing them of the need to participate in planning and developing programs. The very process of program development should be one of collaborative consultation among professionals and members of community groups (Caplan and Grunebaum, 1967; Ruiz et al., 1975:153-154).

Community individuals and groups should participate from the start in developing and appraising social support models, and realistically assessing their abilities to implement these. Ultimately, as government funds diminish, community organizations may need to provide the administrative umbrella for such efforts, funds for operating them, and even personnel to carry them out. While this level of community participation will be difficult to achieve and troublesome for drug program staffs, it may be necessary if new approaches to the social integration of drug abusers are to be initiated under the present funding conditions. Additionally, such participation will lead to a feeling of ownership and responsibility among community members, which will be important for the success of social integration approaches.

Theoretically and empirically supportable approaches exist for establishing supportive networks for the rehabilitation of drug abusers (cf. Ishiyama, 1979; Callan et al., 1975; Stanton, 1978; Ch'ien, 1979; Hawkins and Fraser, 1983). These approaches require new roles of treatment staff and the active participation of community groups and individuals in reaching out to reintegrate former street drug abusers into conventional society. In these approaches, staff act primarily as consultants and supports to community members who are the agents of reintegration, working directly with ex-abusers to provide new networks of social support for maintaining rehabilitated lives.

NOTES

1. Programs were allowed to enter as covariates because respondents differed across the programs on nine of sixty variables tested. Only Program Two, which appears to serve a less deviant population, entered as a control in this analysis.

2. All comparisons of networks over time were tested for significance using a t test for differences between dependent observations.

REFERENCES

Akers, Ronald L., Marvin D. Krohn, Lonn Lanza-Kaduce, and Marcia Radosevich, "Social learning and deviant behavior: A specific test of a general theory." American Sociological Review, 44 (August): 636-655, 1979.

Arling, Greg, "The elderly widow and her family, neighbors, and friends." Journal of Marriage and the Family, November: 757-768, 1976.

Bale, Richard N., "The validity and reliability of self-reported data from heroin addicts: Mailed questionnaires compared with face-to-face interviews." The International Journal of the Addictions, 14(7): 993-1000, 1979.

Bale, Richard N., W. W. Van Stone, J. M. Kuldau, T. M. J. Engelsing, R. M. Elashoff, V. P. Zancone, "Therapeutic communities vs. methadone mainte- nance." Archives of General Psychiatry, 37(February):179-193, 1980.

Bernard, H. R., and P. D. Killworth, "Informant accurancy in social network data II." Human Communication Research, 4(1) 3-18, 1977.

Bloom, W. A., and E. W. Sudderth, "Methadone in New Orleans: Patients, problems, and police." In S. Einstein (Ed.), Methadone Maintenance. New York: Marcel Dekker, 1971.

Brown, B. S., "Introduction." In B. S. Brown (Ed.), Addicts and Aftercare. Beverly Hills, CA: Sage, 11-22, 1979.

Callan, D., J. Garrison, and F. Zerger, "Working with the families and social networks of drug abusers." Journal of Psychedelic Drugs, 7(1): 19-25, 1975.

Caplan, G., and H. Grunebaum, "Perspectives on primary prevention: A review." Archives of General Psychiatry, 17:331-345, 1967.

Ch'ien, James M., "Alumni associations of Hong Kong." In B. S. Brown (Ed.), Addicts and Aftercare. Beverly Hills, CA.: Sage, 155-163, 1979.

Cohen, A. K., Delinquent Boys: The Culture of the Gang. New York: Free Press, 1955.

Curtis, Bill and D. Dwayne Simpson, "Differences in background and drug use history among three types of drug users entering drug therapy programs." Journal of Drug Education, 7(4):369-379, 1977.

Drug Abuse Council, The Facts About "Drug Abuse." New York: Free Press. 1980.

Fararo, T. J., and M. H. Sunshine, A Study of a Biased Friendship Net. Syracuse, N.Y.: Syracuse University Youth Development Center, 1964.

Gove, W., and M. Geerken, "The Effect of children and employment on the mental health or married men and women." Social Forces, 56(1): 66-76, 1977.

Hammer, Muriel, "Predictability of social connections over time." Social Networks, 2:165-180, 1980a.

Hammer, Muriel, "Reply to Killworth and Bernard." Connections, 3(3): 14-15, 1980b.

Hammer, Muriel, "Some comments on the validity of network data." Connections, 3(1):13-15, 1980c.

Hammer, Muriel, "Social supports, social networks, and schizophrenia. Schizophrenia Bulletin, 7(1): 45-57, 1981.

Hammer, M., S. K. Polgar, and K. Salzinger, "Speech predictability and social contact patterns in an informal group." Human Organization, 28:235-242, 1969.

Hawkins, J. David, "Reintegrating street drug abusers: Community roles in continuing care." In B. S. Brown (Ed.), Addicts and Aftercare. Beverly Hills, CA: Sage, 25-79, 1979.

Hawkins, J. David, and Mark W. Fraser, "Social support and the treatment of drug abuse." In James K. Whittaker and James Garbarino (Eds.),

Social Support System in the Human Services. New York: Adline, 355-380, 1983.

Hawkins, J. David, and Norman Wacker, "Verbal performances and addict conversion: An interactionist perpective on therapeutic communities." *Journal of Drug Issues*, 13(2):281-298, 1983.

Hawkins, J. David, and Richard F. Catalano, "Reversing drug abuse: A theory of rehabilitation." Paper presented at the Pacific Sociological Association Meeting, San Francisco, 1980.

Hirschi, T., *Causes of Delinquency*. Berkeley, CA.: University of California Press, 1969.

Ishiyama, Toaru, "Self help models: Implications for drug abuse programming." In B. S. Brown (Ed.), *Addicts and Aftercare*. Beverly Hills, CA: Sage, 117-133, 1979.

Johnson, Bruce D., *Marijuana Users and Drug Subcultures*. New York; Wiley, 1973.

Johnson, Bruce, D., "Toward a theory of drug subcultures." In Dan J. Lettieri, M. Sayers, and H. W. Pearson (Eds.), *Theories on Drug Abuse: Selected Contemporary Perspectives.* Washington, D.C.: U.S. Government Printing Office, 1980.

Kandel, Denise, B. "Adolescent marihuana use: Role of parents and peers." *Science,* 181 (September): 1067-1069, 1973.

Kandel, Denise, B., "Interpersonal influence on adolescent illegal drug use." In Erick Josephson and Eleanor E. Carroll (Eds.), *Drug Use: Epidemiological and Sociological Approaches*. New York John Wiley and Sons, 1974.

Kandel, Denise B., "Developmental stages in adolescent drug involvement." In Dan J. Lettieri, Mollie Sayers, and Helen Wallenstein Pearson (Eds.), *Theories on Drug Abuse*. Washington: U.S. Government Printing Office, 120-127, 1980.

Kandel, Denise, B., Donald Treiman, Richard Faust, and Eric Single, "Adolescent involvement in legal and illegal drug use: A multiple classification analysis." *Social Forces*, 55(2): 438-458, 1976.

Killworth, P., and H. R. Bernard, "Catiji: A new sociometric and its application to a prison living unit." *Human Organization*, 33:335-350, 1974.

Killworth, P., and H. R. Bernard, "Informant accuracy in social network data." *Human Organization*, 35(3):269-286, 1976.

Killworth, P., and H. R. Bernard, "Informant accuracy social network data III." *Social Networks,* 2(1): 19-46, 1979.

King, J., *The Probation and After-Care Service*, London: Butterworth, 1969.

Leinhardt, Samuel, "Social network research: Editor's introduction." *Journal of Mathematical Sociology*, 5:1-4, 1977.

Lewis, David C., and John Sessler, "Heroin treatment: Development, status, outlook." In Drug Abuse Council (Eds.), *The Facts About Drug Abuse*. New York: Free Press, 1980.

Lowinson, Joyce, and John Langrod, "Neighborhood drug treatment centers: Opposition to establishment." *New York State Journal of Medicine* (April) 766-769, 1975.

Mitchell, J. Clyde, *Social Networks in Urban Situations*. Manchester, England: Manchester University Press, 1969.

National Institute on Drug Abuse, "Nonresidential self-help organizations and the drug abuse problem: An exploratory conference." (DHEW No. (ADM) 78-752,) Washington: U.S. Government Printing Office, 1978.

Pattison, E. Mansell, Robert Llamas, and Gary Hurd, "Social network mediation of anxiety." *Psychiatric Annals*, 9(9): 474-482, 1979.

Phillips, Swan, L., "Network characteristics related to the well-being of normals: A comparative base." *Schizophrenia Bulletin* 7(1): 117-123, 1981.

Pink, W. T., and M. F. White (Eds.), Delinquency Prevention: A Conference Perspective on Issues and Directions. Portland, OR: Regional Research Institute, 1973.

Ruiz, Pedro, John Langrod, and Joyce Lowinson, "Resistance to the opening of drug treatment centers: A problem in the community psychiatry." International Journal of the Addictions, 10(1): 149-155, 1975.

Segal, S. P., and U. Aviram, The Mentally Ill in Community-Based Sheltered Care: A Study of Community Care and Social Intergration. New York: Wiley, 1978.

Sokolovsky, Jay, Carl Cohen, Dirk Berger, and Josephine Geiger, "Personal networks of ex-mental patients in a Manhattan SRO hotel." Human Organization 37(1): 5-15, 1978.

Stanton, M. Duncan, "Some outcome results and aspects of structural family therapy with drug addicts." In D. Smith, S. Anderson, M. Buston, T. Chung, N. Gottlieb, and W. Harvery (Eds.), A Multicultural View of Drug Abuse: The Selected Proceedings of the National Drug Abuse Conference. Cambridge, MA: Schenkman, 1977.

Stanton, M. Duncan, "The client as family member: Aspects of continuing treatment." In B. S. Brown (Ed.), Addicts and Aftercare. Beverly Hills, CA: Sage, 1979.

Stanton, M. Duncan, and T. C. Todd, "Structural family therapy with heroin addicts." In E. Kaufman and P. Kaufman (Eds.), The Family Therapy of Drug and Alcohol Abusers. New York: Gardner, 1978.

Stanton, M. Duncan, T. C. Todd, David B. Heard, Sam Kirschner, Jerry I. Kleiman, David T. Mowatt, Paul Riley, Samuel M. Scott, and John M. Van Deusen, "Heroin addiction as a family phenomenon: A new conceptual model." American Journal of Drug and Alcohol Abuse, 5(2):1-25, 1978.

Stephens, R. C. and R. B. Smith, "Copping and caveat emptor: The street addict and consumer." Addictive Diseases, 2(4): 585-600, 1976.

Waldorf, Dan, "Rock Bottom." Chapter 9 in Careers in Dope. Englewood Cliffs, N.J.: Prentice-Hall, 1973.

Waldorf, Dan, and Patrick Biernacki, "The natural recovery from opiate addiction: Some preliminary findings. Journal of Drug Issues, 11(1): 61-74, 1981.

Weinman, B., and R. J. Kleiner, "The impact of community living and community member intervention on the adjustment of the chronic psychotic patient." In L. I. Stein and M. A. Test (Eds.), Alternatives to Mental Hospital Treatment. New York: Plenum, 139-159, 1978.

Wolf, Kenneth, and Douglas, M. Kerr, "Companionship therapy in the treatment of drug dependency." In B. S. Brown (Ed.), Addicts and Aftercare, Beverly Hills, CA: Sage, 1979.

DRUNKENNESS IN THE OLD TESTAMENT:

A CLUE TO THE GREAT JEWISH DRINK MYSTERY[1]

John Maxwell O'Brien and Sheldon C. Seller

Departments of History and Sociology
Queens College
City University of New York
Flushing, New York 11367

This paper is based on a project which is designed to develop a series of tables that catalog and categorize references to alcohol in the sacred scriptures. This report deals with references to alcohol in the Bible -- or if you will -- the Old Testament. This work appeared in the Drinking and Drug Practices Surveyor (18:18-24, 1982), and similar tables relating to the Apocrypha, the Pseudepigrapha, the New Testament, the Koran, and the Talmud will make their appearance over the next few years. Hopefully, by the time we have finished we shall have mined most of the basic scriptural sources within Judaism, Christianity, and Islam.

The primary purpose of each table is to enable researchers to locate references to alcohol in the respective texts. Each of these references is characterized as either positive or negative, that is, as reflecting perceptions of the potential positive or negative effects of alcohol. These references are then assigned to their appropriate realms -- physical, psychological, social, religious, or economic. Finally, each reference is subsumed under a category which epitomizes the general message conveyed.

The following are some examples to illustrate the methodology employed. All translations are taken from the Oxford Study Edition of The New English Bible, New York, 1972.

POSITIVE OR NEGATIVE? Positive	Eccles. 2:3
REALM. Psychological	
CATEGORY. Stimulates the mind	So I sought to stimulate myself with wine, in the hope of finding out what was good for men to do under heaven throughout the brief span of their lives.

Now the Speaker in Ecclesiastes would never find a satisfactory answer to his question, but his musings reveal that he was at one time operating on a premise that wine had real possibilities in this respect. Hence a positive functional perception of alcohol in the psychological realm stimulates the mind.

POSITIVE OR NEGATIVE? Negative	Jer. 48:26
REALM. Social	
CATEGORY. Elicits derision	Make Moab drunk--he has def-ied the LORD--until he over-flows with his vomit and even he becomes a butt for deri-sion.

These tables are designed to call attention to the full range of perceptions of alcohol in each of the texts. We have categorized implicit and sometimes metaphorical messages as well as those which are quite explicit. Is there something to learn of importance in this process of textual codification? We certainly think so.

Not too long ago one of the authors (O'Brien, 1980) constructed a table classifying the attributes of alcohol in the reading of Alexander the Great. (See Appendix.)

Alexander was extremely well read. One can count Homer, Aeschylus, Pindar, Herodotus, Euripides, Sophocles, and Xenophon among the authors with whom he was familiar. We can be fairly sure that perceptions of alcohol in Hellenic culture were well represented in the pages with which that bibulous Macedonian was familiar. The resultant table was startling--especially in terms of the ratio it revealed between positive and negative attributes of alcohol. There were 42 positive categories and only 11 negative categories.

In sharp contrast, the Old Testament, according to our preliminary calculations, contains 36 positive categories and 65 negative categories.

A closer look at those negative categories permits us to offer some tentative observations, the validity of which will have to be determined by further research and evaluation.

- First, the Old Testament is perhaps the richest, quarry of quasi-clinical information in history. It is an encyclopedia of alcohol abuse.

- Second, it may provide the earliest historical evidence of one group differentiating itself from others in terms of its use of alcohol.

Drinking characteristics of habitual drunkards, chronicled in the Old Testament, will be very familiar to "alcoholism" specialists. First of all, it is made abundantly clear, through the text, that some drinking patterns are strikingly different than others. For instance:

Drunkards drink at the wrong time of day and lack self-control:	Eccles. 10:17
	Happy the land when its king is nobly born, and its princes feast at the right time of day, with self-control, and not as drunk-ards.

Drinking can affect one's thought processes:	Hos. 4:12
	New wine and old steal my people's wits.

Drinking can become a full-
time activity:

Shame on you! You who rise
early in the morning to go
in pursuit of liquor and
draw out the evening in-
flamed with wine.

This Isaiah passage suggests physiological dependence, and the question
of addiction is addressed directly a few verses later:

Isa. 28:7

These two are addicted to
wine, clamouring in their
cups; priest and prophet
are addicted to strong
drink and bemused with wine;
clamouring in their cups,
confirmed topers.

The most provocative and illuminating passage (in a clinical sense) is
found in the Wisdom tradition of Proverbs where a convincing profile of the
habitual drunkard, and his attendant problems, was recorded several thousand
years ago:

Prov. 23:29-35

Whose is the misery? whose
    the remorse?
Whose are the quarrels and
    the anxiety?
Who gets the bruises without
    knowing why?
Whose eyes are bloodshot?

Those who linger late over
    the wine,
those who are always trying
    some new spiced liquor.
Do not gulp down the wine,
    the strong red wine,
when the droplets form on
    the side of the cup;
in the end it will bite
    like a snake and sting
    like a cobra.
Then your eyes see strange
sights, your wits and your
speech are confused;
you become like a man toss-
    ing out at sea,
like one who clings to the
    top of the rigging;
you say, 'If it lays me
    flat, what do I care?
If it brings me to the
    ground, what of it?
As soon as I wake up, I
    shall turn to it again.'

A few words about the "Great Jewish Drink Mystery." We are referring to the low incidence of alcohol-related problems among Jews and how that came about. Our work on the Old Testament has revealed clues which might be germane to the problem. Its pages are permeated with the assumption that drunkenness is or should be a non-Jewish phenomenon. Drunkenness signifies spiritual atavism, a regression to heathen polytheism and godlessness. In fact, it reflects a failure to accept God and his commandments.

Deut. 25:18-21

When a man has a son who is disobedient and out of control, and will not obey his father or his mother, or pay attention when they punish him, then his father and mother shall take hold of him and bring him out to the elders of the town, at the town gate. They shall say to the elders of the town, 'This son of ours is disobedient and out of control; he will not obey us, he is a wastrel and a drunkard.' Then all the men of the town shall stone him to death, and you will thereby rid yourselves of this wickedness. All Israel will hear of it and be afraid.

Thus, habitual drunkenness is offered as prima facie evidence of a violation of the commandments and the penalty prescribed is execution.

The Old Testament is replete with grim warnings intoned against the abuse of God's gift, and may very well be the ultimate source of authority for the discouragement of excessive drinking among Jews.

NOTE

1. Some preliminary observations in this paper now appear fully developed in Sheldon C. Seller, "Alcohol abuse in the Old Testament," Alcohol and Alcoholism, 20, 1985.

REFERENCES

O'Brien, J. M., "Alexander and Dionysus: The invisible enemy," Annals of Scholarship, 1:83-105, 1980.

Positive and Negative Attributes of Wine in Alexander the Great's Readings[a,b]

Positive	Homer	Aeschylus	Pindar	Herodotus	Euripides	Sophocles	Xenophon
PHYSICAL							
Revitalizes	Iliad (VI.258-62)•						
Helps generate strength	Iliad (IX.705-06)•						
Warms the body		•			Ion (552-53)•		
Slakes thirst	•					Philoctetes (713-15)•	•
Accompaniment to food	Odyssey (V.165-66)•	•		•		•	•
Pleasant aroma		•			Electra (497-99)•		
Looks attractive	Iliad (V.341)•	•			•		
Pleasing taste	•		•	•		•	
Life giving			Paean (IV.25-26)•	•			
Relaxant	Iliad (XIV.5)•						
Soporific	•			•	Cyclops (573-84)•		
PSYCHOLOGICAL							
Engenders love		•		•	Bacchae (773-75)•		
Brings joy and cheer	•		•		Bacchae (773-75)•	•	•
							Anabasis (VI.4.6)•

(continued)

Positive and Negative Attributes of Wine in Alexander the Great's Readings (Continued)a,b

Positive	Homer	Aeschylus	Pindar	Herodotus	Euripides	Sophocles	Xenophon
Provides bravado	Iliad (XX.83-85) •						•
Alleviates despair			Paean (IV.26) •				
Heals grief		•			•		Symposium (II.24-25) •
Frees truth					•	Oedipus Rex (779-780) •	
Excites you				•			Hellenica (VI.4.8) •
SOCIAL							
Cultivates fellowship	•			•			Anabasis (IV.5.32) •
Solidifies friendship (drinks to health)	•			•			Anabasis (VII.3.27) •
Marks special occasions	•		Olympian (VII.1-6) •		•		
Makes oaths and compacts binding	•			Histories (IV.70) •	•	•	
Inspires Music, singing & dancing	•		Nemean (IX.48-50) •		•	•	
Reward for work	Iliad (XVIII.544-47) •	•		•			
RELIGIOUS							
Communion with God					Bacchae (284-85) •		
Necessary in expiatory rites		Choephori (538-39) •					

ECONOMIC

Medium of exchange	Iliad (VII.472-75)
Valuable resource	Iliad (XVII.248-50)
Drinking vessels & jars valuable commodity	Cyropaedia (VIII.8.18)

NEGATIVE

PHYSICAL

Gets you drunk	Cyclops (427-36)
Drains your strength	Iliad (VI.264-65)

PSYCHOLOGICAL

Affects your memory	Iliad (VI.264-65)
Drives you mad	Histories (VI.75-84)
Fogs your mind	Cyclops (427-36)
Evokes violence	Cyclops (434)
Leads to lust	Phoenissae (21-22)
Is an enemy	Cyclops (678)
Brings out bestiality	Cyclops (678) Cyropaedia (V.2.17)

(continued)

Positive and Negative Attributes of Wine in Alexander the Great's Readings (Continued)[a,b]

Positive	Homer	Aeschylus	Pindar	Herodotus	Euripides	Sophocles	Xenophon
Can leave you in ruin							Socrates' • Defense (31)
RELIGIOUS							
Dionysus can make you do things					Bacchae • (1296) •		

[a] A dot • indicates that one or more statements of the listed messages are found in the author's writings. Source reference indicates a specific example.

[b] Parenthetical references are to the Loeb Classical editions. Line references are to the Greek text.

FACTORS GOVERNING RECRUITMENT TO AND MAINTENANCE OF SMOKING

J. F. Golding

University of Newcastle upon Tyne
Department of Pharmacological Sciences
The Medical School
Newcastle upon Tyne
NE2 4HH, United Kingdom

G. L. Mangan

University of Auckland
Department of Psychology
Auckland, New Zealand

INTRODUCTION

In attempting to identify factors underlying recruitment to and persistence of the smoking habit, an initial problem is that the particular motives identified may be relevant to one or the other phase, or to both. As we shall see in our discussion of theories about smoking, this introduces considerable overlap into an already confused area. A second problem of perhaps greater consequence is that all theories of smoking, with the possible exception of genetic theories, are pleasure seeking/reward theories. Thus, analysis of the nature of the rewards (and punishments) inherent in smoking inevitably leads to one or another of the theoretical models described below. From this point of view, the main difference between models lies in the emphasis given to the social, as opposed to pharmacological, rewards, and the exact nature of such rewards. Insofar as pharmacological reward inconcerned, consider, for example, the view proposed by addiction theory, that smokers gain their reward from relief of nicotine withdrawal symptoms, compared with that proposed by arousal modulation theory, that smokers enjoy positive reward from nicotine, which is employed as a mood control agent to maintain arousal level at a hypothetical optimum. This difference reflects the as yet unresolved debate in the animal learning literatures as to whether relief from punishment is equivalent to reward and is mediated by similar mechanisms (Gray, 1971). While such debates persist, it is obviously difficult to define precisely the role of nicotine in smoking maintenance.

THEORIES ABOUT SMOKING

A number of theories have been proposed concerning the reasons why individuals take up and continue smoking. Each of these has attracted some empirical support, but the most persuasive are undoubtedly pharmacological addiction and arousal modulation theories. These we shall describe in some detail and briefly mention the less-well-substantiated theories such as opponent process, genetic, orality, and social learning theories.

Simple Pharmacological Addiction

Nicotine addiction theories (e.g., Schachter, Silverstein, Kozlowski, Perlick, Herman, and Liebling, 1977) postulate that certain CNS receptors become adapted to nicotine and signal punishment when the nicotine level at these sites falls below a critical level. From this point of view the desire to smoke may be regarded as a consequence of nicotine "hunger," the purpose of smoking being to maintain nicotine homeostasis, just as regular heroin injections maintain opiate homeostasis at CNS opiate receptors. Indeed, suggestions have been made that nicotine from smoking may cause the release of the endogenous "brain opiates" (Jaffe and Kanzler, 1979; Mello et al., 1985; Pomerleau et al., 1983).

There are two main lines of evidence supporting an addiction model: (a) "withdrawal effects" following cessation of smoking, and (b) variations in smoking rates following experimental manipulation of nicotine levels in the smoker.

1. Withdrawal Effects Following Cessation of Smoking

A tobacco withdrawal syndrome has been identified, especially in heavy smokers, and a number of behavioral effects have been described. These include: increases in irritability, appetite, laziness, drowsiness, depression (Ryan, 1973; Schachter et al., 1977; Shiffman, 1979), as well as psychophysiological effects, usually in the direction of decreased arousal, on measures such as dominant EEG frequency (Ulett and Itil, 1969; Knott and Venables, 1977; Herning et al., 1983), evoked potential (Hall, Rappaport, Hopkins, and Griffin, 1973; Vogel et al., 1977), pulse rate and diastolic blood pressure (Knapp, Bliss, and Wells, 1963), and adrenaline and noradrenaline excretion (Myrsten, Elgerot, and Edgren, 1977). One report of hypersensitivity of the evoked response to low-intensity stimuli (Knott and Venables, 1978) may be reflecting the behavioral observation that both symptoms of increased irritability as well as drowsiness are observed after tobacco abstinence.

However, the magnitude of withdrawal effects is not in any way comparable with that reported by alcoholics "drying out" or by heroin addicts going "cold turkey." Additionally, it could be noted that while such withdrawal effects may be a "rebound" phenomenon, particularly for the long-term, heavy smoker, it is equally plausible that this withdrawal syndrome may reflect a return by some smokers to their "normal" constitutional level of functioning. The possibility that smokers may be inherently different from nonsmokers derives from a variety of sources: from genetic studies, longitudinal personality studies of individuals before they start smoking (Cherry and Kiernan, 1978), and comparison of EEG characteristics of heavy and light smokers and ex-smokers and non-smokers (Brown, 1973).

2. Experimental Manipulation of Nicotine Levels

Better evidence in support of nicotine addiction theories derives from studies in which nicotine levels of smokers are artificially manipulated and the effects on smoking rate assessed. These methods have employed both (i) direct pharmacological control of nicotine utilizing intravenous or oral administration of nicotine or lobeline (Lucchesi, Schuster, and Emley, 1967; Kumar, Cooke, Lader, and Russell, 1978), nicotine receptor antagonists such as mecamylamine (Stolerman, Goldfarb, Fink, and Jarvik, 1973) or manipulation of nicotine excretion rates via urinary pH (Schachter et al., 1977; Feyerabend and Russell, 1978); (ii) <u>manipulation of the maximum possible nicotine delivery of a cigarette</u> by means of variations in nicotine content of tobacco, length of cigarette, and cigarette filter efficiency (cf. Stepney, 1980, for review of 16 published experiments utilizing this paradigm; Mangan and Golding, 1984, for a general review of this area of research; and Ando and Yanagita, 1981, for animal models).

Generally speaking, the observed as opposed to the predicted changes in smoking rates, following direct pharmacological or cigarette nicotine delivery manipulation have been relatively small. A number of factors, however, may have depreciated treatment effects. First, regarding direct supply of nicotine to the subject, this may have to be delivered in the form of pulsed microinjections into the carotid, in order to more closely mimic the delivery of nicotine to the brain by inhalation-style cigarette smoking, nicotine being thought to arrive at the brain in the form of a slug or "bolus" of nicotine- enriched blood (Feyerabend and Russell, 1978).

Second, smoking behavior itself may have some degree of "functional autonomy," i.e., the activity of smoking may come to acquire secondary reinforcing properties by virtue of its close temporal connection to the arrival of nicotine at the brain. The sight, smell, taste, physical manipulation, and inhalation associated with each puff may form a compound conditioned stimulus. Given the many thousands of such reinforcements and the short latency of nicotine arrival at the brain ("puff to brain" times of 8-10 seconds, approximately), conditioning theory predicts that smoking behavior would acquire such secondary reinforcing properties.

Third, it may be that the usual method of assessing smoking rate, i.e., numbers of cigarettes consumed, is far too crude an index. The smoker may control nicotine intake, the so-called "self-titration" for nicotine, by varying such smoking style parameters as number, pressure and duration of puffs, and depth of inhalation. Some such explanation is necessary to account for the recent observation that cigarette smokers appear to be able to maintain their plasma nicotine levels within fairly tight limits with only slight changes in cigarette consumption, in the face of large experimental variations in predicted nicotine delivery of cigarettes above and below their habitually smoked medium nicotine cigarettes (Ashton, Stepney, and Thompson, 1979; Ashton and Stepney, 1982).

Arousal Modulation

The "hedonic theory of arousal" (Berlyne, 1971) postulates that increase in arousal from low levels, and decrease in arousal from unpleasantly high levels are reinforcing events for the individual. "Low" and "high" arousal levels are relative to the postulated "optimal arousal level" which may vary both between individuals and as a function of particular environmental circumstances.

This is the rationale for the arousal modulation theory of smoking, which suggests that smoking is an activity which has the function of controlling arousal -- the smoker smokes to increase arousal when bored or fatigued and to reduce arousal when tense (Mangan and Golding, 1978, 1984). When the smoker is neither under- nor over-aroused, smoking continues partly through habit, and partly because it has become a positive reinforcer (cf. "secondary reinforcement" of smoking, discussed earlier). On a more general plane, such a theory explains why individuals who might be expected to be poor at controlling their arousal at the optimal level on the basis of their personality traits, should use any readily available stimulant (e.g., caffeine) or depressant (e.g., alcohol), as well as nicotine (which has both stimulant and depressant properties) in order to maintain homeostatic arousal level.

Thus, at the simplest level of analysis, whereas addiction theories of smoking postulate that smokers seek to maintain <u>nicotine</u> homeostasis, by contrast, arousal modulation theories suggest that smokers attempt to maintain <u>arousal</u> homeostasis. According to this viewpoint, cigarettes are only one of a number of ways by which this is possible, other means being drinking, drug taking, overeating, sports, and yoga.

Other Theories

Less well-supported theories about smoking are the genetic, opponent process, orality, and social learning theories.

Genetic theories, which have again assumed some prominence in the literature, suggest that there is a genetic predisposition underlying recruitment to and persistence of smoking. As yet, however, the evidence is not compelling, largely because of the unavailability of samples of MZ twins, reared apart, which, arguably, is the most critical test of the genetic hypothesis, the only instances recorded being relatively small sample studies reported by Fisher (1958a,b), Juel-Nielsen (1960), Shields (1962), and Holden (1980). On the other hand, there is substantial support for strong genetic involvement in a number of personality and life-style dimensions which have been shown to be correlated with smoking recruitment and maintenance. Arguments for, and evaluation of the various genetic hypotheses have been well reviewed elsewhere (cf. Fisher, 1959; Burch, 1978; Eysenck, 1980; Cederlof et al., 1977; Surgeon-General, 1979).

Opponent process theory postulates that the smoker experiences initially a period in which smoking is "positively" reinforcing, by contrast with relief from unpleasant withdrawal symptoms. The latter type of reinforcement becomes dominant in the subsequent, addictive phase (Surgeon General, 1979, 16: 9). This theory, therefore, is a more sophisticated version of addiction theory.

Orality/psychoanalytic theories claim that the etiology of smoking is a projection of the pleasure derived in childhood from stimulation of the mucous membrane of the mouth. Orality theories range from the rather simple models based on phallic analogies with the shape of the cigarette/cigar, to the more general "oral frustration" theory, which suggests that lack of maternal affection, which results in anxiety and rebelliousness, is the real motive underlying smoking (Freud, 1901; Izard, 1978).

At the other end of the spectrum, we have social learning theories, which propose that smoking is a learned habit, initially acquired under conditions of social reinforcement such as peer pressure. The habit is subsequently maintained, in the absence of primary social reinforcement, by factors which we have already identified in our discussion of addiction and arousal modulation theories. However, at second-order level, "social" factors still influence smoking behavior through the process of ritualization and incorporation of the habit into many everyday activities.

Finally, we might mention the various smoking typologies. Their importance lies in the fact that whereas the smoking theories described above assume that smokers are a homogeneous group, smoking typologies have examined the extent to which smokers are heterogeneous with regard to self-reported motives. A number of smoking typologies have been suggested, for example by Myrsten et al. (1975), Tomkins (1968), Ikard et al. (1969), Mausner and Platt (1971), McKennell (1970, 1973), Russell et al. (1974), and Crumpacker et al. (1979), identifying a wide range of smoking motives, including relaxation, stimulation, social situational, sensory motor/handling, "pleasure," and addictive/craving.

FACTORS DETERMINING RECRUITMENT OF SMOKING

There is little doubt that recruitment occurs mainly during adolescence (Surgeon General, 1979, A-14; Bewley, Day, and Ide, 1974), that boys tend to smoke more than girls (although this sex difference is reducing), that the percentage smoking increases with age toward 30% (boys) and 25% (girls) at age 17 years. These figures vary between regions, ethnic groups, and countries. For example, in Japan there is a much greater sex difference in cigarette smoking (males smoking more than females) than in other developed countries (Todd, 1978).

Most individuals have either tried or had the opportunity of accepting a cigarette during their adolescence or teens. The possible factors which determine whether or not a smoker is produced from the initial encounter with tobacco are discussed below.

The factor of primary importance is social rather than genetic, whether this is exercised by the family, peer group, or by culture through schools and mass media. This is demonstrated by the dramatic increase in smoking by women over the last fifty years, which has followed emancipation in terms of status and jobs. Nobody would suggest that the gene pool has changed over that period of time.

Studies (reviews in Surgeon General, 1979, Section 17; Cherry and Kiernan, 1978) have demonstrated the importance of various influences such as smoking by parents, siblings, and peers, and of personality and behavioral traits such as rebelliousness, extraversion, and neuroticism. Also of interest is the observation that loss of a parent is related to smoking (Holland et al., 1969). Relatively clear-cut evidence suggests that low academic achievement is predictive of smoking (although there may be some confounding effects of SES) which is not a reflection of differences in intelligence (Clarke et al., 1972), thus suggesting that noncognitive factors contribute to both underachievement and to smoking. The influence of school environment (including whether or not teachers smoke, educational antismoking programs), and also the mass media is as yet unclear, although no doubt important.

However, in spite of the mass of largely observational data, only a fraction of which has been cited, relatively few studies have attempted to test a broad battery of questions in order to assess which are the most predictive of smoking. The study reported here was directed to this goal.

General Plan Research

In Phase I of the study (1974-1975), between 450-500 items measuring a variety of biographical, sociocultural, and psychobiological variables were administered over a number of sessions to representative samples of 12- to 13-, 14- to 15-, and 16- to 17-year-old boys and girls (n = 752) attending Oxfordshire comprehensive schools. Internal consistency checks were run and data analyzed by stepwise multiple regression analysis, which identified sets of items predicting number of cigarettes smoked within each age-sex group. Biographical items were concerned with age, sex, structure of family, SES, details of pocket-money, earnings from part-time jobs, information on hobbies, TV preferences, place in class. Personality trait items were drawn from Eysenck and Eysenck's PQ scale, Cattell's HSPQ, Wilson and Patterson's Conversation Scale, Rotter's Locus of Control Inventory, Bett's QMI, Zuckerman's Stimulation Seeking Scale, and measures of "oral dependence" (consumption of sweets, alcohol, tea, coffee, thumb- and pencil-sucking, nail-biting). Sociocultural items were adapted from Robins' (1966) criteria of clinically diagnosed "psychopathy," including aggression, identification with the group, somatic symptoms, quality and stability of family, peer relationships, and attitudes to school and parental authority. Smoking behavior was assessed by items concerning number of cigarettes smoked per day, time, location, style (inhalation) of smoking, reasons for starting smoking, smoking and attitudes to smoking of parents and siblings, exposure and reactions to antismoking propaganda.

In Phase II of the study (1975-1976), these discriminative item sets were employed to produce a series of short, 6- to 30-item questionnaires for different age and sex groups with predictive power for smoking, nonsmoking, and "given-up" groups. These short questionnaires were validated on 1550 school children. Multiple regression and multiple discriminant analyses identified items predicting prevalence rates in still-smoking and given-up groups, and those differentiating still-smoking from given-up and nonsmoking groups in this sample.

Results/Discussion of Smoking Recruitment Study

For the sake of brevity, only the major results are detailed. The percentage of smokers (using the definition of at least 1 cigarette per week) were: 2nd Form (12-13 years), 18% (boys), 13% (girls); 4th Form (14-15 years), 41% (boys), 33% (girls); 6th Form (16-17 years), 18% (boys), 24% (girls). The relatively low proportions of smokers in the 6th Form probably reflects the fact that many pupils have left school before the 6th Form. Given the known relationships between scholastic achievement, SES, and smoking, it is reasonable to suggest that the smoker tends to leave school early. Otherwise the percentages are similar to those in the United Kingdom reported by other authors (Bewley et al., 1974).

Lists of items with predictive value for smoking, which include the "given-up" (GU) as well as still-smoking (SS) groups, are given in Table 1.

It is immediately obvious that there are great similarities between the GU and SS groups, and it is perhaps as well to bear in mind that neither category is fixed. The GU group obviously numbers a proportion of "undecided smokers," just as some of the SS group will no doubt cease smoking; a certain proportion of both groups are "waiverers," likely to cross boundaries. Thus, the item "If you have given up, do you intend to take up smoking again (a) yes, (b) undecided, (c) no," is given and (a)/(b) endorsement by 35% of the 2nd Form and 31% of the 4th Form GU pupils.

Regarding the detailed item content of Table 1, it is apparent that items predictive of smoking for the older age groups are high alcohol consumption, extraversion, and for girls, neuroticism. With the lower age groups, the importance of sibling and peer group influences becomes apparent in the 4th Form, although drinking items are also of importance (N.B. this potentially denotes rule-breaking behavior since the serving of alcohol to teenagers below the age of 18 years old is illegal in the United Kingdom). For the 2nd Form smoking groups (albeit representing a relatively small proportion of that age group), this trend is accentuated; smoking appears to be a marker for a number of critical life events: anxiety, rebellion, psychosexual fears (among boys, "belief in coeducation"). A more detailed picture of one such critical life event, that of parental absence, is given below.

A trend can be observed (Table 2) of increased smoking (SS versus GU versus nonsmoking) with parental absence. Sex differences regarding reason

Table 1

Group	Still smoking	Given up
6th Combined	High alcohol consumption; extraverted; permissive	High alcohol consumption; has used cannabis; permissive
6th Boys	Enthusiastic/happy-go-lucky; steady/ responsible; positive home relationships	Enthusiastic/happy-go-lucky; risk-taker
6th Girls	Poor mood control; values new experiences; tense	Large age gap to next sister; values new experiences; relaxed/composed
4th Combined	Oral dependent/stimulation-seeker; older brother heavy smoker; aggressive/dominant; high P	Oral dependent/stimulation seeker; aggressive/dominant; high P; older brother heavy smoker
4th Boys	High alcohol consumption; socially compliant parents divorced; outgoing/participating; permissive; anxious	High alcohol consumption; outgoing/partici- pating; authoritarian; unconcerned about health; anxious
4th Girls	Older sister smokes heavily; extraverted/ socially compliant; earns from part-time job	Older sister smokes heavily; socially comp- liant; values freedom from worry; receives generous pocket money; anxious
2nd Combined	Anxious; tense/frustrated; rebellious; undemonstrative; high P; high alcohol con- sumption; close friends academically oriented; vivid imagery	Aggressive/dominant; places little value on having fun; friends not academically orien- ted; wants to leave home; unconcerned about health
2nd Boys	Disapproves of coeducation; weak imagery, particularly tactual; older sister smokes heavily; anxious; group dependent/ socially bold	Disapproves of co-education; poor imagery, particularly olfactory; strong tactual imagery; father rated as not understanding; parent absent through death or divorce; group dependent/socially bold
2nd Girls	Threat-sensitive/anxious; wants to leave home; internally restrained; number of younger brothers at home; seeks advice from teachers	Poor imagery; parent absent from home through death or divorce; authoritarian; group- oriented

Table 2. Parental Absence in Still-Smoking, Given-Up,
and Nonsmoking groups -- 2nd Year Boys and Girls
(12-13 Years of Age)

| Reason for paren-tal absence | % Parental absence | | | | | |
| | Still smoking | | Given-up | | Nonsmoking | |
	♂	♀	♂	♀	♂	♀
Death	7.1	4	3.1	6.7	5.7	2.0
Divorce	9.5	32	0	6.7	5.7	4.9
Separation	11.9	16	9.2	5.0	2.9	2.0
Father works	23.8	8	24.6	11.6	11.6	8.8
All reasons combined	52.3	60	36.9	30	25.7	17.7

for parental absence are apparent -- divorce and separation having a great effect on girls and absence of the father more effect on boys.

It is perhaps instructive to compare the items predictive of smoking among teenagers with those found from a comprehensive study of the adult smoking population (Cederlof et al., 1977). A great many of the items predicting smoking among the teenagers obviously carry over into the adult population -- excessive alcohol drinking, anxiety, extraversion, and poor home relationships, for example. For the young adult population who are smoking, drug-taking may be added to this constellation of interrelated behaviors (Golding et al., 1983).

Smoking Maintenance

The factors maintaining smoking are obviously very diverse. The relative importance of "addictive" as opposed to "arousal modulation" factors may vary between individuals -- as suggested by smoking typologies -- and over time -- as suggested by opponent process theory. Certainly, the various animal studies utilizing acute and chronic nicotine administration indicate that both viewpoints have some validity insofar as the effects of nicotine on physiological and performance measures are concerned (Russell, 1976). In addition, human smoking becomes enmeshed in a variety of everyday activities, such as social drinking, which throws into prominence those factors emphasized by social learning theories.

We espouse an arousal modulation theory of smoking maintenance, ignoring for the moment the question of whether gains in mood control and performance efficiency from smoking are more a consequence of relief from nicotine withdrawal symptoms than true "above baseline" effects. This model seems to have greater explanatory power in accounting for the observed effects and for some of the apparent inconsistencies and contradictions reported in the smoking literature.

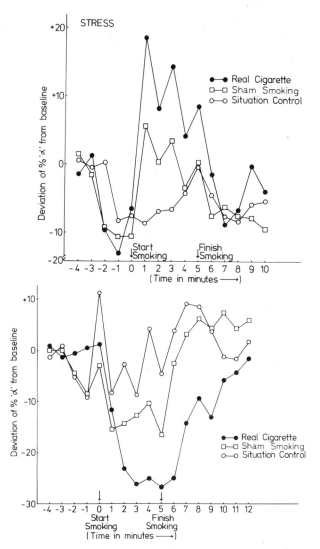

Figure 1. Effects of cigarette smoking on EEG alpha during stress (top) and relaxation/sensory isolation (bottom) conditions (n = 8, 8, 8 for each experimental group: real smoking; activity control/sham smoking/situation control).

In the context of a general consensus that nicotine is the primary reinforcer for smoking, we have to account for the clearly demonstrated biphasic, stimulant/depressant, dose-related effects of nicotine, small doses being stimulant, high doses depressant (Armitage et al., 1969; Ashton et al., 1974, 1978). This has been clearly shown in animal studies; nicotine can both facilitate and impair the performance of animals on a variety of tasks, including bar pressing for food reward, maze learning, active avoidance of shock, and so on (Russell, 1976, for review). Whether the effect is facilitatory or depressive is largely a function of dosage, which, of course, is under experimenter control.

Insofar as human smoking is concerned, nicotine reinforcement has two quite distinct aspects -- psychological, in the sense of mood control, and behavioral, in the sense of performance enhancement, although the latter may be in part a consequence of the former.

The role of nicotine as a mood control agent is clearly demonstrated in the Golding and Mangan (1982a) study. Details are presented in Figure 1. From the figure it is clear that, under stressful conditions, cigarette smoking decreases arousal (i.e., promotes relaxation) and produces alpha-blocking (i.e., increases arousal) under conditions of sensory isolation.

Such effects may help explain, in terms of the well-known "inverted U curve" relationship of cortical efficiency to arousal, the improved performance of smokers in a variety of tasks, including vigilance (Wesnes and Warburton, 1978; Mangan, 1982), driving simulation (Heimstra, Bancroft, and De Kock, 1967), Stroop (Wesnes and Warburton, 1978), reaction time (Myrsten and Andersson, 1978). Consolidation of memory traces, as measured by subsequent retest, is also generally improved (Andersson, 1975; Mangan, 1983; Mangan and Golding, 1983).

While the acquisition of simple verbal rote learning (Andersson, 1975) and immediate memory for digits (Williams, 1980) may be slightly impaired or unaffected by cigarette smoking, the data reported by Mangan and Golding (1978) and Mangan (1983) suggest that, with the inclusion of more "irrelevant" material in the learning task (thus making the task more difficult), the major effect of smoking is to focus the subject's attention (cf. Johnston, 1965, 1966) on the immediate task, and to prevent him from attending to irrelevant stimuli. A similar type of effect on "incidental learning" has been demonstrated (Andersson and Hockey, 1977), and is congruent with suggestions made, on the basis of the observation of increased habituation rate to irrelevant or irritating stimuli under cigarette smoking (Friedman et al., 1974; Golding and Mangan, 1982a,b) that smoking can help "screen out" irritating or distracting stimuli.

Clearly, in the case of human smoking, dosage is, to a degree, after the control of the smoker. There is considerable evidence to suggest that differential effects, i.e., stimulant or depressant, can be obtained through the manner in which the cigarette is smoked, as demonstrated in Figure 1. It has been observed that smokers in high-arousal situations tend to puff more frequently, more vigorously, and to inhale to a slightly greater extent than when in low-arousal situations (Golding, 1980). These parameters of "smoking style" can be measured accurately using plumbed holders to assess puff numbers, pressure, and duration and measures of inhalation depth (thoracic Hg strain gauge) (Figure 2).

The smoker's ability to rapidly alter his or her nicotine intake by varying the vigor with which the cigarette is smoked is indicated by the close similarity of physiological effect of smoking cigarettes through selective filters delivering a predicted 1.8 mg (high nicotine) as compared

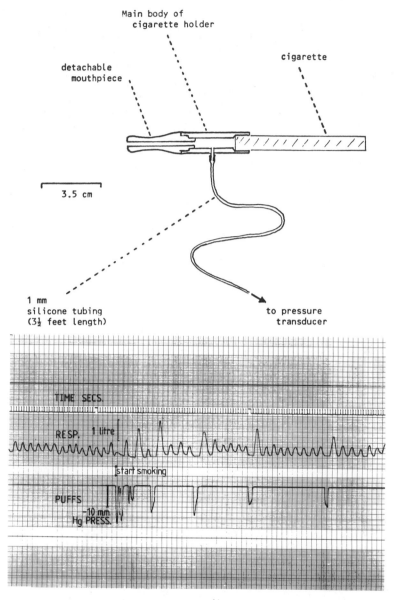

Figure 2. Plumbed cigarette holder (top) and specimen output together with associated inhalation during smoking.

to 0.4 mg (low nicotine), these values being "predicted" on the basis of "standard smoking" by a smoking machine.

As can be seen from Figure 3, however, the differential effects on heart rate, skin conductance level (SCL), and EEG alpha of smoking high-

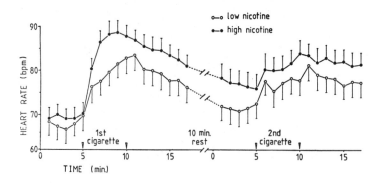

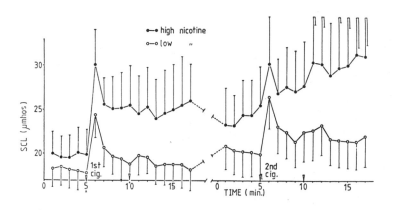

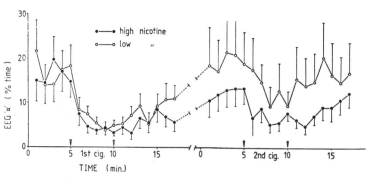

Figure 3. Effects of smoking high- and low-delivery-nicotine cigarettes on top: heart rate (mean, S.E. bars), middle: skin conductance level (mean, S.E. bars), and bottom: EEG alpha (mean, S.E. bars) (% time > criterion).

versus low-nicotine-delivery cigarettes was much less than might be expected, most differences being nonsignificant (Golding, 1980). However, analysis of smoking style and residual butts indicated that subjects smoked the low-nicotine-delivery cigarettes more vigorously, which no doubt reduced differences on physiological measures. These particular results are, of course, compatible with addiction theory also (cf. Herning et al., 1981).

Such "self-titration" for nicotine by the smoker may have important consequences not only for the understanding of the dynamics of smoking, but also in terms of health. If smokers switching to low tar/nicotine/carbon monoxide cigarettes simply puff harder, then the postulated health advantages will be reduced (Russell, 1980).

Consideration of all these effects of cigarette smoking on arousal and performance suggests that the number of factors maintaining smoking is considerable and indicates why some smokers should find it difficult to quit the habit. In those individuals who might be expected to be inherently poor at arousal control (neurotic and anxious heavy smokers, for example), it is found that they do, in fact, have greater difficulty in giving up (Cherry and Kiernan, 1978). For these individuals, alternative mood control agents, for example yoga, biofeedback, may be of great help. On the other hand, for those individuals whose problem is mainly underarousal, that is the "stimulation smoker" described in various typologies, alternative "substitute" activities should be of a more arousing nature.

Of course, it is possible that, for some smokers, giving up may lead to increased use of other mood control agents such as alcohol, a point underlined in a recent reanalysis of changing mortality patterns in British doctors who have given up smoking (Lee, 1979). In younger age groups, this "drug need" may reemerge in different forms, for example as solvent abuse.

CONCLUSION

Since the 1950s the prevalence of cigarette smoking has reduced significantly from about 70% to less than 40% (approx.) in the United States, for example. However, 95% of this reduction has been as a result of smokers' own efforts, due to any one or combination of a number of factors, such as cost or general public health education (Surgeon-General, 1979). In spite of a proliferation of specific intervention programs utilizing a variety of techniques, e.g., aversive conditioning, rapid smoking, contractual management, individual and group counseling, there is still no unequivocal evidence that treatment-induced cessation rates are superior to no-treatment control rates, when the target groups are adequately observed over 6- to 12-month periods (Leventhal and Cleary, 1980), although some reviewers strike a more optimistic note (Orleans et al., 1981a,b). Of course, it can be argued that the smokers presenting themselves for these treatments represent a highly self-selected "hard core" who have difficulty in quitting. Nonetheless, it would appear from the research evidence currently reviewed that the failure of specific intervention programs may be due in part to a lack of appreciation of the varying motives for the maintenance of cigarette smoking, and consequently the adoption of a "one suit fits all" approach to cessation programs. Future research could be directed toward more accurate identification of particular subgroups of smokers (cf. Best and Hakstian, 1978), a better understanding of the reinforcing nature of cigarette smoking, and increased use of objective biological assays for smoking (e.g., serum thiocyanate, exhaled carbon monoxide) rather than self-report, which in this context we would expect to be unreliable.

In the long term some of the money and effort currently expended on persuading people to stop smoking might be more usefully employed in efforts to dissuade young people from taking up the habit, for example by increased health education in schools. It is surely a truism that prevention is better than cure.

REFERENCES

Andersson, K. "Effects of cigarette smoking on learning and retention." Psychopharmacologia, 41: 1-5, 1975.

Andersson, K., and Hockey, G. R. S. "Effects of cigarette smoking on incidental memory." Psychopharmacologia, 52: 223-226, 1977.

Ando, K., and Yanagita, T. "Cigarette smoking in Rhesus monkeys." Psychopharmacology. 72: 117-127, 1981.

Armitage, A. K., Hall, G. H., and Sellers, C. M. "Effects of nicotine on electrocortical activity and acetylcholine release from the cat cerebral cortex." Brit. J. Pharmacol., 35: 152-160, 1969.

Ashton, H., and Stepney, R. (1982). Smoking, Psychology and Pharmacology. Tavistock, London, U. K.

Ashton, H., Stepney, R., and Thompson, J. W. "Self-titration by cigarette smokers." Brit. Med. J. 2: 357-360, 1979.

Ashton, H., Millman, J. E., Telford, R., and Thompson, J. W. "The effects of caffeine, nitrazepam and cigarette smoking on the contingent negative variation in man." EEG & Clin. Neurophysiol. 37: 59-71, 1974.

Ashton, H., Marsh, V. R., Millman, J. E., Rawlins, M. D., Telford, R., and Thompson, J. W. "The use of event-related slow potentials of the brain as a means to analyse the effects of cigarette smoking and nicotine in humans." In R. E. Thornton (Ed.), Smoking Behaviour: Physiological and Psychological Influences, pp. 54-68. Edinburgh: Churchill Livingstone, 1978.

Berlyne, D. E. Aesthetics and Psychobiology. New York: Appleton, 1971.

Best, J. A., and Hakstian, A. R. "A situation-specific model for smoking behavior." Addictive Behaviours, 3: 79-92, 1978.

Bewley, B. R., Day, I., and Ide, L. Smoking by Children in Great Britain, A Review of the Literature. London: Social Science Research Council, Medical Research Council, 1974.

Brown, B. B. "Additional characteristic EEG differences between smokers and non-smokers." In W. L. Dunn (Ed.), Smoking Behavior: Motives and Incentives, pp. 67-81. Washington, D. C.: Winston, 1973.

Burch, P. R. J. "Smoking and lung cancer: The problem of inferring cause." J. Roy. Statist. Soc. A 141: 437-477, 1978.

Cederlof, R., Floderus, B., and Friberg, L. "Cancer in MZ and DZ twins." Acta Geneticae Medicae et Gemellologiae, 19: 69-74, 1970.

Cederlof, R., Friberg, L., and Lundman, T. (1977). "The interactions of smoking, environment and heredity and their implications for disease etiology. A report of the epidemiological studies on the Swedish twin registries." Acta Medica Scandinavica, Supplement No. 612.

Cherry, N., and Kiernan, K. E. "A longitudinal study of smoking and personality." In R. E. Thornton (Ed.), Smoking Behaviour: Physiological and Psychological Influences, pp. 12-18. Edinburgh: Churchill Livingstone, 1978.

Clarke, R. V. G., Eyles, H. J., and Evans, M. "The incidence and correlates of smoking among delinquent boys committed for residential training." Brit. J.Addict. 67: 65-7, 1972.

Crumpacker, D. W., Cederlof, R., Friberg, L., Kimberling, W. J., Sorenson, S., Vandenberg, S. G., Williams, J. S., McClean, G. E., Grever, B., Iyer, H., Krier, M.J., Pedersen, N. L., Price, R. A., and Roulette, I. "A twin methodology for the study of genetic and environmental control of variation in human smoking behaviour." Acta Genet. Med. G. 28:173-195, 1979.

Eysenck, H. J. (with contributions by L. J. Eaves). The Causes and Effects of Smoking. London: Maurice Temple Smith Limited, 1980

Feyerabend, C., and Russell, M. A. H. "Effect of urinary pH and nicotine secretion rate on plasma nicotine during cigarette smoking and chewing nicotine gum." Brit. J. Clin. Pharmacol. 5:293-297, 1978.

Fisher, R. A. "Lung cancer and cigarettes." Nature 182: 108, 1958a.

Fisher, R. A. "Cancer and smoking." Nature 182: 596, 1958b.

Fisher, R. A. Smoking. The Cancer Controversy. Edinburgh & London: Oliver & Boyd, 1959.

Freud, S. Complete psychological works of Sigmund Freud. In J. Strachey (Ed. and trans.), "Infantile Sexuality II." Vol. VII. "Auto-Eroticism," p. 182. London: Hogarth Press, 1953.

Friedman, J., Horvath, T., and Meares, R. "Smoking and a stimulus barrier." Nature 248: 455-456, 1974.

Golding, J. R. Short-Term Effects of Cigarette Smoking. Unpublished Ph.D Thesis, Bodleian Library, Oxford, 1980.

Golding, J. F., and Mangan, G. L. "Arousing and de-arousing effects of cigarette smoking under conditions of stress and mild sensory isolation." Psychophysiology. 19:449-456, 1982a.

Golding, J. F., and Mangan, G. L. "Effects of cigarette smoking on measures of arousal response suppression and excitation/inhibition balance." Int. J. Addict. 17:793-804, 1982b.

Golding, J. F., Harpur, T., and Brent-Smith, H. "Personality drinking and drug-taking correlates of cigarette smoking." Personality and Individual Differences 4: 703-6, 1983.

Gray, J. A. The Psychology of Fear and Stress. London: Weidenfeld and Nicolson, 1971.

Hall, R. A. M., Rappaport, H. K., Hopkins, H. K., and Griffin, R. "Tobacco and evoked potential." Science 180: 212-214, 1973.

Herning, R. I., Jones, R. T., Bachman, J., and Mines, A.H. "Puff colume increases when low-nicotine cigarettes are smoked." Brit. Med. J. 283: 187-189, 1981.

Holden, C. "Identical twins reared apart." Science 207: 1323-1328, 1980.

Holland, W. W., Halil, T., Bennett, A. E., and Elliott, A. "Indications for measures to be taken in childhood to prevent chronic respiratory disease." Millbank Mem. Quart., Vol. XLVII, 215-227, 1969.

Ikard, F.F., Green, D. E., and Horn, D. A. "A scale to differentiate between types of smoking as related to management of affect." Int. J. Addict. 4: 649-669, 1969.

Izard, C. "Neuropsychology and tobacco." In R. E. Thornton (Ed.), Smoking Behaviour: Physiological and psychological influences, pp. 44-53. Edinburgh: Churchill Livingstone, 1978.

Jaffe, J. H., and Kanzler, M. "Smoking as an addictive disorder." Nat. Inst. Drug Abuse Res. Monogr. Ser. 23: 4-23. Rockville, Md.: N.I.D.A., 1979.

Johnston, D. M. "Preliminary report on the effect of smoking on the size of visual fields." Life Sciences 4: 2215-2211, 1965.

Johnston, D. M. "Effect of smoking on visual search performance." Percept. & Mot. Skills 22: 619-622, 1966.

Juel-Nielsen, N. (Pers. comm. quoted by E. Raaschou-Nielsen). "Smoking habits in twins." Danish Med. Bull. 7: 82-88, 1960.

Knapp, P. H., Bliss, C. M., and Wells, H. "Addictive aspects of heavy cigarette smoking." Am. J. Psychiat. 119: 966-972, 1963.

Knott, V. J., and Venables, P. H. "EEG alpha correlates of non-smokers, smokers, smoking and smoking deprivation." Psychophysiology 14: 150-156, 1977.

Knott, V. J., and Venables, P. H. "Stimulus intensity control and cortical evoked response in smokers and non-smokers." Psychophysiology 15: 186-192, 1978.

Kumar, R., Cooke, E. C., Lader, M. H., and Russell, M. A. H. "Is tobacco smoking a form of nicotine dependence?" In R. E. Thornton (Ed.), Smoking Behaviour: Physiological and Psychological Influences, pp. 244-458. Edinburgh: Churchill Livingstone, 1978.

Lee, P. N. "Has the mortality of male doctors improved with the reductions in their cigarette smoking?" Brit. Med. J. 2: 1538-1540, 1979.

Leventhal, H., and Cleary, P. D. "The smoking problem: A review of the
research and theory in behavioral risk modification." Psychol. Bull.
88: 370-405, 1980.

Lucchesi, B. R., Schuster, C. R., and Emley, G. S. "The role of nicotine as
a determinant of cigarette smoking frequency in man with observations
of certain cardiovascular effects associated with the tobacco
alkaloid." Clin. Pharmac. & Therapeutics 8: 789-796, 1967.

Mangan, G. L. "The effects of cigarette smoking on vigilance performance."
J. Gen. Psychol. 106: 77-83, 1982.

Mangan, G. L. "The effects of cigarette smoking on verbal learning and
retention." J. Gen. Psychol. 108: 203-210, 1983.

Mangan, G. L., and Golding, J. F. "An `enhancement' model of smoking
maintenance?" In R. E. Thornton (Ed.), Smoking Behaviour:
Physiological and Psychological Influences, pp. 87-114. Edinburgh,
Churchill Livingstone, 1978.

Mangan, G. L., and Golding, J. F. The Psychopharmacology of Smoking.
Cambridge, U. K.: Cambridge University Press, 1984.

Mausner, B., and Platt, E. S. Smoking: A Behavioral Analysis. New York:
Pergamon Press, 1971.

Mello, N. K., Lukas, S. E., and Mendelson, J. H. "Buprenorphine effects on
cigarette smoking." Psychopharmacology 86: 417-425, 1985.

Myrsten, A-L., Elgerot, A., and Edgren, B. "Effects of abstinence from
tobacco smoking on physiological and psychological arousal levels in
habitual smokers." Psychosom. Med. 39: 25-38, 1977.

Myrsten, A-L., Andersson, K., Frankenhaeuser, M., and Elgerot, A.
"Immediate effects of cigarette smoking as related to different smoking
habits." Percept. & Mot. Skills 40: 515-523, 1975.

McKennell, A. C. "Smoking motivation factors." J. Soc. Clin. Psychol 9: 8,
1970.

McKennel, A. C. A comparison of two smoking typologies. Research Paper 12,
London: Tobacco Research Council, 1973.

Orleans, C. T., Shipley, R. H., Williams, C., and Haac, L. A. "Behavioral
approaches to smoking cessation - I. A decade of research progress
1969-1979." J. Behav. Ther. & Exp. Psychiat. 12: 125-129, 1981a.

Orleans, C. T., Shipley, R. H., Williams, C., and Haac, L. A. "Behavioral
approaches to smoking cessation - II. Topical Bibliography 1969-1979."
J. Behav. Ther. & Exp. Psychiat. 12: 131-144, 1981b.

Pomerleau, O. F., Fertig, J. B., Seyler, L. E., and Jaffe, J.
"Neuroendocrine reactivity to nicotine in smokers." Psychopharmacology
81: 61-67, 1983.

Raaschou-Nielsen, E. "Smoking habits in twins." Danish Med. Bull. 7:
82-88, 1960.

Robins, L. N. Deviant Children Grown Up. Baltimore: William & Witkines,
1966.

Russell, M. A. H. "Tobacco smoking and nicotine dependence." In R. J.
Gibbons, Y. Israel, H. Kalant, R. E. Popham, W. Schmidt, and R. G.
Smart (Eds.), Research Advances in Alcohol and Drug Problems, Vol. III,
pp. 1-47. New York: Wiley, 1976.

Russell, M. A. H. "The case for medium-nicotine, low-tar, low-carbon
monoxide cigarettes." In G. B. Gori and F. G. Bock (Eds.), Banbury
Report: A Safe Cigarette, pp. 297-310. New York: Cold Spring Harbor
Laboratory, 1980.

Russell, M. A. H., Peto, J., and Patel, V. A. "The classification of
smoking by factorial structure of motives." J. Roy. Statistical
Society (A) 137: 313-46, 1974.

Ryan, F. J. "Cold turkey in Greenfield, Iowa: A follow-up study." In W.
L. Dunn (Ed.), Smoking Behavior: Motives and Incentives, pp. 231-241.
New York: Wiley, 1973.

Schachter, S., Silverstein, B., Kozlowski, L. T., Perlick, D., Herman, C.
P., and Liebling, B. "Studies of the interaction of psychological and

pharmacological determinants of smoking." J. Exper. Psychol. (Gen.) 106: 3-40, 1977.

Shields, J. Monozygotic Twins Brought Up Apart and Brought Up Together. London & New York: Oxford University Press, 1962.

Shiffman, S. M. "The tobacco withdrawal syndrome." NIDA Monogr. Ser., 23: Rockville, Md.: N.I.D.A., pp. 158-184, 1979.

Stepney, R. "Consumption of cigarettes of reduced tar and nicotine delivery." Brit. J. Addiction 75: 81-88, 1980.

Stolerman, I. P., Goldfarb, T., Fink, R., and Jarvik, M. E. "Influencing cigarette smoking with nicotine antagonists." Psychopharmacologia 28: 247-259, 1973.

Surgeon-General. Smoking and Health: A Report of the Surgeon General, U. S. Department of Health, Education and Welfare. Washington, D. C.: U.S. Government Printing Office, 1979.

Todd, G. F. "Cigarette consumption per adult of each sex in various countries." J. Epidemiol. & Community Health 32: 289-293, 1978.

Tomkins, S. S. "A modified model of smoking behavior." In E. F. Borgatta and R. Evans (Eds.), Smoking, Health and Behavior. Chicago: Aldine, 1968, pp. 165-186.

Vogel, W., Broverman, D., and Klaiber, E. L. "Electroencephalographic responses to photic stimulation in habitual smokers and non-smokers." J. Comp. Physiol. Psychol. 91: 418-422, 1977.

Ulett, J. A., and Itil, T. M. "Qualitative electroencephalogram in smoking and smoking deprivation." Science 164: 969-970, 1969.

Wesnes, K., and Warburton, D. M. "The effects of cigarette smoking and nicotine tablets upon human attention." In R. E. Thornton (Ed.), Smoking Behaviour: Physiological and Psychological Influences, pp. 131-147. Edinburgh: Churchill Livingstone, 1978.

Williams, D. G. "Effects of cigarette smoking on immediate memory and performance in different kinds of smoker." Brit. J. Psychol. 71: 83-90, 1980.

SOCIAL PHARMACOLOGY FROM A MACROSYSTEMATIC PERSPECTIVE

Romuald K. Schicke

Medizinische Hochschule Hannover
D-3000 Hannover 72

INTRODUCTION

Prior to analyzing systemic aspects of social pharmacology a definition of social pharmacology itself may be indicated. Social pharmacology can be termed a social science which deals with the multifaceted relationships of medicines in society. More specifically it embraces:

1. The manifold relationships between the individual-medicines-and society

2. The role of medicines in a patient-physician relationship

3. The relationship of the physician to medicines

4. The role and position of the pharmaceutical industry

5. The economy and efficacy of the medication system

6. The ensuing information and communication system

7. The behavioral patterns of providers and consumers with respect to medicines

It deals, in summary, with the social ecology of the drugs or medicines and their multiple relationships in society (Schicke, 1976). Such, of course, requires a multidisciplinary approach (Venulet, 1974).

Needless to say, social pharmacology may be viewed in its macrosystemic or microsystemic dimensions; either at the society's level or at that of individual actors in their transactional processes and relationships. This analytic article proposes to focus upon the macrosystemic aspects.

In this connection the definition of systems refers mainly toward an integration of social and economic systems (Parsons and Smelser, 1965). The system itself is considered a body of an organized set of ideas and facts arranged to form a regular interaction or interdependence and to constitute an integrated whole. This approach denotes a holistic one. It centers upon the role and position of the medication or drug in society which, in turn, may lend itself to indicate the constituent elements or subsystems from cultural, historical, social, and socioeconomic perspectives.

THE CULTURAL-HISTORICAL SYSTEM PERSPECTIVE

The historical and cultural approach may help to identify some broad
stages:

1.   Up to the 17th century -- the prevalence of mystically ascribed
      properties to substances;

2.   During the 18th century -- efforts to isolate qualities and
      properties by experiments and empirical methods;

3.   In the 19th century -- extraction and analysis of "useful"
      medicaments;

4.   And in the 20th century -- synthesis of effective and potent
      substances and their mass production and distribution. (Schicke,
      1974)

Such developments have been accompanied by a conceptual path leading
from experimental to social pharmacology (Venulet, 1974, 1978).

No extensive attempts should be made within this context to expand upon
century-long developments in the search for remedies or cures of human
ailments. It must be pointed out, however, that the road from mainly herbal
folk- and house-remedies to the modern synthetic pharmaceuticals has been
paved with obstacles of aberrations in human mind and beliefs. These
included the everlasting search for miracle drugs and instant cures. The
latter range from such doctrines as F. A. Mesmer's (1734-1815) magnetism,
wearing of copper bracelets to remedy rheumatic diseases, using Laetrile to
cure cancer, or consuming high dosages of ascorbid acid to arrest same, up
to "fresh-cell" therapy of Dr. P. Niehans (1882-1971) to ensure rejuvenation
and increase vitality.

The faith in the healing powers of nature (herbs, balneology,
climatology) especially doctrinized in the 19th century Continental Europe,
has markedly influenced the healing sciences, medication production, and
consumption right up to the present. These include, for instance, the wide
usage of herb-extracts with high alcohol concentrations to remedy a variety
of heterogeneous ailments and discomforts, not seldom by the very same
preparation; S. Kneipp's (1821-1897) water therapy for curing various
conditions, and S. Hahnemann's (1755-1843) treatment by homeopathic
remedies, currently in use, with unproved efficacy.

MEDICAMENTS AND DRUGS AND SOCIOECONOMIC SUBSYSTEMS

A further systemic dimension in viewing medicines and drugs may be
offered from the social and economic perspective. If one turns to the
classification of social markets of therapeutics one may find that the
sociological and economic aspects may be viewed congruently. In this
connection, we may distinguish between three basic types of markets:

1.   Illicit drugs of nonmedical application

2.   The parapharmaceuticals or over-the-counter (OTC) drugs --
      predominantly self-administered

3.   The ethical or prescription drugs (see Table 1)

As far as illicit drugs are concerned, the drug problem may be
associated with acculturation processes and factors, so that in each society
problems may arise in coming to grips with unknown drugs or substances. At

Table 1

Type of drug or medicament (and its mode of administration)	Group	Social		Market
		Subsystem	Control	
1. Drugs for non-medical use (self-administered)	opiates, hashish, marijuana, LSD, etc.	subculture (anticulture)	reference and peer groups	illegal restricted
2. Parapharmaceuticals (predominantly self-prescribed)	digestive, analgetics, antipruretics, laxatives, etc.	lay-system (sub- or para-professional)	primary groups	legitimate unrestricted
3. Pharmaceuticals-ethical drugs ("other prescribed," in part, self-administered)	antibiotics, cardiacs, psychotropics, etc.	professional	physician	legal restricted

Source: Schicke, 1974.

the social level one can link those problems with disturbed socialization processes of the drug users and the influence of the peer groups. When transferred to the microsociological level of the individual, drug abuse may progress through several stages, i.e., from acculturation, psychosomatic dependence with deviant behavior, to economic dependence and delinquency. Satisfaction of curiosity, with the dominant influence of normative reference -- and peer-groups, plays a rather significant role in the use of nonmedical and illicit drugs, as demonstrated by several studies (Wertz, 1972; Berger et al., 1980).

Following the category of illicit drugs, the use and misuse of parapharmaceuticals or OTC remedies may be considered. Here the lay-system and the sub- and paraprofessionals play an important facilitating role, next to the primary groups. The main impetus for self-treatment and self-medication stems from primary groups (spouse, parents, etc.). The extent of self-medication may, in a wider perspective, be culturally determined depending upon the stage of societal development. The latter relates to the relationships of myth/belief medicine, folk medicine, and the level of scientific pharmacological industrial development and general education (see Figure 1).

To what extent the total consumption level or the composite of preference is influenced by sociocultural factors, on the one hand, or possibly by shortages in supplies and the spectrum of choices available on the other, cannot easily be quantified. The latter aspects, however, may be of interest in examining the role of the OTC medications allotted within the entire market of medications. Such an analysis may be quantified in some monetary terms. From this perspective, in some industrialized countries (e.g., West Germany, the United States, England), OTC medications may range between 15-30% of total medicinal sales. When viewing the total expenditures for drugs and medications and the relationship between the OTC and ethical preparations, one has also to consider the constraint imposed by the given health insurance schemes and problems of an operationalized definition of a medicine covered by insurance which may vary internationally (Schicke, 1974, 1976). Thus one could consider, for example, the aspirin or oral contraceptives, which could be dispensed in different markets (see Figure 2).

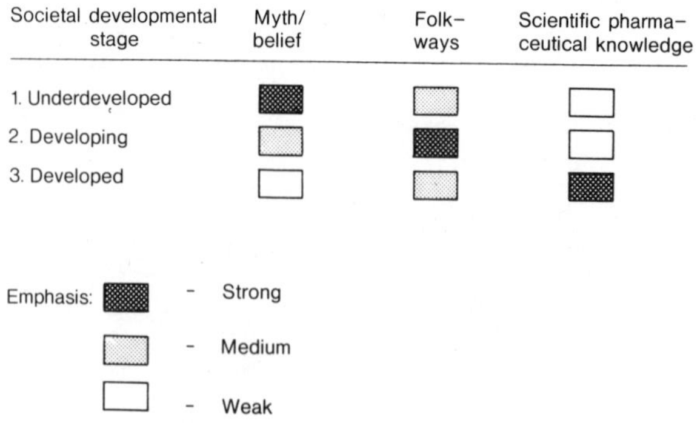

Societal developmental stage	Myth/belief	Folk-ways	Scientific pharma-ceutical knowledge
1. Underdeveloped	▓ (strong)	▒ (medium)	☐ (weak)
2. Developing	▒ (medium)	▓ (strong)	☐ (weak)
3. Developed	☐ (weak)	▒ (medium)	▓ (strong)

Emphasis: ▓ – Strong

☐ – Medium

☐ – Weak

Figure 1. Societal development stages in the pharmaceuticals' and remedies' diffusion process.

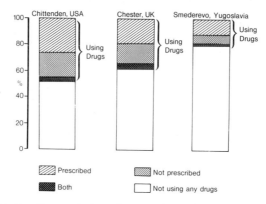

Figure 2. Drugs taken in previous two days, adults.
3 selected countries.

Moreover, in examining the OTC consumption of medications one must duly consider the level of a patient's perception of symptoms, discomfort, pain, and expectation which may be socially, culturally, subculturally, or personality differentiated (Zborowski, 1952).

In analyzing the actual and/or potential impact of the ethical drug sector one must consider the following parameters:

1. The systematic aspects of professional control

2. Determination of need

3. The resultant transactional process between the patient and the physician

4. The physician's attitudes

5. The patient's expectations and anxiety in terms of and/or compared to the severity of the case.

At the dispensing physician's level one must also take note of:

1.   The physician's knowledge, experience, and attitudes to drugs

2.   His customary way of practice, which would also reflect his customary prescription pattern

3.   The information level on new drugs

4.   His attitude to innovation

5.   Monitoring adverse reactions and interactions

6.   Assessment of effectiveness

7.   Refraining from polypharmacy

A further aspect constitutes the patient's attitude toward medications' intake and his readiness to comply with the medication regimen. One thus could speak of malconsumption which considers both under- and over-medication, both at the prescriber's and the consumer's level. In self-administration of drugs various errors may be committed, which reduce the effectiveness of drug consumption. This causes, at times, economic waste, aside from side-effects or other serious health hazards. These issues will not be elaborated upon in this paper.

When medications are administered by others, mostly during confinement in an institution, errors of commission or omission may be made by nurses and allied personnel. These events should be or may be duly monitored. Some studies in the United States indicate an error rate in hospitals of 15% (Barker et al. 1968) and in nursing homes ranging between 20-40% (Senate Special Committee on Aging, 1975).

SYSTEMS OF PRODUCTION, DISTRIBUTION, AND CONSUMPTION OF MEDICINES

Another systemic aspect represents the production of effective drugs by the pharmaceutical industry and adequate quality control. Here self-control may be most important and desirable. Depending upon society, different control mechanisms have been employed (e.g., The Federal Drug Administration in the USA) or control left to the industry (e.g., Switzerland, West Germany). Overwhelmingly, however, OTC "medications" and homeopathic remedies are exempt from proof of efficacy and safety.

Different modes of quality control may obviously have different and safeguarding effects. They should not constitute a bureaucratic delay function in the proliferation of new effective drugs. Their primary role is to curb dispensation of "medications" of unknown efficacy and safety already on the market.

SOCIOECONOMIC SYSTEMS OF PHARMACEUTICAL MARKETS

At another level of macroanalysis the given national structures of drug markets could be examined in its ogranizational and social dimensions including their economic aspects.

This allusion to various types of markets from a socioeconomic point of view may stimulate an attempt toward a broad typology, which would embrace five types of models such as:

1.   The East European type, where the production and distribution of pharmaceuticals is fully socialized within a state-monopolistic central planning structure, competing with other heterogeneous items within the state budget, in which consumer goods do not rank extremely high.

2.   The Scandinavian system which basically preserves a quasi-free market production sector with state controlled distrubution at the wholesale level (e.g., "Apoteksbolaget" in Sweden and state-owned distribution in Norway).

3.   The British system where the National Health Service occupies, vis-a-vis the pharmaceutical industry, the role of quasi-monopolist and represents a "countervailing power" on the demand side, whose endeavors are aided by the "Pharmaceutical Price Regulation Scheme."

4.   The North American system (especially in the USA) where relatively strict controls (Federal Drug Administration) are centered upon quality aspects at the manufacturer's level, and at the distribution level; a diversified mechanism permits more price competition than that which is possible in the Federal Republic of Germany.

5.   The West European system (e.g., West Germany and Switzerland.) where quality and effectiveness controls are predominantly left to the industry itself, in part, with a resultant greater abundance of medications on the market (Schicke, 1973). Since a concept of basically administrative (fixed) prices still traditionally prevails (carried forward since the Middle Ages).

REFERENCES

Barker, K. N. et al. A Study of Medication Errors in a Hospital, 2nd ed.
    Oxford, Mississippi: University of Mississippi, March 1968.
Berger, H. et al. Wege in die Heroinabhängigkeit; Zur Entwicklung
    abweichender Karrieren. Munich: Juventa, 1980.
Parsons, T., and Smelser, N. J., Economy and Society: A Study in the
    Integration of Economic and Social Theory. New York: The Free Press,
    1965.
Schicke, R. K. "The pharmaceutical market and prescription drugs in the
    Federal Republic of Germany: cross-national comparisons",
    International Journal of Health Services, 3: 223-236, 1973.
Schicke, R. K. "Arzneimittelmärkte aus sozio-ökonomischer Sicht",
    Pharmazeutische Zeitung 119: 1415-1420, 1974.
Schicke, R.K. "Soziookonomische Aspekte der Arneimittelversorgungs-systeme",
    Niedersächsisches Ärzteblatt, 47: 766-780, 1974.
Schicke, R. K. Sozialpharmakologie, Stuttgart-Berlin-Köln-Mainz: W.
    Kohlhammer Verlag, 1976.
Senate Special Committee on Aging, Subcommittee on Long-term Care, 94th
    Cong., 1st Sess., Nursing Homes: Misuse, high costs and kickbacks,
    Washington, D.C.: G.P.O., United States Government Printing Office,
    1975.
Venulet, J. "Aspects of social pharmacology", Intern. J. Clin. Pharmacol.
    10: 203-205, 1974.
Venulet, J. "Aspects of social pharmacology", Progress in Drug Research, 22:
    9-25, 1978.
Wetz, R. Jugendliche und Rauschmittel, ed. Bundeszentrale für
    gesundheitliche Aufklärung, Cologne, 1972.
Zborowski, M. "Cultural components in response to pain", J. Soc. Issues, 8:
    16-30, 1952.

COOPERATION BETWEEN THE POLICE AND SOCIAL CARE IN THE TREATMENT

OF ALCOHOL AND FAMILY VIOLENCE PROBLEMS IN FINLAND

Teuvo Peltoniemi

The Finnish Foundation for Alcohol Studies
SF-00100 Helsinki 10
Finland

BACKGROUND

Fifteen years ago the following letters were exchanged between the Helsinki Police Department and the Social Welfare Department of Helsinki:

Police Department: "The police authorities can provide the social care system with more than enough alcoholics in need of care. The Police Department hopes that the social care will be greatly intensified, expecially with regard to compulsory care ... measures must be taken efficiently and rapidly."

The Social Welfare Department: "It should be mentioned that the inauguration of compulsory measures concerning alcoholics in the first stage depends on the activities of police authorities. If the police do not arrest an alcoholic and bring him to the PAVI office (alcohol care unit) of the Social Welfare Department after he has been sobered up there are often no compulsory measures taken."

The issues relating to both the cooperation as well as the division of activities between different authorities is by no means a rare phenomena. Many organizations like to send their unwelcomed clients to be handled by other agencies. Often a group of persons remains whom no organization will take care of.

This paper will focus on the problems of cooperation between two authorities in Helsinki, the police and social welfare, in connection with the treatment of alcoholism and family violence problems. The data are drawn from studies carried out between 1976-1978 in the Helsinki Police Department and the PAVI office (charged with the care of alcoholics) of the Helsinki Social Welfare Department.

Data were collected by the observation method over two and a half years in the PAVI office. During the 1977 a survey was conducted on contacts with all organizations working partly or totally with "alcoholic" clients. Domestic disturbance calls were investigated by observation and questionnaires during the autumn of 1977 in the Helsinki Police Department (Peltoniemi, 1980, and this volume, Chapter 15).

FRAMES OF REFERENCE

The distinction between police, social welfare authorities, and health care is often problematic. Theoretically it can be illustrated by a triangle where the corners consist of the three general functions of these types of organizations (Peltoniemi, this volume). These functions can be categorized as <u>Control, Care, and Treatment</u> (Figure 1). These functions -- which are organized and goal oriented processes--can be operationally defined in the following ways:

- Control is the limitation of nonacceptable behavior (e.g., aggressive behavior) even by means of compulsory measures.

- Care is the satisfaction of the basic needs of man (e.g., food or human relations).

- Treatment is the influencing of the behavior of man (e.g., alcoholism through internal changes.

It might appear as if these functions simply are the same as the three organizations mentioned above. In practice we can seldom find an organization which is active with only one of these functions. Usually it is a question of a combination of all three. Graphically, organizations usually are situated somewhere inside of the triangle rather than at its corners.

Because the functions of the different organizations overlap, difficulties in cooperation automatically follow. This is also clearly demonstrated by the data.

The Police

Traditionally it has been said that the police are responsible for the control functions of law and order. In practice it has been found that a great deal of police work is the giving of information and guidance and the solving of many types of problems. It has even been estimated that more time is given by the police to these activities than to traditional law and order work (Banton, 1964). In this sense, the two different roles in police work (1) control and (2) mediation (Manning, 1977, 112), explain why police work comes close to both the social welfare and treatment aspects of intervention.

One reason for the variety of the police tasks is to be found in external factors. The police are one of the few authorities that work round the clock and are easy to contact. Therefore, at night and on weekends the

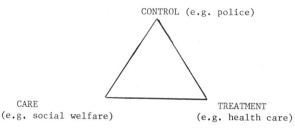

Figure 1. Control, care, and therapy in relation to the police, social, and health care in the treatment of alcohol and family violence problems.

police provide services usually provided by the other authorities, especially that of social welfare.

The different functions of police work lead to problems because police training is more directed to control function roles, even if in the actual work knowledge of care and even of treatment is needed. When working in fields related to social care, the police role changes from a professional into an amateur one. (Cumming et al, 1965, 277). That is probably why so many policeman consider these tasks outside of their "proper" role. A policeman of the Helsinki Police Department stated:

> We are not family consultants but the safekeepers of public law and order. It is not our task to give therapy. There are other institutions and workers than the police who provide these services.

A discrepancy exists between the real activities and the self-concept of the police.

The Police and the PAVI-office

Close cooperation exists regarding alcoholism and related issues between the Care Section of the Police Department and the PAVI office. The Care Section is the part of the police which handles vagrants and alcoholics. The usual uniformed police branch handles the bulk of the arrests of drunks. The PAVI office is one of the agencies of the City of Helsinki Welfare Department with the primary task of taking care of clients with serious alcohol problems (Peltoniemi, this volume, Chapter 15).

Both of these organizations deal with vagrants. The Care Section of the police helps the PAVI office by bringing clients who do not come in to answer to the statutory "invitation" letter of the PAVI office. In 1977 the Care Section took 1190 persons to the PAVI office.

The PAVI office thus is a very important cooperative partner of the Care Section of the Helsinki Police Department. Up to 70% of the outside contacts of the Care Section were directed to the PAVI-office. For the latter, however, this police agency is a quantitatively small partner which received only 4% of all the contacts of the PAVI office. The contents of the contacts between the PAVI-office and the Care Section are quite routine. The police bring clients to the PAVI office and the agency checks whether the police are interested in their clients for police reasons.

In evaluating the contacts, the Care Section of the Helsinki Police Department felt that the many contacts with the PAVI office could be somewhat reduced. More contacts were wished with other alcohol care organizations as well as with child care. Theoretically, this means that the establishment of more contacts with the treatment corner of the triangle was desired.

The treatment side, on the contrary, was not very interested in expanding contacts with either the Care Section or the PAVI office. Treatment agencies considered both of the latter agencies to be control-type organizations and wished to refer only those clients with the greatest problems to them. In practice typical mental patients with only slight or no alcohol problem were forced to become clients of the PAVI office or the Care Section instead of a mental hospital, for instance. These two organizations were apparently serving the role of the "last resort". (cf. Wiseman, 1970).

The interim cooperation of the PAVI office and the Care Section is illustrated by the fact that the PAVI office sometimes changed its normal policies because of a request from the police. During the European Security Conference in 1975 police arrested more drunks than usual and requested that the PAVI office adopt a harder line as well. This was done and the PAVI office referred many more clients to alcoholism institutions elsewhere in the country away from public sight in the center of Helsinki.

In many cases the PAVI office and the Care Section had different views concerning the use of statutory compulsory treatment; the police representing a harder line than did the PAVI office. In spite of these principal differences of opinion, cooperation between these two organizations was quite routine and worked well in comparision to health care.

FAMILY VIOLENCE

The other area where cooperation was studied was in the treatment of family violence. This phenomena came into public discussion during the late 1970s. It was a problem long known to the PAVI office, where it is registered under the label "dangerous alcoholics". Concerning these cases, very close cooperation with the Care Section exists. Most of the family violence, however, is taken care of by the ordinary uniformed police branch. In 1977, the police in Helsinki handled 8,221 domestic disturbance calls

When a police patrol deals with a domestic disturbance call it enters a private area, the home. It is no wonder that the police do not consider this task very pleasant, as is shown by the comment of the policeman cited earlier. In these situations the police are under a strong stress. The police expected to act as a kind of umpire in the conflict. In American studies it has also been shown how domestic distrubance calls are dangerous and more policemen get wounded or killed in dealing with those calls than in other tasks (Thompson and Gilby, unpublished). Even if that is not the case in Finland (where most police deaths at work are due to traffic accidents) domestic disturbance calls are regarded as risky in Helsinki.

The police would certainly like to see some other organization take care of the treatment of family violence. There are many factors which explain why this is not done. One reason is the lack of outside resources. Domestic crisis seem to be concentrated late at night and over weekends when the police are the authorities easiest to reach. And of course, there is the threat of physical violence in these situations which makes the police the natural authority to turn to. The staff of other authorities usually have no training in dealing with a violent situation. On the other hand, the police do not have much training in dealing with the crisis behind the conflict. Thus in fact no organization feels very much at home in regard to these cases (Thompson and Gilby, unpublished). This is apparent in the fact that even if the police seem to value the mediation of domestic disturbance cases, in practice very concrete measures dominate the treatment (Table 1). Mediation accounts for only 15% of all measures taken. About two of three cases were handled in a concrete way by either evicting somebody from the house or by arresting somebody (most often because of drunkenness) (Peltoniemi, 1980).

Since the police regard themselves as a bad choice in dealing with family violence, one would expect them to often refer these cases to other authorities after the initial disturbance has been calmed down. Often, however, that does not appear to be the case. In only 13% of the cases did the police refer some participant in the conflict to other agencies. Most often the police brought a party into contact with the criminal

Table 1. Measures Taken by the Police in Response
to Domestic Disturbance Calls in Helsinki during
the Fall of 1977

Measures	Number	%
No measures needed as:		
-the disturbing part has left		
the residence before the police entered	20	14
-the fight calmed down once the police		
entered	10	7
Measures taken:		
-mediating with the disputing partners	21	15
-evicting the disturbing partner from		
the residence	42	30
-taking the disturbing party to jail	49	4
Total	142	100
(no information on 7 cases)		

investigation branch of the police, to whom the victim has to make a report
in order for nonserious private violent acts to be prosecuted as a crime.
More frequently (in 38% of the cases) the police merely suggested to some of
the participants of the conflict that they contact some other agency. Most
often, the suggested agency was once again the criminal investigation brance
of the police. In addition, a hospital was often suggested in order to
ensure evidence in a possible lawsuit.

Family violence seems to be strongly associated with the heavy use of
alcohol. In the domestic disturbance calls it was often the case that all
the partners were under the influence of alcohol. This ties the treatment
of family violence more closely to alcohol care.

In 1977, however, only 162 cases ended in the PAVI office through the
Care Section. A part of them (17%) were even withdrawn before the process
in the PAVI office. A policeman in the Care Section noted that:

> In the police we are fed up with the women who don't know what
> they want. A wife makes a claim and the husband is taken into the
> custody here. Soon the wife, however, feels sorry and comes here
> again to withdraw the claim and take her husband back home before
> there is any hearing of the case in the PAVI office. The wives
> seem to be prepared to understand the behavior of their
> husbands to such extremes -- sometimes it is incomprehensible.

One could assume from this perspective that if the police are not
satisfied with their possibilities in the treatment of family violence they
should refer some of the cases to other organizations. Certainly it would
be possible to improve the cooperation here. The study, however, revealed
one barrier from the point of view of the police. Most of the people whose
behavior leads to the domestic disturbance calls to the police already are
clients of alcohol and social welfare care. The police are well aware of
that and seemingly do not believe that a renewed contact with those agencies
would change anything. The police also complained that social care is not
especially interested in these cases. These people are not easy clients for
the organizations and therefore are no quite welcomed by any.

CRISIS INTERVENTION

Similar experiences in the treatment of family violence in North America have led to the establishment of different crisis intervention systems. An attempt has been made to combine the family violence-related activities of the police and social and health care in a new way. This is designed to avoid the present situation where no agency feels very capable of handling and willing to handle these cases.

One crisis intervention system is found in London, Ontario, Canada (Jaffe and Thompson, 1979). In the London Police Force there is a special Family Consultant Service. It is situated on the premises of the police force, but, even so, works independently. The police and health and social care agencies and the church are represented on the office board.

The telephone calls from the public continue to be received at the police radio desk. A police patrol is still sent to the house if needed. The difference is that after the acute conflict is calmed down the policeman asks over the radio that a member of the Family Consultant Service come to the house to take care of the situation.

The Family Consultants helped the crisis by discussion and reasonsing in one of every three conflicts. Most cases referred to some other agency in social or health or law services. In many conflicts this meant a renewing of a previous or present client relations. Good results were found especially in cases where an appointment could be made immediately for the following morning (Jaffe and Thompson, 1979).

The police seem to view the Family Consultant Service very positively. This is very understandable, even theoretically, since the unit takes care of tasks where the police feel themselves to be amateurs. The police retained control over the part of family violence which comes closer to its traditional "proper" role, the law and order side.

Similar systems have been tested also elsewhere in Canada and in the United States (Levens, 1977). The experiences in general seem to be good. Many problems of cooperation have been settled and the police often seem to be satisfied with them.

The crisis intervention systems have been criticized with the argument that they complicate the organization and increase costs (Freeman, 1979). It has been found that in many cases the clients are not especially interested in any help after the acute violent crisis has been settled. They would rather prefer to take care of their problems by themselves. There has not been much evidence that the crisis intervention systems would result in less family violence. Feminists have stressed the danger that because of these systems not even those cases which seemingly would merit prosecution are entering the courts (Freeman, 1979).

From the theoretical point of view, however, the cooperation problems concerning family violence and alcoholism are easy to predict. Because of many organizations -- not only the police used here as an example -- seem to have a nonrealistic view of their objectives and everday activity, the problems seem to remain unsettled in practice. That is why some kind of crisis intervention systems would help in eliminating problems of cooperation even if it is not shown that they would give good results in eliminating the problem of family violence. The London, Ontario's model, where the system is closely connected to the police, is of course not the only possibility. In the Nordic countries (where the roles of social and health care are different from the North American situation) it would be possible to connect the crisis intervention systems to social or health care of, for example, to the shelter system.

The examples discussed here show, however, that theoretically predictable problems of cooperation do exist also in practice. Border areas always exist and there is no way of drawing a clear demarcation line between the tasks of the different organizations. Sometimes activities overlap. Where some groups of persons are left without help from any organization, the situation must be considered as unacceptable, however, and organization cooperation must be reformed.

REFERENCES

Banton, M. The Policeman in the Community. London: Tavistock, 1964.

Cumming, E., Cumming, I., & Edell, L. "Policeman as philosopher, guide and friend." Social Problems, 12:276-286, 1965.

Freeman, M. D. A. Violence in the Home. Hampshire: Saxon House, 1979.

Jaffe, P., & Thompson, J. "Family consultant service with the London police force- A description." Canadian Police College Journal, 3:115-125, 1979.

Levens, B. R. A Literature Review of Domestic Dispute Intervention Training Programs. Ottawa: Department of the Solicitor General, Monograph 3, 1977.

Manning, P. K. Police Work: The Social Organization of Policing. Cambridge, MA: MIT Press, 1977.

Peltoniemi, T. "Family violence: Police House Calls in Helsinki, Finland in 1977." Victimology, 5:213-224, 1980.

Thompson, J., & Gilby, R. "Correlates of domestic violence and the role of police agencies," unpublished.

Wiseman, J. P. Stations of the Lost: The Treatment of Skid Row Alcoholics. Englewood Cliffs, NJ: Prentice-Hall, 1970.

THE PARENT'S MOVEMENT: AN AMERICAN GRASSROOTS PHENOMENON

Carol Marcus

Pacific Institute for Research and Evaluation
Bethesda, Maryland 20814

A remarkable and unusual social movement in the area of drug abuse prevention has arisen in the United States over the past three years. Since 1977, parents in unprecedented numbers across the country have been organizing neighborhood parent groups for drug-free youth. This movement, a truly grassroots nonprofessional effort, has begun to have significant impact on the drug abuse field.

To date, there have been no substantive studies of the movement itself, or its impact on the drug using behavior of the children it was designed to support and protect. The parents themselves, however, numbering in the thousands, claim the movement allows them to reclaim the lives of their children from the dangers of drug use and abuse.

This paper describes the development and growth of the parents movement, suggests some of the reasons for its attraction to parents, its impact on existing systems, its implications for the field of prevention, and its potential as a model of prevention activity for other nations.

Pacific Institute for Research and Evaluation in collaboration with Albert Einstein Medical College and Pride, Inc., a parent training organization, has undertaken a study of the characteristics of parent group members. In a survey conducted in the spring of 1981, one hundred and twenty-five active parent group members answered questions about the reasons for their initial involvement, the perceived changes in the behavior and attitudes of their children, and the perceived changes in their own attitudes. The survey asked questions about the parents' values concerning parental control, youth independence and decision-making, and drugs and the drug culture. This study is expected to be completed this fall.

The Parent's Movement as a national phenomenon began rather inauspiciously with a letter from an angry parent to the Director of the National Institute on Drug Abuse (NIDA) in Washington, D.C. This parent, an Atlanta mother of three, had discovered with shock and outrage, that her thirteen-year-old daughter was deeply involved in a peer group that regularly used marijuana. Up to that point, the Schuchard family had only an uneasy sense that this formerly affable and outgoing child was becoming sullen, aloof, and somehow different. When they discovered that marijuana smoking, not an adolescent "phase," was causing the problem, they decided to take dramatic steps to stop drug use by their daughter and her friends. In the process they discovered many things that parents across the country were experiencing with growing resentment and frustration:

- that drugs and alcohol were easily available to their children.;

- that they as parents knew little about drugs, the drug culture, and the dangers involved for their children;

- that the traditional agencies established to do something about the problem were not having great impact in their communities;

- that schools were as likely to deny having a drug problem as to do something about it;

- that popular media were glamorizing the use of drugs;

- that marijuana, which parents saw as a serious threat to their children, was considered "harmless" by many;

- and finally, that they as parents had been passive and uninvolved, feeling overwhelmed, ignorant, and powerless to take any action.

In her letters to NIDA, this mother, Keith Schuchard, described her anger at these discoveries, and her determination to take matters into her own hands. She outlined a process she and her neighbors had undertaken to stop drug use by their children. This process has become a widely replicated national model that includes:

- organizing the parents of her children's peer group, creating a "Parent Peer Group";

- closely supervising and setting curfews for all children in the peer group and neighborhood;

- chaperoning all activities involving groups of teens and preteens, and establishing firm ground rules against any kind of drug use;

- researching the effects of marijuana and other drugs, and informing parents throughout the community about the dangers of these substances;

- developing more effective drug education programs with the school which include parents as well as their children;

- working to enact legislation limiting the sale and distribution of drug-related paraphernalia, and lobbying against the decriminalization of marijuana; and

- publishing parent-oriented newsletters and information about drugs and youth.

NIDA's interest in the letter, and its subsequent earnest and supportive response to it, was enhanced by a variety of factors. First, professionals in the drug abuse field have remained uncomfortable with the concept of prevention. Although common sense, some early research data, and clinical observation have pointed toward prevention as a cost-effective and practical strategy for confronting drug abuse, solid evidence of its effectiveness had yet to be forthcoming. Federal officials have hesitated to take risks with untested programs, and consequently prevention budgets have reflected a low level of trust and priority.

Second, recent surveys of drug use by teenagers had confirmed that, far from fading into oblivion in the late 1960s and early 1970s, the patterns of drug use indicated far more alarming trends for young people than anything they had observed a decade earlier. A 1978 survey showed that:

- While 47% of the class of 1975 used marijuana at least once during their lifetime, fully 59% of high school seniors reported having used some form of illicit drug.

- More than 5% of those surveyed used alcohol on a daily basis.

- Perhaps most alarming, especially to parents, the number of respondents using marijuana on a daily basis increased from 6% in 1975 to nearly 11% in 1978. (Institute for Social Research, 1979).

Thus, NIDA expressed great interest in the Atlanta mother's program and sent representatives to visit the Atlanta parents, provided access to drug abuse technical assistance to the parents, and published Keith Schuchard's book detailing her experience entitled, Parents, Peers, and Pot (NIDA, 1979.) The encouragement and support of the Federal government contributed to the early rapid strengthening and expansion of the Parents Movement.

But at its base and at heart, the Parents Movement is neither a government funded nor directed effort. It is a private local effort which today claims more than 1,000 active parent groups across the United States.

As the Atlanta group was developing its program model and sharing it with neighboring counties, other isolated groups of parents across the country were responding with anger and frustration to the same unnerving circumstances and seeking some direction. The Atlanta model, disseminated by the media and by its leaders traveling across the country, became the catalyst and unifying force for a nationwide network of parents. In effect, drug use and abuse had entered the homes of the American white middle class through its youngest members and had sparked activity by these normally nonpolitical, nonactivist citizens.

In talking with parent group members from across the country over the past few years, we have observed a number of common factors which motivated these traditional, mainstream American parents, mostly mothers, to take to their neighborhood streets. In addition to their shock at the pervasiveness of the drug problem, and at the younger and younger ages of children becoming involved with drugs, these parents expressed great concern about their own broader roles as parents and as family members. They reported that:

- the drug issue brought home to them their former reluctance and fear in the face of this problem;

- they felt powerless and inadequate to act alone to change the worrisome behavior of their children in regard to dating, curfews, driving cars, alcohol use, and school grades, as well as drugs; and

- they were tired of taking all the blame and feeling guilty for the problems which their children were facing in a changing world.

These parents, who feel they are in danger of losing control over their children, their neighborhoods, and the quality and direction of their lives, are representative of many American parents today.

A recent Gallup poll found that 45% of Americans think family life has gotten worse in the last 15 years. And when asked to identify the three most harmful influences on family life, 65% of those surveyed listed drug abuse and 50% named alcohol abuse. When asked what three items have the most negative effects on family life, they listed the high cost of living (63%), fear of crime (36%), and declining religious, moral, or social standards (The Gallup Organization, 1980).

Recent United States census figures reveal a deep change in the structure of the American family:

- Between 1970 and 1979 there was a 79% increase in one-parent households.

- 19% of all families with children living at home are single-parent families.

- 38% of one-parent families are headed by divorced women compared with 7% in 1970.

- Between 1960 and 1978 the number of wives in the labor force increased from 12.6% to 40.9% (U. S. Department of Commerce, 1979).

The pressures and changes which confront families today may be unprecedented. Unemployment and inflation, changing social mores, challenges to the traditional notions of family, and isolation from traditional supports have combined to cause anxiety and bewilderment for many families.

The Parents Movement against drugs may be a response to broader concerns of parents offering them an opportunity to band together to recreate some structures, control, and social supports for raising and protecting their children.

The Parents Movement does not lend itself to a structured description of its activities. The spontaneous development of these groups responding to local needs in widely varying communities has spawned a myriad of programs often as different from one another as their communities and neighborhoods are different. However, the initial actions and early development of many groups follow a pattern similar to the Atlanta model.

Parents often seek each other out initially because of a drug incident in their neighborhood or family. Often the first target for action is an external force which parents see as contributing to the problem. Closing drug paraphernalia shops and urging crackdowns on neighborhood drug trafficking are common first steps. Surveillance and control of their children's activities to stop or prevent any use of drugs is usually the first internal activity of the group. These initial actions tend to attract the attention of more parents, the media, and the community. Parents then develop common codes of conduct for parents and children in an attempt to create a new "community" of families.

As the groups work together, they begin to set broader goals and undertake activities such as parent drug education, parenting and communication skill-building, school and community-based alternatives programs, health promotion, legislative and legal action, and adolescent health.

Parents are becoming a formidable social and political force in the drug abuse field in the United States. In 1980, the National Federation of

Parents for Drug Free Youth was formed to represent parent group interests on a national scale. Their combined impact has been hardhitting:

- they have influenced the direction of Federally funded research;

- they have influenced national prevention programs;

- they have influenced the passing of legislation on State and Federal levels;

- they have lobbied to place a drug abuse official in the White House who is a long-time supporter of the Parents Movement;

- they have been written about in The New York Times, the Washington Post, and Reader's Digest;

- they have appeared many times on national television news and talk programs;

- they are receiving support from many State substance abuse agencies;

- and they continue to grow.

While the parents have found many supporters, they have also run into barriers: the pro-marijuana lobby, the courts, reluctant parents, and disinterested schools, drug abuse and family counselors with differing views, and perhaps most seriously, drug abuse prevention and treatment professionals.

Prevention professionals in the United States have made some important strides in programing and research over the past five years. They have worked against great odds facing small budgets, a medically oriented treatment establishment, little public support, and in recent years, a vocal anti-affective education lobby. These professionals initially viewed the active parents as overzealous, too drug specific in their orientation, too law enforcement oriented, manipulative of research findings to support their anti-drug stance, insensitive to broader needs of adolescents, misguided in their opposition to many prevention strategies, and consequently in danger of setting the field back ten years.

Parents, on the other hand, have complained that many professionals were too cavalier in their attitudes toward drugs, too accepting of drug experimentation as a rite of teenage passage, and too blaming of and insensitive to parents.

Over the past year, some of the misunderstandings and problems between parents and professionals have been resolved, although a great deal of wariness and mistrust still exists. Resolution of the current tensions will, we hope, protect the important work of the professionals and maintain the enthusiam of the parents.

The future direction of the Parents Movement is unclear. According to some leaders, many groups are in transition facing decisions about "what to do next." Having accomplished their initial goals over the past two or more years, they want to move beyond the neighborhood parent meetings and community seminars to new roles and strategies. Many groups have already done so and are involved in active support of comprehensive community health initiatives. Groups are working with social and service clubs, criminal justice and other service agencies, and schools and local businesses. It would seem in this time of restricted social service funding that this nationwide network of active parents can be of critical importance to health and service providers.

The phenomenon of the Parents Movement has been an interesting and enlightening experience for many in the drug abuse prevention field. Because the parents have generated an enormous amount of public awareness about the harmful effects of marijuana through the media, locally and nationally, and because they have urged thousands of parents to say "no" to drug use by their children, it is possible that their efforts have already begun to effect the patterns of drug use of many young people.

A 1980 survey of high school seniors indicates a decrease in peer acceptance of smoking tobacco, and a long-term increase in young people's concerns about health. These factors combined with the recent publicity about the ill effects of marijuana may have figured in the drop of daily marijuana use to 9.1% following a period of dramatic increases. High school seniors who began to think that there was "great risk" to regular marijuana use rose from 35% to 50% (Institute for Social Research, 1981). The 1980 survey also reveals a slight drop in annual (down 2%) and monthly (down 3%) marijuana use.

This survey data is very encouraging to parents because it indicated a downturn for the first time in years. Yet they fear that public officials and citizens may become complacent in the light of the new trends and fail to continue to work actively to lower what is still an unacceptable and tragically high rate of use.

REFERENCES

Institute for Social Research, University of Michigan. Highlights from Drugs and the Class of '78: Behaviors, Attitudes and Recent National Trends. Rockville, Maryland: National Clearinghouse for Drug Abuse Information, 1979.
Institute for Social Research, University of Michigan. Highlights from Student Drug Use in America 1975-1980. Rockville, Maryland: National Clearinghouse for Drug Abuse Information, 1981. pp. 27-32.
National Institute on Drug Abuse. Parents, Peers, and Pot. Washington, D.C.: Superintendent of Documents, U. S. Government Printing Office, 1979.
The Gallup Organization and White House Conference on Families. American Families, 1980. Princeton, New Jersey: The Gallup Organization, 1980.
U. S. Department of Commerce, Bureau of the Census. Current Population Reports: Population Characteristics, Household and Family Characteristics and Statistical Abstract of the United States Washington D.C.: The Bureau of the Census, 1979.

SELF-IDENTIFIED PRESENTING PROBLEMS OF

DRUG ABUSERS ENTERING RESIDENTIAL TREATMENT

L. Ball*, D. Morgan, and G. Small

Alcohol and Drug Programs, Ministry of Health
British Columbia, Canada

BACKGROUND

There are various methods which have been employed in psychology and other disciplines to determine the focus of treatment. Psychometric tests, personal inventories, clinical interviews, and behavior ratings have all found favor in different times and places. Most frequently it is the report of others, whether relative, teacher, case worker or counselor which determines the nature and form of therapy. All too often the patient's answer to the query, "What changes do you wish to make?" is ignored.

Walls et al., in Cone and Hawkins (1977), review some 166 checklists but specifically do not include scales that call for client response. Thus, Wolpe's and Lang's Fear Inventory (1964) and the Self-Rating Behavior Scale (Upper and Cautela, 1975) are excluded, as are most of the techniques/forms presented by Cautela (1977). Still, Wolpe (1973) recommends that a fairly large amount of information should be obtained about the individual in several areas of functioning, including the patient's own view of difficulties being faced. Riem and Adams employed a Specific Needs Assessment Inventory in their study of individuals entering drug treatment (Riem and Adams, 1979).

The purpose of the present study was to compare the results of three approaches to treatment planning: (1) assessment battery, (2) behavioral observations, and (3) self-ratings. The subjects for the study were 116 drug abuse clients (average age 30.8 years, 74% male, 26% female) who were referred for residential treatment at a 45-bed Residential Treatment Center in British Columbia. The major drugs of abuse of these clients included 55% opiates, 19% alcohol, 10% barbiturates, 6% benzodiazepines, 3% stimulants, 3% psychedelics, 4% other drugs. Table 1 contains demographic detail about the clients. The study was completed in 1980-81.

The sources of assessment data are described individually in the following assessment battery. All clients who were referred to the Residential Treatment Center, (R.T.C.) had their initial contact with the treatment system at the Province's outpatient units. In order to determine the client's need for treatment, level of treatment (outpatient vs. inpatient), and focus of treatment, all clients were exposed to a complete assessment battery. This battery was composed of the following elements: medical assessment, social history, and psychological testing. The

---

* Nanaimo Treatment Unit, Alcohol and Drug Programs, Ministry of Health, Nanaimo, British Columbia, Canada

Table 1. Background Characteristics of Subjects

Current legal status

1. No legal status	41.4%
2. Probation	10.6
3. Bail	11.6
4. Outstanding warrants	4.0
5. Court diversion	2.5
6. Parole	4.6
7. Other	25.3

Current marital status

1. Single	41.3%
2. Married	13.0
3. Common law	17.4
4. Divorced	9.4
5. Separated	17.4
6. Widowed	1.5

Highest level of education

$\bar{X}$ = 10.34, SD = 2.19

Current employment status

Yes	18.3%
No	81.7

Usual employment status

1. Full-time employment	48.0%
2. Part-time employment	7.8
3. Seasonal	8.3
4. Unemployed	21.3
5. Welfare	14.6

Claimed to be drug dependent

Yes	67.7%
No	19.3
Not sure	13.0

When first used narcotic drugs

$\bar{X}$ = 18.86 years, SD = 4.18

Age when started using regularly (i.e., developed pattern)

$\bar{X}$ = 19.76 years, SD = 3.93

Ever attempted to end own life

Yes	40.0%
No	60.0

psychological testing component of the battery was most comprehensive and included the following measures: Wide Range Achievement Test (reading only), Bender Gestalt with Background Interference Procedure, Weshcler Adult Intelligence Scale, Minnesota Multiphasic Personality Inventory, Revised Bogardus Scale, S-Roles Anomie Scale, Beck Depression Inventory, Levenson Locus of Control, Rotter Locus of Control, Sachs Sentence Completion Test, Self-Image Scale, Life Events Inventory, Hopkins Symptom Check List, Michigan Alcohol Screening Test, and the Drug Abuse Screening Test. Table 2 lists the tests used in the assessment battery by title, areas sampled, skills required reference, and interpretation, and took an average of 11-1/2 hours. This was supplemented by 12-1/2 hours devoted to medical examinations, intake procedures, orientation to programs, paneling, and debriefing. This assessment battery was, in most cases, completed within 72 hours of initial client contact. The completeness of the battery and the rigorous time frame for completion was influenced by the possibility of legal status being attached to being "in need of treatment." The treatment planning data from the assessment was generally written up in terms of the eight major objective codes for the drug treatment program. The objectives included:

1. Immediate needs
2. Drug/alcohol abuse
3. Physical health
4. Family/social adjustment
5. Emotional development
6. Psychological health
7. Educational concerns
8. Recreational leisure time activities
9. Vocational concerns

Behavioral Observation

Once the client was assissed with the initial battery and the decision was made for referral to residential treatment, the client was then transferred to the R.T.C. Most were in need of detoxification. Length of detoxification varied depending upon the needs of the client, averaging 13 days, but ranging from one day for those admitted in a drug-free state (merely to confirm this fact) up to six weeks for those rare cases admitted while using opiates and other substances extremely heavily. Detoxification from opiates was usually affected through a withdrawal regime employing methadone, Valium or Serax, chloral hydrate, Dalmane, and/or paraldahyde. This time period provided as excellent opportunity to observe the client in a stressful environment. Following detoxification the client was transferred to the drug-free area for orientation and treatment planning. This provided further opportunities for behavioral observation prior to entering the treatment phase of the program. Of the three sources of treatment planning data, the behavioral date was distinctly least structured and was based on the client's free-form behavior with no observational guidelines being imposed. Sources of observations routinely included those from: physician, nurses, health care workers, social workers, arts and crafts instructors, physical fitness instructors and, occasionally, from the resident psychologists. This data source proved to be very rich due to the fairly diverse observations.

Self-Ratings

On entering the orientation-treatment planning phase, the client trained in emergency self-control procedures (thought-stopping, clearing, deep breathing, etc.). The training and orientation period usually lasted 7-10 days, thereby allowing the client to begin to acclimate to a drug-free environment. Toward the end of this period of time, the client was asked to complete the R.T.C. Behavior Problem Check-List (see Appendix I). This check-list was divided into the eight major areas of concern identified above as objectives in treatment. Each area was further subdivided into more specific areas of concern. For example, under Immediate Needs the checklist included medical problems, residence, hygiene, clothing, finances, legal, social and psychological. These specific areas were again further qualified; for example, under residence, to include rent to pay? plants to water? things to move? In the other major areas of concern, breakdown into specifics was similar but restricted to one level, e.g., under attainment of good physical health such items as learn proper diet, develop capacity to relax, development of good hygiene habits, development of good physical health. An additional section was provided for other areas not assessed so that the patient could insert matters of concern which were not otherwise listed. A total of 86 specific behaviors or problems was identified. For each, if a positive response was elicited, the patient was requested to set a priority on that item of from 1 to 10, with the latter indicative of a most severe problem. Finally, the patient was asked to summarize the problem areas and give an overall priority to each. Thus an overall statement from the client was obtained that identified specific behaviors important to the client.

In the utilization of the three sources of data for treatment planning, the client's data was first listed in order of priorities, the assessment battery data was then added, and finally the behavorial observation data was included. The three sources were then combined in terms of commonalities and priorities established based on the perceived needs of the client. This approach guaranteed a focused mandate for treatment, with all sources of data included. Appendix II outlines the procedures used and the process followed to attain this end.

METHOD OF ANALYSIS

The data from three sources were analyzed through Spearman rank-order correlation coefficients.

To prepare for the correlational analysis, the data were first ranked in terms of the top 25 most commonly mentioned areas of concern across all three areas. Thus three lists of priorities were developed: the first based on the recommendations from the assessment battery, the second from the behavioral observations, and the third from the client ratings. In forming these lists the low frequency items in each of the areas were collapsed into larger areas, e.g., "meeting straight people" and "making new friends" were combined with the larger category of communication skills; and "self-control" and "relaxation" were included with stress management skills. The correlation coefficients were calculated between the assessment battery recommendations and the client ratings, and the assessment battery recommendations and the recommendations based on the behavioral observations.

RESULTS

The results of the rank ordering across the three sources of data are presented in Table 3. As can be noted from Table 3, there was a fairly high degree of assessment between the three sources; the following correlational coefficients were obtained:

- Assessment recommendations and client ratings r_s = .72 (p < .05);

- Behavioral observation recommendations and client ratings r_s = .80 (p < .05);

- Assessment battery recommendations and the recommendations from the behavioral observation r_s = .91 (p < .05).

DISCUSSION

In presenting the discussion, three areas are included: the relative fit between the three sources of treatment planning data; the overall ranking and its relationship to program goals; and the functional necessity

Table 2. Details of Assessment Battery

---

1. <u>Wide Range Achievement Test (WRAT) - Reading Only</u>
   The subject reads a list of words to determine his reading level. A fourth grade reading level is a requirement to complete the assessment battery. It is a quick (10 minute) standardized way of determining reading level. Reference: Jastak, J. F. & Jastak, S. R., <u>Wide Range Achievement Test Manual of Instructions</u>, 1976 Edition, Wilmington, Del.: Guidance Assoc. of Delaware Inc., 1976.

2. <u>Canter Background Interference Procedure for the Bender Gestalt Test</u> (BG-BIP)

3. <u>Wechsler Adult Intelligence Scale</u> (WAIS)
   The subject must perform a variety of verbal and performance (motor) tasks in order to generate verbal and performance IQ scores. This test measures intellectual functioning, emotional functioning, and neurological involvement. Reference: Wechsler, D., <u>Manual for the Wechsler Adult Intelligence Scale</u>, New York, New York: Psychological Corporation, 1955.

4. <u>Minnesota Multiphasic Personality Inventory (MMPI)</u>
   The subject must answer true or false to 566 questions. It taps motivation level, personality variables (stress, social adjustment, past family adjustment), and psychological variables (defense mechanisms used, response patterns). Reference: Hathaway, S. R. & McKinley, J. C., <u>Manual. Minnesota Multiphasic Personality Inventory</u>, New York, New York: Psychological Corporation, 1967.

5. <u>Beck Depression Inventory</u>
   This scale consists of 21 items, each consisting of 3 to 5 statements. The subject must choose the statement which best applies to him. It discriminates people with regard to degree of depression and reflects change in depression over time. It takes approximately 2 to 10 minutes to administer. Reference: Beck, A. T. & Beamsderfer, A., "Assessment of depression: The depression inventory." In P. Pichot (ed), <u>Modern Problems in Pharmopsychiatry, Volume 7</u>, Basel, Switzerland: Karger, 1974.

6. <u>Rotter Internal-External (I-E) Locus of Control Scale</u>
   This test consists of 29 items, each having two statements from which the subject must choose one. It measures the degree to which a person feels that life is controlled by outside circumstances or by one's self--whether or not a person feels they can control what happens to them. Reference: Rotter, J. B. "Generalized expectancies for internal versus external control of reinforcement." <u>Psychological Monograph</u>, 80, 80, Whole 609, 1966.

7. <u>Life Events Inventory (LEI)</u>
   This test measures the amount of stress in a subject's environment in the preceding year. Those experienced are indicated from a list of 55 possible disruptive factors. It is a good check on the history of the past year and taps psychological and physical health factors. It takes approximately 30 minutes to administer. Reference: Cochrane, R. & Robertson, A., "The Life Events Inventory: A measure of the relative severity of psycho-social stresses." <u>J. Psychosomatic Research</u>, 17, 1973.

8. <u>Hopkins Symptom Checklist (HSCL)</u>
   This test has 58 items representative of symptom configurations seen in outpatients. Subjects indicate which items apply to them. It taps psychological adjustment (somatization, obsessive-compulsive, interpersonal sensitivity, depression, anxiety) and takes approximately

(continued)

Table 2. Details of Assessment Battery (Continued)

15 minutes to administer. Reference: Leonard, R., et al. Behav. Science, 17, 1974.

9. Michigan Alcoholism Screening Test (MAST)
This test consists of 25 items which taps alcohol abuse. It takes approximately 10 to 30 minutes to administer. Reference: Selzer, M. "The Michigan Alcohol Screening Test: the quest for a new diagnostic instrument." American Journal of Psychiatry 133, 1971.

10. This is a test consisting of 20 questions related to social adjustment and drug use. It takes approximately five minutes to administer. Reference: Khavari, K. A. & Douglass, F. M., IV. "The Polydrug Assessment Scale: A Psychometric technique for the indirect measurement of drug use." J. Consult & Clinical Psychology, 46, 6, 1978.

11. Drug Abuse Screening Test (DAST)
A research questionnaire developed for the treatment program by Dr. Harvey Skinner and consisting of 28 questions concerning involvement and abuse of drugs.

12. Levenson Internal-External Locus of Control Scale
The subject must indicate whether or not he agrees with 24 statements. The test measures the degree to which a person feels life is controlled by outside circumstances or by one's self, similar to the Rotter scale. It takes approximately 25 minutes to complete and was included to tap anxiety and psychological adjustment. Reference: Levenson, Hanna. "Distinctions within the concept of internal-external control: Development of a new scale." Proceedings of the Annual Convention of the American Psychological Association, 1972, Vol. 7 (Pt. 1), 261-262.

13. Revised Bogardus Scale
This scale measures the degree of alienation from community and has been used specifically in drug research. It takes less than 5 minutes to complete; the subject checks a situation in which he feels the average man would be most comfortable. Reference: N/A.

14. S Roles - S Roles Anomie Scale
This test measures hopelessness and social disorganization. Subjects check true or falise to five items. Administration time is less than 5 minutes. Reference: N/A.

15. Sachs Sentence Completion Test = Revised (SSCT-R)
The subject must complete phrases to make complete sentences. It taps overall adjustment and takes approximately 40 minutes to complete. Reference: N/A.

16. Self=Image Scale (SIS)
This test is very short (less than 5 minutes) and measures the degree of addict-subject's satisfaction with self. It taps overall adjustment and can be used as an indicator of motivation. Subjects indicate which of four statements apply to themselves. Reference: N/A.

of extensive assessment batteries in the development of treatment plans. Each of these areas is discussed individually.

The relative degree of fit between the three sources of data is quite high. One might expect a good level of fit between the assessment and observational recommendations, since both are being made by professional staff, and this was obtained. However, such a good fit with the client ratings was surprising. This result can likely be attributed to the very specific nature of the items on the client checklist. A far different

Table 3. Rank Ordering of Problem Areas by Assessment,
Recommendation, Observational Recommendations, and
Client Recommendations (N = 116)

| Problem area* | Rank order | | |
	Assessment recommendations	Observation recommendations	Client recommendations
1. Stress Management	2	1	1
2. Vocational/academic	1	2	3
3. Communications skills	6	3	2
4. Life skills	3	8	4
5. Leisure skills	10	4	5
6. Health restoration	9	5	6
7. Drug-free alternatives	4	7	10
8. Self-concept development	7	6	12
9. Relationships, family therapy	8	9	9
10. Dealing with depression	11	11	8
11. Medical/dental concerns	12	12	7
12. Learn to share, trust	14	13	10
13. More assessment data	5	12	24
14. Assertiveness training	13	12	17
15. Legal concerns	17	14	14
16. Dealing with drug urges	16	16	13
17. Employment concerns	15	19	15
18. Financial concerns	19	17	15
19. Develop responsibility	18	17	18
20. Pain control	20	13	21
21. Motivation for treatment	15	17	25
22. Control mood swings	16	18	23
23. Sleep problems	20	18	20
24. Immediate needs	20	20	21
25. Deal with guilt	18	15	19

*Problem areas are listed based on combined ranks across the three areas.

result might have been obtained had a less specific client measure been employed. On this point it is also important to note that when the client completed the checklists, it was subsequently reviewed by a residential counselor, and ambiguous areas and priorities were clarified by the client in consultation with the counselor. The most common difficulty encountered in this process was the assignment of priority ratings, with clients tending to rate all problem areas equally high. The tendency can be understood as a part of the addiction syndrome, with addicts being unwilling to make appropriate distinctions with regard to the delay of gratification. Such an unwillingness or lack of skill is also typically demonstrated in budgeting and time-management areas. The present results argue strongly for the expanded use of behaviorally specific client input in the treatment planning process. The use of the client priorities was also found to be highly useful in generating a mandate for change.

Table 4. A Comparison of Program Goals and Priority Treatment Areas

Program goals	Treatment areas
1. Immediate needs	1. Stress management
2. Drug/alcohol abuse	2. Vocational/academic
3. Health restoration	3. Communication skills
4. Family/social adjustment	4. Life skills
5. Emotional/psychological health	5. Leisure skills
6. Education	6. Health restoration
7. Recreation/leisure	7. Drug-free alternatives
8. Vocational	8. Self-concept development
	9. Relationship/family therapy
	10. Dealing with depression

It is interesting to note the relationships between the overall ratings and the espoused program goals. By looking at the top ten overall priority ratings and the eight program goals, such a comparison is possible.

As displayed in Table 4, there was found to be a good general fit between the program goals and the treatment areas rated as highest by clients and professional staff, with the exception of immediate needs. In most cases the client's immediate needs would already have been looked after by completion of detoxification. In collapsing the treatment areas into the program goal areas, the following result was obtained:

1. Immediate Needs – Accomplished prior to drug-free inpatient treatment.

2. Drug/Alcohol Abuse – Drug-free.alternatives, stress management.

3. Health Restoration – Health restoration.

4. Family/Social Adjustment – Communication skills, life skills, relationship/family therapy.

5. <u>Emotional/Psychological Health</u> - Stress management, self-concept development, dealing with depression.

6. <u>Education</u> - Vocational/academic upgrading.

7. <u>Recreational/Leisure</u> - Leisure skills, health restoration.

8. <u>Vocational</u> - Vocational/academic upgrading.

As will be noted, the areas of greatest concentration for treatment fall into goals 4 and 5: Family/social adjustment and emotional/psychological health.

The assessment battery was introduced to assist panels in determining whether or not an individual should be committed to treatment under law. The current results indicate only a marginal need for an extensive assessment battery for treatment planning in a residential setting. While arguments may be raised in favor of such assessments in deciding the type of treatment (inpatient vs. outpatient), one still has to question the validity of test scores that may be biased, or invalidated, as a result of drug impairment. When consideration is given to the cost of these assessments, the questionable validity of the test results prior to detoxification, and the findings of the above study, it would appear that there is little support for the need of an extensive psychological assessment battery prior to treatment.

This study will assist outpatient units when conducting followup to continue appropriate treatment modalities and enable long-term assessment of effectiveness of the methods employed. Success or failure of treatment is not dichotomous but rather is assessed in terms of achievement of program goals, with followup being carried out six months following discharge from the Residential Treatment Center. It will also assist the Ministry to select nurses, psychologists, social workers, health care workers, recreational and occupational therapists who have prior training and experience in treatment which addresses the problem identified by clients.

Future research in this area should focus on the utility of complex assessment batteries in treatment planning for a variety of disorders in addition to chemical addictions and should include comprehensive longitudinal follow-up studies to ascertain that program goals have been met.

REFERENCES

Cautela, J. R. <u>Behavior Analysis Forms for Clinical Intervention</u>. Champain, Ill.: Research Press Company, 1977.
Reim, Karl, F., and Adams, Roderick, E. "Concerns of drug abusers entering an outpatient treatment program." <u>J. Drug Education</u>, 9:151-161, 1979.
Upper, D., and Cautela, J. R., The process of individual behavior therapy. In: M. Hersen, R. M. Eisler, P. M. Miller (Ed.), <u>Progress in Behavior Modification</u>, Vol. 1 New York: Academic Press, 1975.
Walls, Richard, T., Werner, Thomas, J., Bacon, Ansley, and Zane Thomas Behavior checklists. In: John D. Cone, and Robert P. Hawkins, (Ed.), <u>Behavioral Assessment</u> New York: Brunner/Mazel, Inc., 1977, pp. 77-146.
Wolpe, J., <u>The Practice of Behavior Therapy</u>. New York: Pergamon, 1973.
Wolpe, J., and Lang, P. J. "A fear survey schedule for use in behavior therapy." <u>Behavior Research and Therapy, 2</u>:27-30, 1964.

Client's Name_____ Date_____

Counselor's Name_____

Problem Behavior Checklist

The purpose of this checklist is to help identify problem areas that clients want to work on while in treatment. To achieve optimal results, a staff member could assist clients who may be unsure and are experiencing difficulty identifying areas to work on.

I. IMMEDIATE NEEDS

		If yes	Priority (1-10)
A.	MEDICAL PROBLEMS		
	1. Current pain	_____	_____
	2. Drug-associated pain	_____	_____
	3. Prescription medications	_____	_____
	4. Medical number	_____	_____
	5. Medical appointments	_____	_____
	6. Lab tests	_____	_____
B.	RESIDENCE		
	1. Rent to pay?	_____	_____
	2. Things to move?	_____	_____
	3. Alternative living location/ accommodations?	_____	_____
	4. Plants to water/pets to be cared for?	_____	_____
C.	HYGIENE		
	1. Soap available	_____	_____
	2. Toothbrush	_____	_____
	3. Razor available	_____	_____
D.	CLOTHING		
	1. Sufficient clothing	_____	_____
	2. Shoes, boots, work clothing	_____	---------------
	3. Mending to do	_____	_____
E.	FINANCES		
	1. Does client have money now	_____	_____
	2. Job available	_____	_____
	3. Debts to pay	_____	_____
	4. Legal fees	_____	_____
	5. Can he/she budget	_____	_____
F.	LEGAL		
	1. Upcoming court dates	_____	_____
	2. Lawyer	_____	_____
	3. Statement from agency	_____	_____
	4. Friend in court	_____	_____
G.	SOCIAL		
	1. No drug-free friends	_____	_____
	2. Can he/she talk to drug-free people comfortably	_____	_____
	3. Participate in drug-free activities	_____	_____

H. PSYCHOLOGICAL
1. Anxiety _____ _____
2. Depression _____ _____
3. Disorientation _____ _____
4. High drug urges _____ _____
5. Sleep problems _____ _____
6. Current life crisis _____ _____

II. MOVING TOWARD FREEDOM OF DRUG ABUSE

A. NONDRUG ALTERNATIVES TO FEEL GOOD _____ _____
B. FREQUENT DRUG URGES _____ _____
C. DO NOT KNOW DRUG-FREE PEOPLE _____ _____
D. DIFFICULTY RELATING TO STRAIGHT
   PEOPLE _____ _____
E. ANXIOUS AROUND PEOPLE IF NOT STONED _____ _____
F. DO NOT FEEL GOOD IF NOT INTOXICATED _____ _____
G. DO NOT KNOW ABOUT SELF-CONTROL
   METHODS _____ _____

III. ACQUISITION OF IMPROVED PHYSICAL HEALTH

A. ACQUIRE DIETARY AND NUTRITIONAL
   KNOWLEDGE _____ _____
B. DEVELOPMENT OF HYGIENE HABITS _____ _____
C. DEVELOP CAPACITY TO RELAX _____ _____
D. DEVELOPMENT OF PHYSICAL HEALTH
   PROGRAM _____ _____

IV. MOVING TOWARD EMOTIONAL MATURITY AND PSYCHOLOGICAL WELL-BEING

A. DEVELOP SELF-CONTROL SKILLS _____ _____
B. INCREASE ABILITY TO BE RESPONSIBLE _____ _____
C. IMPROVE PROBLEM-SOLVING SKILLS _____ _____
D. LEARN TO UNDERSTAND WHY WE DO WHAT
   WE DO _____ _____
E. DEVELOPMENT OF AWARENESS OF THE
   HUMAN POTENTIAL _____ _____
F. DO MORE MEANINGFUL ACTIVITIES _____ _____
G. IMPROVE CAPACITY TO TRUST _____ _____
H. BECOME MORE ASSERTIVE, LEARNING TO
   SAY NO AND YES WHEN APPROPRIATE _____ _____

V. ACQUISITION OF RECREATIONAL AND LEISURE TIME SKILLS

A. ABILITY TO HAVE FUN WITH PEOPLE _____ _____
B. LEARN NEW WAYS TO PASS TIME _____ _____
C. LEARN PHYSICAL GAME SKILLS _____ _____
D. APPRECIATION OF THE ARTS _____ _____
E. ABILITY TO ACCEPT ASSISTANCE IN
   LEARNING NEW ACTIVITIES _____ _____
F. ABILITY TO ENJOY TIME SPENT ALONE _____ _____

VI. ACQUISITION OF IMPROVED LEARNING/EDUCATION SKILLS

A. ASSESSMENT OF CAPACITY TO LEARN _____ _____
B. AWARENESS OF PRESENT ACADEMIC LEVEL _____ _____
C. IDENTIFICATION OF LEARNING INTERESTS_____ _____

```
       D.    WILLINGNESS TO LEARN NEW THINGS
       E.    WILLINGNESS TO ACCEPT SUPPORT        ____  _____
             IN LEARNING
       F.    EXCITEMENT ABOUT LEARNING            ____  _____
       G.    UNDERSTANDING LEARNING AS A LIFELONG
             EVENT
                                                  ____  _____
```

VII. ACQUISITION OF WORK COMPETENCY SKILLS

```
       A.    ASSESSMENT OF CURRENT SKILL LEVEL    ____  _____
       B.    DEVELOPMENT OF WORK-READINESS SKILLS ____  _____
       C.    ABILITY TO WORK -- READINESS SKILLS  ____  _____
       D.    JOB SEARCH SKILLS                    ____  _____
       E.    FINDING HELPERS IN GETTING A JOB     ____  _____
       F.    RESUME WRITING                       ____  _____
       G.    LOOKING GOOD IN A JOB INTERVIEW      ____  _____
       H.    HOW TO KEEP JOBS YOU'VE GO           ____  _____
       I.    FINDING MORE EXCITING JOBS           ____  _____
```

VIII. MOVING TOWARD A MATURE AND REWARDING LEVEL OF FAMILY AND SOCIAL
      ADJUSTMENT

```
       A.    IMPROVEMENT OF COMMUNICATION SKILLS  ____  _____
       B.    IMPROVED UNDERSTANDING OF YOUR
             WANTS/NEEDS AND OTHERS
       C.    RESOLVING CONFLICT                   ____  _____
       D.    IMPROVEMENT OF SKILLS IN DEALING     ____  _____
             WITH GROUPS OF PEOPLE
       E.    HELPING OTHERS SUCCEED               ____  _____
       F.    GREATER UNDERSTANDING OF
             INTERPERSONAL RELATIONSHIPS          ____  _____
       G.    IMPROVED ABILITY TO SHARE AND TRUST  ____  _____
       H.    GREATER ABILITY TO BE ASSERTIVE      ____  _____
       I.    DEALING WITH ANGER                   ____  _____
       J.    REVEALING GUILT FROM THE PAST        ____  _____
       K.    MEETING NEW PEOPLE                   ____  _____
       L.    GETTING ALONG WITH STRAIGHT PEOPLE   ____  _____
```

IX. OTHER AREAS NOT ASSESSED

```
       A.
       B.
       C.
       D.
       E.
       F.
       G.
       H.
       I.
       J.
```

Summary of Problem Areas

Problem Areas (be as specific as possible   Overall Priority
in stating the problem areas)
       1.
       2.

3.
4.
5.
6.
7.

APPENDIX II.   Toward Implementing   Behaviorally Based Treatment Plans at the
               R.T.C. (Guidelines and Procedures*)

The general model to be used can be conceptualized thus:

I. BEHAVIORAL ASSESSMENT

    A.   IDENTIFY PROBLEM AREAS
         1.   Involvement of client is essential.**
         2.   Introduce as much data as possible (e.g., ACC assessment
               data, probation reports, medical and psychological records).
         3.   Input from treatment staff.
    B.   SET PRIORITIES
         1.   Use client priorities where possible.**
         2.   Develop short-term intermediate, and long-term goals.
         3.   What can be done at the R.T.C., what can be done in the
               clinic?
    C.   SELECT PROBLEMS FOR IMMEDIATE INTERVENTION
         1.   Don't take on too much; 1-3 areas is a maximum.
         2.   Use the "kiss model"; keep it simple, shared.
    D.   ANALYZE THE SELECTED PROBLEM AREAS** (RAC-S)
         1.   Problematic behavioral responses (R)
         2.   Antecedent conditions (A)
         3.   Consequences (C)
               a)   Positive
               b)   Negative
         4.   Response Strength (S)
               a)   Frequency
               b)   Intensity
               c)   Duration
               d)   Latency

II. TREATMENT PLANNING

    A.   ESTABLISH BEHAVIORAL CRITERIA
         1.   What behaviors will indicate progress in the problem areas?
         2.   Quantify successive approximations of the criterior behavior
               (e.g., 1 = not drug free, 2 = 20% drug free, 3 = 40% drug
               free,
               4 = 80% drug free, 5 = 100% drug free).
         3.   Set expectations against baseline data.
    B.   IDENTIFY INTERVENTIONS
         1.   Anxiety and stress management skills
         2.   Occupational therapy
         3.   Assertiveness training
         4.   Life skills training
         5.   Vocational upgrading
         6.   AA
         7.   N.A.
         8.   etc.

    C.   IDENTIFY REINFORCER FOR COMPLYING WITH INTERVENTIONS**
    D.   ASSIGN RESPONSIBILITIES
         1.   Who is responsible for what?
         2.   Who will be supporting what interventions?
         3.   How will the team work together?
    E.   ESTABLISH HOW AND WHEN THE INTERVENTIONS WILL BE EVALUATED
         1.   Set a date and keep to it.
         2.   Determine exactly how the criteria will be used to evaluate
              the intervention before the intervention is implemented.
    F.   SET REVIEW DATES TO MONITOR THE PROGRAM
         1.   At first the program should be monitored at least on a weekly
              basis.
         2.   After the program is well in place, hold monitoring sessions
              less frequently.
    G.   WRITE UP THE TREATMENT PLAN AS A TEAM CONTRACT
         1.   Specify all of the above points.
         2.   Ensure that responsibilities are clearly stated.
         3.   Sign the plan.
              a)   Team members
              b)   The client must sign.

III. FIELD TREATMENT PLAN

IV. MONITOR PROGRESS

 V. EVALUATE PLAN

    A.   EVALUATION SHOULD FOLLOW PREDECIDED PLAN
    B.   IDENTIFY STRENGTHS AND WEAKNESSES OF PLAN
    C.   REFOCUS IF REQUIRED
    D.   FIELD NEW PLAN
    E.   FURTHER EVALUATION

APPENDIX III.   Procedures for Linking Specific Treatment Planning Processes
                to the General Model

General Model   R.T.C. Procedures

  I.   BEHAVIORAL ASSESSMENT

       A.   IDENTIFY PROBLEM AREAS

            1. Client Involvement   Residence worker completes
                                        problem behavior checklist

            2. Introduce Data   Review assessment binder
                                        presented at IPP by Case
                                        Coordinator)

            3. Input from treatment staff IPP Meeting

---

*This builds on the Behavioral Assessment and Treatment Planning Symposium
on May 7, 1980.
**The primary role here falls to the residence worker to complete the
problem behavior checklist and TT11, TT12 with the client.

B. SET PRIORITIES

    1. Use client priorities   Residence workers' respons-
                                              ibility to present at IPP
                                              meeting

    2. Short-term, intermediate,   Prioritize the areas at IPP
        long-term goals   meeting

    3. R.T.C., clinic goals   Select IPP top areas at IPP
                                              meeting

C. SELECT PROBLEMS FOR   MINI TEAM RESPONSIBLE (Overall
    IMMEDIATE INTERVENTION   Responsibilities Fall to Case
                                          Coordinator)

    1. 1 to 3 areas   Mini Team responsible

    2. Keep it simple, shared   Mini Team responsible

D. ANALYZE SELECTED PROBLEM AREAS HIGH CLIENT INVOLVEMENT (Client
                                          must join meeting)

    1. Problem behavioral re-   Mini Team responsible
        sponses

    2. Antecedent conditions   Mini Team responsible

    3. Consequences   Mini Team responsible

    4. Response strength   Mini Team responsible

II. TREATMENT PLANNING   OVERALL RESPONSIBILITIES FALL
                                          TO CASE COORDINATOR

A. ESTABLISH BEHAVIORAL CRITERIA MINI TEAM RESPONSIBLE (OPERA-
                                          TIONAL OBJECTIVES)

B. IDENTIFY INTERVENTIONS   MINI TEAM RESPONSIBLE (WORK
                                          FROM GENERAL TO SPECIFIC
                                          INTERVENTIONS)

C. IDENTIFY REINFORCERS   RESIDENCE WORKERS RESPONSI-
                                          BILITY (TT11, TT12)

D. ASSIGN RESPONSIBILITIES   CASE COORDINATORS ROLE

E. EVALUATION TIMES   CASE COORDINATORS RESPONSIBIL-
                                          ITY

F. REVIEW PROCESS   CASE COORDINATORS RESPONSIBIL-
                                          ITY

G. WRITE UP AND SIGN   HIGH CLIENT INVOLVEMENT
    TREATMENT PLAN   (CLIENT AND STAFF SIGN TREATMENT
                                          CONTRACT)

III. FIELD TREATMENT PLAN   MINI TEAM RESPONSIBILITY

IV. MONITOR PROGRESS   CASE COORDINATORS RESPONSIBIL-
                                               ITY (PRESENTED BY CASE COORDIN-
                                               ATOR AT CASE REVIEW)

V.   EVALUATE PLAN   CASE COORDINATOR HAS OVERALL
                                               RESPONSIBILITY

     A. FOLLOW PREDETERMINED SCHEME   MINI TEAM AND CLIENT INVOLVED

     B. IDENTIFY STRENGTHS AND   MINI TEAM AND CLIENT INVOLVED
         WEAKNESSES

     C. REFOCUS IF REQUIRED   REFER BACK TO IPP PRIORITIES

     D. FIELD NEW PLAN   ENSURE CLIENT INVOLVEMENT

     ... CONTINUE   MINI TEAM RESPONSIBLE

STAGES IN THE MULTIFACETED TREATMENT OF THE MULTIPROBLEM DRUG ABUSER

Sam Lison, Ruti Saenger, and Ilana Vinder

Jerusalem Center for Drug Misuse Information--
Ezrath Nashim in Derech Beit Lechem
Jerusalem, Israel

INTRODUCTION

Professionals and the public at large have continued to report and believe that drug abusers are both difficult to treat and to treat successfully (Brecher, 1972; Brill and Lieberman, 1972; Sells, 1974). Neither the medical model, which is considered by some to exacerbate the problem (Cummings, 1979), nor traditional psychotherapy, which too narrowly embraces the diversity of problems associated with drug abuse (Hughes, 1977), have been reported as being effective. The few optimistic reports of treatment success have often been obtained by the use of idiosyncratic methods (Cummings, 1979). This situation has necessitated the development of a broadly applicable, multifaceted approach to the problem as recommended by Glatt (1974) and which has recently been reported as existing in greater numbers (Steer, 1979). That is, several programs now exist where a wide range of coordinated therapeutic interventions are applied over a period of time by a network of professionals in various settings and agencies. The caseworker or therapist from the primary drug-treatment facility is responsible for the treatment program. What has been lacking in this approach, though, is a coherent description of the multifaceted therapeutic process. This paper presents such a description based on the authors' experience at the Jerusalem Center for Drug Misuse Intervention.

JERUSALEM CENTER FOR DRUG MISUSE INTERVENTION (JCDMI)

The JCDMI is the only ambulatory drug-free clinic serving clients from the Jerusalem catchment area, a population of 400,000 people. Its voluntary clients are either self-referred or referred from welfare, correctional, and medical agencies. The decision to stay in treatment is that of the client alone. The clinic has been staffed by a multidisciplinary team of psychologists, social workers, vocational rehabilitation counselors, criminologists, occupational therapists, a doctor, a nurse, and a lawyer on retainer (Einstein, 1981). The director is a psychologist. The clinic, affiliated with a private psychiatric hospital, was initiated in 1976 at the request of the Ministry of Health. All the staff have responsibility for the treatment of specific drug-using clients and their families while at the same time they work in their specialty areas (i.e., nursing services, occupational therapy, etc.). The center has developed into a multifaceted facility in the sense that it offers a broad range of treatment interventions and that it has developed a network of contact with support systems and community agencies, introduced into the treatment program at opportune times.

Table 1

Stage	Client	Therapist	Support systems
1 - Trust	1. Testing by acting out 2. Suspends mistrust 3. Accepts minimal conditions for treatment	1. Accepting; warm but setting limits 2. Ready for acting out 3. Develops program	Crisis intervention agencies
2 - Symptom reduction	1. Improved behavioral functioning 2. Cooperative in program	1. Directive 2. Supportive of client initiative 3. Moderates high expectations	Conditionally provide long-term medical and rehabilitation services
3 - Two world conflict	1. Choice between: a) health - illness b) conventionality-delinquency 2. Testing both worlds 3. Beginning successful experience	1. Assesses client's real strength 2. Encourages and pushes for positive conflict resolution 3. Acts as role model	Continued provision of long-term services Greater involvement of family support system Contact with conventional community services and institutions
4 - Social values internalization	1. Developing positive self-image 2. Functions according to real potential 3. More acceptance of world and self within it 4. Drug abstinence	1. Work on deeper psychological issues 2. Reflects issues as problems in living and not as result of deviance 3. Encourages more varied experiences	1. Separation from distress and welfare institutions 2. Continued contact with conventional institutions 3. Continued involvement of family
5 - Individuation and termination	1. Greater emotional and functional maturity and independence 2. Starting self-realization 3. Possible regression in face of cutting therapeutic bond	1. Accepts having to let go 2. Emphasizes and summarizes achievements 3. Termination is not rejection	1. Independent client - environment contact 2. Conventional institutional contact

The vast majority of clients are polydrug users (opiates, barbiturates, amphetamines, tranquilizers, hashish, cocaine, psychedelics, and alcohol). Long- or short-term heroin users are rare. The majority are of Sephardi (Middle Eastern and North American) origin, the minority are Ashkenazi (Western). The socioeconomic backgrounds vary, but are primarily toward the low range of the scale. One finds considerable pathology in all of the clients' families of origin. The nature of the family problems are different. Sometimes the client is the only "identified patient" in the family, that is the "black sheep," and sometimes he comes from a multiproblem family. Almost all of the clients have a long history of adjustment difficulties and have long been considered problematic and deviant by their families and other agencies, as well as by the community at large. They are typically highly impulsive and have difficulty tolerating deferred gratification. In short they are functioning poorly in most conventional life areas and can thus be considered "multiproblem." For these clients the therapeutic process is inevitably long, involving a number of stages and life experiences before it brings them to achieve their goals. These goals are typically multiple, in keeping with the clients deteriorated functioning in most life areas, as well as with the center's ideology and treatment model.

STAGES IN THE TREATMENT PROCESS

Table 1 illustrates the stages through which the client progresses within the multifaceted treatment program and outlines the functions of the therapist and support system over the treatment period.

Stage 1: The Development of Trust

During the initial phases of treatment the client naturally approaches both the treatment process and techniques with trepidation and doubt. As a result, the focus of the first stage of treatment, as in all therapeutic systems, becomes the establishment of rapport and basic trust. The client typically tests the therapist in a number of ways: by acting out, by questioning, by disappearing. The therapist is aware that this is a universal tendency with these clients. Cognizant of the client's suffering, the therapist responds by being consistent, by not taking the acting out personally, and by reaching out when necessary. S/he accepts the client as s/he is, and simultaneously sets the limits or boundaries that the client must accept. This constitutes the minimal conditions for treatment to take place. When the client is prepared to suspend his mistrust of the therapist's and the agency's intentions, actions, and demands, and at the same time signals his intention of misunderstanding and abiding by the minimal conditions for treatment—not being too drugged at the center, coming on time to sessions, curtailing destructive tendencies—a therapeutic alliance can be established.

One example of the establishment of trust by reaching out is that of R who came to one initial interview, but missed his second appointment. In reply to the therapist's letter he came for another meeting but then disappeared again. He responded later to numerous calls to his home, started coming regularly to sessions, and eventually settled down in treatment.

Initial and long term treatment goals are agreed upon following a thorough review of the client's strengths and weaknesses in various conventional and deviant/criminal life areas. Immediate special and critical problems might need solution by the timely intervention of outside agency support systems such as casualty and emergency rooms at hospitals and welfare. But the main treatment site at this time is the drug abuse treatment facility where the client needs to settle down in to the treatment-oriented routine. During this phase contact needs to be made with the staff of the community agencies who are partners to the treatment effort. These workers tend to be unenthusiastic about assisting the distressed client because of actual or heard about failed past treatment interventions (often resulting in stereotypes and stigmas) and because of a tendency to assume that the client's requests for help are part of a plan to manipulate special favors. The therapist therefore need to work at this stage to improve the relationship between the involved and/or rejecting institutions and the client.

This first phase is a difficult and in many ways the most critical one. Its positive resolution sees the client more relaxed and ready to cooperate in the joint development of the treatment program with his/her therapist. Failure at this stage may result in a temporary, as well as a permanent, end to treatment.

Stage 2: Symptom Reduction and General Situation Improvement

The second phase is characterized by sympton reduction and an improvement in the general situation. Now that the client is more familiar with his therapeutic surroundings and more accepting of the ground rules of the therapy s/he is more ready to take an active part in treatment. Personal levels of functioning improve, s/he begins to show some initiative and starts to carry out the treatment plan. Consequently we often see a considerable reduction in symptomatology: drug usage is reduced and criminal activity may stop altogether. The client begins to see the possibility of a different future. All to often this is a source of a new problem. Notwithstanding the reality that most often the gap between his new found dreams and his current abilities is vast, the patient wants to see quick results. At the same time his hostility to the outside world, which he perceives as responsible for his difficulties, is more directly expressed. The therapist at this point attempt to moderate unrealistic high hopes, encourages optimism, and legitimizes and defuses extreme feelings of anger, hostility, and despair. The therapist acts as a source of support, a provider of information, and a role model.

Many clients have individual needs which ought to be dealt with at this juncture if the conditions for the continuation of treatment progress are to be maintained. Some patients may be candidates for methadone maintenance or for inpatient detoxification services. Others may require special housing or welfare assistance. It may be necessary to refer clients to various educational, vocational, and rehabilitational courses. These may eventually prove to be very significant actions. Further, remedies may be sought for previously neglected medical problems. Thus, at this stage of treatment many important therapeutic events take place outside the physical area of the drug treatment facility, but the therapist remains in the picture, mediating these developing client-support system relationships.

Although there are clear signs of improvement these are unfortunately only superficial. The clients have started at a very low level of functioning and it is not too surprising to have seen relatively rapid movement. Neither the internal (psychic) nor external (family support) systems have been put under much strain, and it is largely for this reason that the relative advancement has continued unhindered. Some clients, who have attained certain limited, usually material, goals, are satisfied to leave treatment at this time. The others who remain in treatment progress to the next important phase of therapy.

Stage 3: Conflict between Two Worlds:

The third stage of treatment brings with it a critical conflict. The general progress that occurred during the previous stage is the first sign that real change is possible. The client finds himself now on the brink of the unfamiliar or that which has been "forgotten" from the past. S/he is faced with what s/he has and does not have in life. In a sense s/he has to decide whether it is worthwhile to cross over into the unknown -- to choose between health or pathology, between conventionality or delinquency. That is, the possibility of, attaining those goals that at the outset of therapy seemed distant and unrealistic suddenly becomes feasible.

At the same time, key support systems, and in particular the family, sense the coming change, and are unconsciously concerned, that change in the family homeostatis will be a threat to them. For example, a wife who is used to having her husband functioning in the house as "another child" may be threatened by her husband beginning to show assertiveness and to hint at his intentions of taking on more appropriate roles at home.

The therapist has to realistically assess client and system flexibility and strength, and according to his evaluation will work intensively towards encouraging a positive resolution to the conflict.

Frequently there is a regressive acting out, such as a return to drug and alcohol use and a return to prostitution or to criminal activities. The client is returning to the familiar as a way of resolving the conflict. In fact, s/he may be hesitant or even afraid to relinquish the old for the new home and may surrender to previous nonconventional activities. Alternately, the clients may find they no longer identify with their previous interests, be more willing to embrace the values represented in therapy, and consequently rely on the therapeutic facility to push them ahead.

S. D., a female client, was getting along very well in therapy, started a course of study, and improved her living conditions. However, one day she received an offer she could not refuse and hopped on a plane for Amsterdam and the Red Light District. Z. S., notwithstanding her making considerable advances in therapy, decided that she wanted to go out on the street one night. She, however, told her therapist afterward that it just was not the same, that she was disgusted with herself and thus would not (and did not) return to prostitution.

A client who has made repeated attempts at treatment but who has not managed to positively resolve this critical phase can be considered "chronic."

During the third phase of treatment the support systems ought to remain stable. There appears to be a tendency amongst welfare services to stop assistance following the advances of the second stage. At this point the therapist must make it clear to the welfare facility that the client is now entering a stage where he needs to have confidence in their continued support and encouragement because he is precisely at the point where he is uncertain about his future direction. It is also appropriate at this time to broaden the therapeutic interventions to formally include the family, the closest support system. It is important that the family be a partner to the client's venture so as to enable it to be sufficiently flexible to allow the client's progress to continue. Thus much of the important therapeutic work is done at the drug treatment center where the responsible therapist endeavors to maintain the conditions that facilitate further advancement.

The client who advances through this stage successfully develops a much more positive and realistic self-image, and begins to have more successful "conventional" experiences. This is a time of true breakthrough.

Stage 4: Social Values Internalization and General Stabilization

The following phase in treatment is typified by both the client's internalization of social values and an easier identification with greater society. S/he increasingly fulfills conventional roles-- spouse, parent, friend, worker/apprentice. As the successful experiences increase there is more to lose by regressing. However, the clients still struggle to find their real places in society. They are often not allowed to forget the past, and usually some stigma remains. For example, clients may be barred from government work, or be ineligible for a driver's license. Together with this, they find that a problem or crisis in one area of life need not result in a feeling of total helplessness. Clients differentiate between achievements in different life areas and are more capable of expressing their feelings on these matters. The therapist relies more on the clients abilities and judgments, and turns to prod them to examine remaining problematic psychological issues more closely. Together they evaluate the changes that have taken place and identify future goals that will enable the clients to give fuller expression to their individual potentials.

Group therapy is useful at this time as the clients can further develop their new interpersonal skills. They can learn positively from each others past and current experiences, and encourage each other to multiply their varied activities. Outside of the therapy as such, the client is more in contact with conventional community agencies such as community centers, places of work, and banks. Welfare assistance may be reduced to a minimum and include more counseling than material help. Most clients totally abstain from drug use.

Stage 5: Individuation and Termination

In due course, treatment, which has continued for at least a couple of years, reaches the stage where the client is clearly capable of functioning autonomously. Self-confidence based on an awareness of internal strengths has been developed. The client grapples effectively with daily problems in living. Satisfying interpersonal contacts replace drugs as a source of gratification. There is a beginning of self-realization. A new difficulty arises: Acknowledging the need to separate from therapy. The final statement of individuation has to be made. The therapist must be willing to let go. He emphasizes to the client that this is being done not because of rejection (and thus replaying for the client a primary fear) but because the client no longer needs the kind of support he has become used to receiving in therapy. Some clients may well capitalize on this and spurt forward. Others may first tend to regress. The therapist interprets the client's regression as an attempt to perpetuate the dependence relationship, and this enables the client to confront ambivalent feelings over termination more directly.

The support systems that have been helping the client therapeutically are phased out. The client has achieved many of the goals of treatment; he has progressed in many life areas, he can fulfill his healthy dependency needs in his relationships with people and not with drugs; he gives more adequate expression to his feelings; he has plans for the future. The therapist and client review the latter's achievements, and thereafter, the painful but necessary separation is effected.

DISCUSSION

The treatment stages in multifaceted treatment programs for drug abuse which have been described are based on the limited experience of the authors in an ambulatory facility in a developing country. The treatment process has, of necessity, been described in an abbreviated fashion, and simplified by having to set aside the unique circumstances that each client brings to therapy. Nevertheless, numerous important implications are suggested by the model and much research is required to the further knowledge of these issues.

The authors' experience has been that the treatment process described is broadly applicable to drug abuse treatment. In many cases this may be so regardless of both the therapist's orientation and style and the client's differential diagnosis. More specifically, the treatment process seems to have a dynamic of its own whereby the first, third, and fifth stages are critical phases of trust, internal conflict, and separation, respectively, and the second and fourth are essentially consolidation and stabilization stages that follow the positive resolution of the previous stage. Further, a client who repeatedly fails to reach or positively resolve the critical third stage can be considered to be a "chronic" patient. The above impressions appear to have considerable significance and are in need of further investigation and verification.

Two methods of evaluating the treatment process, and thereby the treatment program, are relevant:

1. The first method is an observational one whereby the therapist evaluates the clients progress by estimating the stage of the treatment process that the client has reached.

2. The second method is more formal and is achieved by administering a quantative evaluation questionaire at appropriate times during the treatment process.

Multifaceted treatment planning requires a realistic concern about staff and their training. It is our opinion that training in a mental health helping profession is essential for those who have direct responsibility for treatment management. Although emphasis is not necessarily placed on interpretation during the treatment process it is important that the responsible therapist have an understanding of the underlying issues and of the expected therapeutic processes. Training in crisis intervention and in organizations and systems are also very important since multifaceted treatment demands the involvement of other agencies. It is useful, too, that paraprofessionals and workers from such agencies, who are involved in the treatment, be actively trained in the role they have to play in the treatment of drug users.

Funding agencies when considering the allocation of resources for drug abuse treatment are clearly aware that the multifaceted treatment program is conceptually different from and rather more expensive, than a unimodal program (i.e., a methadone maintenance depot). The duration of treatment, the relatively high staff-client ratio, staff training, and the utilization of various community resources add to the accumulated treatment costs. However the actual and/or potential benefit to society of this greater investment should be considered sufficient to make it worthwhile. This remains to be demonstrated.

Further research and investigation is necessary, therefore, to provide the needed guidelines whereby the treatment process described here can be most effectively applied, from all points of view. Wider knowlege and application of this treatment process will serve to demystify drug abuse treatment, reduce the anxiety of professionals who come into contact with drug abusers, and consequently will improve treatment success rates.

REFERENCES

Brecher, E. M. Licit and Illicit Drugs. Boston: Little, Brown & Co., 1972.
Brill, L., & Lieberman, L. (eds.) Major Modalities in the Treatment of Drug Abuse. New York: Behavioral Publications, 1972.
Cummings, N. A. "Turning bread into stones: our modern antimiracle. American Psychologist, 34 (12), 1119-1129, 1979.
Einstein, S. Annual Report: Jerusalem Center for Drug Misuse Intervention. Jerusalem, 1981 (Hebrew).

Glass, M. M. A Guide to Addiction and its Treatment. Lancaster, England: Medical and Technical Publishing, 1974.

Hughes, P. H. Behind the Wall of Respect. Chicago: University of Chicago Press, 1977.

Sells, S. B. The Effectiveness of Drug Abuse Treatment, Vol. 1. Cambridge, Mass.: Ballinger Publications, 1974.

Steer, R. A. "Differences in heroin addicts seeking inpatient detoxification, ambulatory detoxification, or methadone maintenance." Drug and Alcohol Dependence, 4, 399-406, 1979.

STAFF AND TREATMENT IN A STATUTORY OUTPATIENT CARE AGENCY

FOR ALCOHOLICS IN HELSINKI, FINLAND*

Teuvo Peltoniemi

The Finnish Foundation for Alcohol Studies
SF-00100 Helsinki 10
Finland

BACKGROUND

One may discern two main orientations in research in the treatment of alcoholics. On the one hand, there are client studies and, on the other, there has been research into the results of treatment. To a lesser extent attention has also been paid to the way in which organization providing treatment actually operates. Among other things, Bruun points out that "unfortunately the content of the treatment and the actual working procedure have virtually been left unstudied. Research has been directed towards clients; we know a lot about their childhoods and the results of the Rorschach tests, but really very little of the staff and its methods of operating (Bruun, 1972, p. 310).

If justification is needed for conducting research into actual treatment organizations and their personnel, we have but to look at the general pessimism of evaluation studies. No one method of treatment has proved to be more efficacious than any other. Indeed, one cannot even demonstrate that providing treatment produces better results than giving no treatment at all (cf., e.g., Ogborne; 1979, Edwards et al., 1977). But evaluation studies may well contain a basic, significant error. The contents of treatment may not have been actually ascertained. Treatment centers may be thought of as "black boxes" with supposedly familiar contents. In a Finnish evaluation study, which was unable to demonstrate any difference between two forms of treatment, Bruun and Markkanen (1961, p. 79) wrote: "During the course of our account we showed that the methods of treatment of the A-clinic and the Hesperia Clinic were formally greatly different from one another. We presumed that this disparity would culminate in differing treatment results. Uniformity appears to be greater than the formal divergence, and this uniformity in basic attitudes might well explain the similar treatment results."

The Research Target

The above points form the background of the present discourse. It examines the philosophies underlying the functioning of PAVI office and its relationship to other agencies. This will be done by analyzing the

---

*This agency is called the PAVI office, by its Finnish name.

background of staff and the way in which they operate and by looking at the prevalence of various treatment ideologies.

This research is directed toward practical, concrete work. There is a great divergence in the statutory care of alcoholics in particular. Some cities, in practice, provide no PAVI treatment at all. Conversely, PAVI treatment is considered of great importance in other places, for example, in Helsinki.

The Finnish Foundation for Alcohol Studies decided to instigate a research project which would center on Finland's largest unit for the statutory outpatient care of alcoholics, the Helsinki PAVI office. The office is part of the Finnish system of caring for alcoholics which inclines to social welfare work. In many countries the treatment of alcoholics is divorced from the general health system. In Finland, Sweden (SOU, 1967, p. 36), and Norway (Christie, 1965) the treatment of alcoholics is rooted in vagrancy care and social welfare. The system includes objectives for the provision of control and support. Later, the development of the treatment of alcoholics was influenced mainly by therapeutic treatment philosophies (in the 1950s, e.g., by Yale and the AA movement). This has strengthened social welfare's general trend toward a more therapeutic orientation and greater professionalism (cf. Wilmot and Ogborne, 1977, p. 1).

A PAVI office that is part of Helsinki's communal statutory care of alcoholics was chosen as the research target. The office has the power to use compulsion when providing care and it may also give financial aid. There are 25 social workers, a doctor, and a nurse working at the office, and ten administrative staff. There were 38,866 client visits at the office in 1977. Opening hours are from 8:30 a.m. till 1 p.m. The PAVI office is situated in the city administration center building where there is also an A-clinic (voluntary treatment only). The welfare police operate from the premises. The organization of the PAVI system is illustrated in Fig. 1.

Previous Studies

There has been little research into the black boxes of treatment organization. Examples of the research that has been conducted are the Wilmot and Ogborne study (1977) of a Canadian halfway house, and Siren's (1977) study of the Helsinki youth clinic. Wiseman's study (1970) may also be thought of as being in this category, although the whole of the study is actually a thoroughgoing analysis of the lives and treatment of skid-row alcoholics. The scarcity of research might well be tied to the fact that the organizations are not generally eager to allow outside interests an opportunity to investigate internal operations. It may also be difficult to see the immediate usefulness of such research from the point of view of the organization. (This is certainly equally true of research concentrating on clients.)

Traditional research on organizations cannot provide the present study with any direct benefit since its orientation is aimed at drawing up general laws. The perspective is often narrow: The study of information, remuneration, etc. (Heydenbrand, 1973). Applied research also has its roots in Finland. School homes (Siren, 1965) and institutions for alcoholics (Saila, 1967) were investigated in the 1960s.

The applied research approach naturally has its own limitations. The most important of these are the partial diminishing of the historical background and the tendency to divorce the research target from its surroundings to some extent (cf., Caudill, 1958, p. 27). Detailed microlevel analysis is, however, such wide-ranging research that one can easily accept the above limitations. The present research approach enables one to view the research target from a different perspective as compared to conventional organizational research. Conventional research generally proceeds from an administrative basis and from the interests of decision-makers (Becker, 1971, p. 19). This paper deliberately proceeds from the point of view of field workers and to a certain extent, it also pays attention to the perspective of the client. It must be stated, though, that departing from the conventional perspective might result in problems regarding evaluation. Becker (1971) gave a detailed account of this problem in his classic paper "Whose Side Are We On?"

Date

In 1976, the author conducted a preliminary study at the PAVI office of Oulu in Northern Finland (Peltoniemi, 1977). The collection of data in Helsinki took place between 1976 and 1978. The author worked in a room at the PAVI office provided by the administration. The office viewed the research work favorably and for the most part, the data were easily gathered. Only three persons partially refused to take part. They were, however, still prepared to be as observed. During the two and a half year period, the author became an ex-officio member of the office and also took part in unofficial activities. His role as a researcher was nonetheless emphasized the whole time. In particular, the author systematically refused to take a stand on controversial matters at the office (cf. Trice 1971, p. 79).

The office personnel elected seven representatives to a cooperative body which met with the researcher about once a month. The cooperative group reviewed practical and ideological problems connected with the study. The Finnish Foundation for Alcohol Studies maintained a back-up group. In addition to university professors, the director of the Social Welfare Board was part of the group. This body met twice a year.

A questionnaire was sent to the personnel of the office during the early stages of the study. The questionnaire collated their backgrounds and attitudes toward their work and alcoholism. Sociometric analysis were also conducted at this point. The main method for collecting data was that of observation. The researcher (or a research assistant) observed "typical" personnel chosen on the basis of sociometric analysis and recorded all discussions and events. In addition to making observations, the author also held three discussions where a systematic appraisal of the objectives of the PAVI office took place (cf. Blum, 1971, P. 89). Individual personnel and clients were interviewed throughout the entire study period.

Owing to the lengthy study period and perhaps because there were often trainees watching the work of the office, the observation work did not seem to have much effect on either the personnel or the clientele.

The author gained the impression that the research relationship was positive throughout. This was partly due to the formal methods which enabled the office personnel to participate in the study. They often thought that their work held problems and they hoped that the research results might help them. When the research was completed, the personnel were allowed to familiarize themselves with the preliminary version of the research report. A seminar lasting one day was held and the matter was dealt with during that time. The final report was amended according to observations made at the seminar when the observations and remarks had to do with clear errors. The opinions of the office personnel on the implications of the report differed widely. It is interesting to note that the opinions expressed at the meeting reflected contradictions recorded in the study and, thus, actually supported the research findings. (The validity and reliability of the study were investigated with the aid of a seminar; cf. Bryun, 1971).

RESULTS

Personnel

The PAVI personnel group are: welfare inspectors, social workers (half of whom has a university-level education), PAVI advisors (who have not had formal education and are active in the AA movement), a doctor and a nurse (mostly for medical inspection for entering an institution as stipulated by law), a director; and an assistant director (one of whom is a person active in the AA movement). (See Figure 1.)

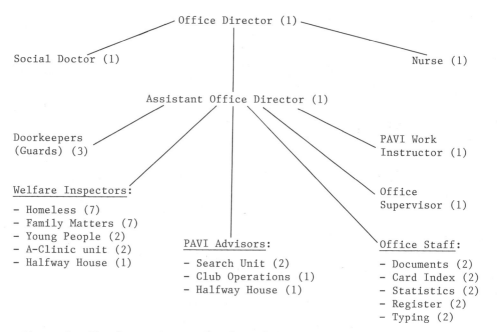

Figure 1. The internal organization of the Helsinki PAVI office. The numerals in parentheses indicate the number of personnel working in a specific area.

Those working at the office are coordinated according to their sex. The majority of workers who look after homeless clients are men. The tasks below were attended to by women in the following proportions in late 1977:

inspectors dealing with the homeless	11%
inspectors dealing with young people	50%
inspectors dealing with family matters	86%

More women inspectors have received formal education in social work than their male counterparts. Women are also more therapeutically oriented. This breakdown of the inspectors serves to illustrate the internal allotment of duties and ideology in the PAVI system.

Working Climate

The author's first encounter with the PAVI system, through initial interviews, brought him into contact with a highly acerbated climate. The relationship between the personnel and the directorship was one of tension. Turnover was high. The atmosphere improved somewhat during the course of the study. Turnover fell, partly because of a worsening employment situation for social workers. Absence due to sickness, which the personnel thought was frequent, proved to be no higher than, for instance, the corresponding rate in industry (a third of the personnel had been absent for at least one day during the previous two-month period; cf. Nyman and Raitasalo, 1978). But is was symptomatic that absence was greater among personnel who actually worked with the clientele (36%) than among the administrative staff (23%).

There were 14 questions on job satisfaction from which an index was formed. This ranged form 0 (low) to 4. Welfare inspectors and office workers were found to be more dissatisfied than the other employee groupings (cf. tabulation).

Office workers	2.1
Welfare inspectors and nurse	1.7
PAVI advisors and door keepers	2.7
Supervisors and social doctor	2.8

Differences were also found in the reasons for dissatisfaction. Office workers expressed the dissatisfaction experienced by people with poorly paid, routine jobs. Welfare inspectors were dissatisfied in relation to the influence that they exerted, with their pay in comparison to the responsibility they bore and with the executive-subordinate relationship. This dissatisfaction was further reflected by the extent to which they were isolated from trade union activity and by the retreating into a passive role of bureaucrat.

Employee Types

Several studies have paid attention to various types of social workers. In his study of American welfare agencies, Kroeger (1975, pp. 90-91) divides social workers into the client-oriented and the agency-oriented types. He finds clear differences in the principles on which these two types work.

Scott (1969, pp. 90-91) divides social workers into four types on the basis of their eduction and reference groups:

1. Professionals (university-educated and sensitive to outside professional reference groups)

2. Only reference groups (lacking university education but sensitive to outside professional reference groups)

3. Only education (university-educated but sensitive to the organization's internal reference groups)

4. Bureaucrats (lacking university education but sensitive to the organization's internal reference groups.

Glaser's (1964) typology proceeds from the duties of social workers and makes a classification on the basis of the aid offered and of control. Glaser outlines four types of criminal welfare personnel:

1. Not much control and a great deal of aid offered--the welfare type

2. Both a great deal of control and aid offered—the paternal type

3. A great deal of control and not much aid offered—the sanction type

4. Not much control and not much aid offered—the passive type

Guttormsen and Höigard (1977, pp. 13-147) use three factors as criteria: attitudes to social welfare, treatment of clients, and ability to resolve internal contradictions in the system. The labels which they employ, however, denote there having been more criteria actually used (Professionalist, Career conscious, Christian, Rebel, Controller, etc.)

Allardt and Littunen's (1972, p. 189) typology has in its background that general attitudes to social policy may in principle be divided into three subfactors: moral, pragmatic, and authoritarian attitudes. Their typology have been applied to the public, to social workers, and to clients. Allardt and Littunen's social worker types are:

1. Performing work of a vocational nature

2. Performing work piously but in a business-like fashion.

3. Severity within the framework of the rules

Allardt and Littunen consider their classification to be of a general nature and do not think that it can be directly applied to individual employees. Nonetheless, they do believe that the typology covers the major attitudes rather well (Allardt and Littunen, 1972, p. 189)

The criteria most often in use in typologies are factors of education, attitudes toward the client, and the organization and attitudes toward social work or social policy itself. At bottom, the question is one of deep differences in operating philosophies.

There are different groups among the PAVI personnel, as indicated by the sociometric analyses. For this reason a grouping was carried out on the basis of ideas formed from the observation material. The following factors of the background of groupings were investigated:

1. Education and attitudes toward it

2. Interest in work and in various work procedures

3. Definition of work objectives (complete abstinence, opportunities for work, causes of alcoholism)

4. Attitudes to the organization

5. Attitudes to outside organizations, to AA and A-clinics in particular

6. Attitudes to social policy (the scope of care, welfare, and control

The groupings were further examined by allowing a few employees to divide the office's field workers into four groups. The names given to these groups attempted to describe the basic characteristics of the groups. No other information on the groups was given. This experimental classification was found to be largely in agreement with the classification made by the researcher, and this would seem to demonstrate the accuracy of the typology.

Four groups were drawn up and were assigned the following names: therapists, social workers, bureaucrats, and controllers. This terminology is not connected with outward official positions or work roles. Instead, the names are meant to bring to mind a picture of the fundamental characteristics of each type through verbal connotations.

The size of the groups were:

Therapist	5 persons
Social workers	11 persons
Bureaucrats	8 persons
Controllers	8 persons

The therapists are those members of the office who are very interested in their work and in therapeutic procedures. The therapists emphasize family work and often have social-work training. They hold the philosophy of A-clinics close to their hearts and do not consider complete abstinence necessary. They work instead toward improving living conditions in general. Clients tend to be a little suspicious of them and to avoid profound discussions.

Social workers are interested in their work, but unlike the therapists, keep their leisure and working lives separate. The level of education of social workers varies. They have a favorable attitude toward treatment but are of the opinion that the resources of the PAVI system are not sufficient to allow treatment to be practiced. They do not expect too much from the results of their work. However, social workers think of themselves as on the side of the client, as helping the "underdog." Clients think of the social worker as a friend and ask after inspectors they know if they do not meet them as they expect.

Bureaucrats are not particularly interested in their work. They refuse to talk about their work, except jokingly, during meal breaks and in their free time. This grouping has formal education, but education is not particularly prized. Experience is thought to be more important. Client visits are attended to in a routine manner. No empathy is felt toward the client. Bureaucrats think that their work provides virtually no opportunities for results. Clients warn newcomers about these inspectors and let each other into the secrets of the game.

Controllers are interested in their work. They do not have a high level of education. Their attitudes toward clients and work reflect that controllers think that they bear responsibility for society, or, at least, for the family. This group forms the executive of the office. The controllers aim a complete abstinence and think that this objective can be reached when clients' recalcitrance is ended, using coercion if need be. Controllers esteem the AA movement and agree with AA in viewing the client as being solely responsible for his problems. They view treatment as a fad which does not correspond to reality. Some clients curse controllers from the bottom of their hearts. On the other hand, there were a significant number of clients who felt that this employee group had helped them.

One should bear in mind the groupings described here are ideal types. It is difficult to find a "pure" type of this kind. Furthermore, these types are tied to client relationship. One client might consider an employee to be largely a therapist while another client might consider that the same employee is largely a social worker. Typology proved to be fruitful when further analyses were made, however. This typology also illustrates the great variety of work which can be carried out within the PAVI system. PAVI is not a single entity; rather, it is a blanket organization for many different kinds of operations.

Treatment Ideologies

Shain postulates that an organization must be in complete agreement on its aims if it is to function efficiently. He lists the following conditions (Shain, 1971, pp. 2-3):

1. The goals of the agency must be clearly stated and must be internally consistent. This means that the various levels of staff must at least to some extent endorse the goals of the agency.

2. The goals must be appropriate in relation to the needs of the population served.

3. The manner in which power or the decision-making process is handled must be consistent with the goals (e.g., autocratic means are unlikely to serve democratic ends).

4. The manner in which communication is handled must be consistent with goals.

5. The values and behavioral capacities of staff must be consistent with goals.

6. The goals of the agency must to some extent be congruent with the demands of the community within which the facility operates.

This view of the necessity for unanimity may not quite hold true. Treatment of alcoholics in Finland, for instance, is characterized by various combinations of the treatment philosophies of the AA movement and the A-clinics (for an account of the treatment philosophies of the AA movement and the A-clinics (see Kiviranta, 1969, pp. 17-24). These combinations seem to work well, at least on the surface. It has also been argued that a treatment center which offers diverse treatment philosophies will be able to care for several different types of client (M. Puro, unpublished).

The philosophies underlying the functioning of the PAVI system are not particularly clear-cut. They are not discussed. Two main orientations may, however, be distinguished. They will be referred to here as the treatment approach and the control approach. The control approach stems from traditional vagrancy care and the traditions of social welfare. The treatment approach is a newer product derived from the development of treatment techniques and professionalism. The control approach rests to a large extent, on the philosophy of AA. The therapists' orientation is based on the A-clinic philosophy. Controllers and bureaucrats lean toward the control approach while therapists and social workers tend toward the treatment orientation.

The control approach is the official treatment philosophy of the Helsinki PAVI office. The therapy orientation has become important, however, because of the general trend of social welfare. The majority of personnel support treatment. Within PAVI, the control approach has emphasized the underlying philosophy by, among other things, placing untrained AA people on the office payroll. Those employees who support the A-clinic system think that these AA advocates are their professional rivals.

Table 1. Work Objectives and Methods by Employee Types

Subsector	Control approach	Treatment approach
Advocates	Controllers and bureaucrats	Therapists and social workers
Underlying philosophy	AA movement	A-clinics
Learning the job	Experience/practice	Formal training
Attitude to PAVI legislation	Follow closely	As lax as possible
Operational basis	Official measures	The client
Use of coercion	More coercion	More voluntariness
Treatment goal	Complete abstinence	Improving relations between people
Characteristics of alcoholism	Disease and irresponsibility	Disease and weakness
Role of alcohol	Cause	Effect
Recovery from alcoholism	Pessimistic view	Optimistic view

From the point of view of the PAVI system, the fact that the executive has completely gone over to the control approach produces problems. It means that the <u>contradictions between the philosophies underlying treatment and the differences between the executive and personnel overlap</u>. This is the background of the exacerbated atmosphere which prevailed in the office from time to time. This complex contradiction is reflected in the following examples. The personnel thought of training as being aimed at increasing professional competence whereas the office executive considered training measures as increasing the unity of the office.

At the general office meeting the personnel put forward their training objectives. These were concerned with treatment and specific topics and lecturers. The executive put forward aims such as "legal aid office operations." The supervisor also thought that there might be a need for specific written instructions on coercive action as "practice in this. field has become confused." The supervisor's proposals were later put into practice.

Ends and Means

Employee types are clearly different in regard to their work objectives and to the methods which they advocate. Table 1 tries to present some of the most important differences.

One important distinguishing factor is attitudes toward therapy. Even though some PAVI personnel set a high value on treatment, most of those who work within the PAVI system would much prefer not to be bothered with the concept at all. This is demonstrated by the following excerpt from a personnel group discussion:

Inspector 1 (a bureaucrat)   "The whole concept of treatment seems odd to me. Do we even know how to give therapy here?"

Inspector 2 (social worker)   "What do you mean by treatment? Let's not talk rubbish."

Work Instructor (controller)	"It's nonsense to say that you can always apply treatment."
Inspector 2 (social worker)	"It all depends on what you mean by treatment."
Work Instructor (a bureaucrat)	"Yeah, how do you understand it and how do I take it?"

Only the therapist type was able to use the work naturally:

| Inspector 3 (therapist) | "Therapeutic relationships break down whenever there's question of economic power." |

Aside from the issue of efficiency, attitudes toward treatment are also tied up with various views about people. Advocates of treatment believe that it is generally possible to help clients. Those who support the control approach tend to view alcoholics with suspicion. When a long-term client relationship was broken off because the client gave up, the office leadership told the therapist-type inspector, "Remember that he took you for a ride. You can't help an alcoholic with normal social work procedures."

Employees who advocate the control approach are often suspicious of the client's wish to recover and his chances of doing so. The following tabulation demonstrates this:

"Alcoholics consist of incurable, hopeless cases":

Therapists	0.8
Social workers	0.9
Bureaucrats	2.1
Controllers	2.3

[N = 31, the point score ranges from 0 (disagree completely) to 4]

The main difference is one of different philosophies underlying operations. There is not much difference between therapists and social workers, on the one hand, or between bureaucrats and controllers on the other.

Compulsory Treatment

Attitudes toward compulsory treatment demonstrate treatment philosophies in a clear way. During October 1978, the personnel made an estimate of whether each single client stood in need of compulsory treatment at a clinic. With regard to homeless people, inspectors gauged that 29% of the clientele stood in need of compulsory treatment. Inspectors thought that the corresponding percentage for family matters (young people, A-treatment unit) was a mere 6%. But compulsory treatment is not only a question of different kinds of clients. Therapist and social worker-type personnel thought that a mere 10% of the clientele needed compulsory treatment. Controllers and bureaucrats estimated 36%. This clear disparity is not explained by different kinds of clients. The same difference is also seen if one classified homeless clients according to the personnel types responsible for looking after the homeless. Among those working with homeless clients, bureaucrats and controllers estimated that 40% of clients needed compulsory treatment whereas social workers and therapists estimated 14%.

Table 2. Reason for Visiting PAVI (%)

	N	%
In need of financial support	22	56
To obtain treatment	11	28
Accommodation matters	3	8
Office summons, supervision	3	8
Total	39*	100

*The N figures are greater than the number of respondents because more than one alternative was sometimes chosen by one respondent.

Table 3. Clients' Opinions of the Characteristics of a Good Inspector (%)

	N	%
One you can talk with	14	26
Treats you as a person	9	17
Trusts and supports you	8	15
Businesslike	4	8
Generous with money	3	6
Arranges practical matters well .	5	9
Inspector (a good one) mentioned by name	10	19
Total	53*	100%

*The N figures are greater than the number of respondents because more than one alternative was sometimes chosen by one respondent.

The Client's Perspective

Thirty clients were interviewed during one autumn day in 1977. The clients felt that they were different from each other. Half of them regarded themselves as alcoholics. The other half opposed the use of the term. This state of affairs is tied up with the clients' backgrounds. Family-related clients consider themselves a typical PAVI clients. Homeless clients, on the other hand, thought that they were typical cases. The PAVI system has thus become labeled by its clients as an institution which mainly treats chronic alcoholics. The clientele thought that the PAVI system served them best when it arranged financial support and directed them to clinics (see Table 2).

Clients thought of the PAVI system as a provider of material aid. But when we asked the question "What is a good inspector like?," responses tended toward "a good person" who is willing to talk and who treated people in an egalitarian fashion (see Table 3).

Even though clients mainly come to PAVI for material aid, the clients definitions of inspectors reflect the need which clients have for human contact. This observation was made by, among others, Wiseman (1970); many skid-row alcoholics are lonely people, and their ties with officialdom may form virtually their only contact with other people.

If we relate clients to the control and treatment approach of personnel, the philosophy of the clientele might be termed the support approach. Clients want support, both material and spiritual. They view treatment with suspicion, and few of them are fond of control.

Discussion

This article examines the prevailing philosophies underlying the functioning of a program providing out-patient care for alcoholics, the Helsinki PAVI office. There are three opposing main treatment philosophies in PAVI, the personnel's therapeutic approach, the control approach, and the clients' support approach. These orientations divide the work of the office in many ways. Personnel may be classified in four ways according to how they view their work. These four categories are termed therapists, social workers, bureaucrats, and controllers.

It is clear that the unity of objectives stipulated by Shain (1971) does not prevail in the Helsinki PAVI-office. Do these different philosophies work to the office's advantage or disadvantage?

The first impression is that these opposing treatment philosophies divide personnel into categories which are overlapping with other divisions—the organization's own divisions in the main. This means that "normal" contradictions are bolstered and the general working climate exacerbated.

Could these differences serve to profit PAVI work? The philosophies do correspond in some measure to the organization's own divisions and this might support such as assumption. Those who advocate the control and official approach are thus mainly found among staff who deal with the homeless. Personnel who deal with family matters incline toward the therapist and social worker types. The classification is, however, by no means water tight. There are exceptions on both sides of the fence. Nor is there a comparable division among the office leadership. Executives advocate the control approach unanimously. Consequently, the PAVI system has not yet formed distinct "therapeutic" units or control units. After the clientele have been separated according to the main criteria (age and accomodation criteria), further classification is of a random nature. This means that selection which might bring together clients and personnel well suited to each other does not take place on any systematic basis.

The abundance of treatment philosophies can, therefore, hardly be said to promote PAVI work. On the contrary, tensions inevitably arise which demand intervention (time, energy, manpower, etc.) in order to limit, neutralize, or prevent them.

It is probably not appropriate to suggest that the hiring of social workers should be limited to those with certain psychological and work-oriented features. In fact, no stand has been taken as far as the ability of each type of social worker to fulfill their work duties. That would initially demand a clear goal choice of the treatment ideologies and then the different types could be viewed against that choice.

It is easy to see, however, that there might be certain workers and certain clients which are more suitable to each other than other combinations. The intake of clients now works on a random basis. One might assume that a more selective system would be helpful. The clients are now received by non-educated workers whose task is quite technical. It would be more functional if a social worker was the intake-screener who would be able to evaluate who is the most suitable treatment agent.

This is not a study of treatment outcome. It should be obvious to the reader that the heterogeneity of treatment could strongly bias an outcome study. It is therefore suggested that in any outcome study, the actual procedures of the organization concerned should be more clearly studied to avoid such possible source(s) of bias.

REFERENCES

Allardt, E., & Littunen, Y. Sosiologia. Porvoo: WSOY, 1972.
Becker, H. S. "Getting individuals to give information to the outsider," in W. J. Filstead (ed.), Qualitative Methodology - Firsthand Involvement with the Social World. Chicago: Markham, 1971.
Blum, F. H. "Getting in

Allardt, E., & Littunen, Y. Sosiologia. (Sociology) Porvoo: WSOY, 1972.
Becker, H. S. "Whose side are we on?" In W. J. Filstead (ed.), Qualitative Methodology - Firsthand Involvement with the Social World. Chicago, Markham, 1971.
Blum, F. H. "Getting individuals to give information to the outsider," In W. J. Filstead (ed.), Qualitative Methodology - Firsthand Involvement with the Social World. Chicago: Markham, 1971.
Bruun, K. Alkoholi: kaytto, vaikutukset ja kontrolli. (Alcohol: Use, Effects and Control). Helsinki: Tammi, 1972.
Bruun, K., & Markkanen, T. Onko alkoholismi parannettavissa? (Is It Possible to Cure Alcoholism) Helsinki: The Finnish Foundation for Alcohol Studies, 1961.
Bryun, S. T. "The methodology of participant observation." In W. J. Filstead (ed.), Qualitative Methodology - Firsthand Involvement with the Social World. Chicago: Markham, 1971.
Caudill, W. The Psychiatric Hospital as a Small Society. Cambridge: Harvard University Press, 1958.
Christie, N. "Temperance boards and interinstitutional dilemmas: A case study of welfare law," Social Problems, 12:415-428, 1965.
Edwards, G., Orford, J., Egert, S., Guthrie, S., Hawker, A., Hensman, C., Micheson, M., Oppenheimer, E., & Taylor, C. "Alcoholism: A controlled trial of treatment and advice," Journal of Studies of Alcohol, 38:1004-1031, 1977.
Etzioni, A. (ed.), The Semi-Professions and Their Organization (Teachers, Nurses, Social Workers). New York: The Free Press, 1969.
Filstead, W. J. (ed.), Qualitative Methodology - Firsthand Involvement with the Social World. Chicago: Markham, 1971.
Glaser, F. The Effectiveness of a Prison and Parole System. Indianapolis, 1964.
Guttormsen, G., & Hoigard, C. fattigdom i en velstandskommune. En undersokelse av sosialomsorgen i Baerum. (Poverty in a Welfare Commune. A Study of the Welfare Agency in Baerum) Oslo: Universitetsforlaget, 1977.
Heydenbrand, W. V. (ed.), Comparative Organizations - The Results of Empirical Research. Englewood Cliffs, NJ: Prentice-Hall, 1973.
Kiviranta, P. Alcoholism Syndrome in Finland. Helsinki: The Finnish Foundation for Alcohol Studies, 1969.
Kroeger, N. "Bureaucracy, social exchange and benefits received in a public assistance agency," Social Problems, 23:182-196, 1975.
Nyman, K., & Raitasalo, R. Tyosta poissaolot ja niihin vaikuttavat tekijat Suomessa. (Nonattendance of Work and Factors Influencing to It in Finland) Helsinki: Reports from the National Pensions Institution, A14, 1978.
Ogborne, A. C. "Towards a systematic approach to helping people with drinking problems," Toronto: Addiction Research Foundation, unpublished.

Peltoniemi, T. "Contents and coordination in statutory outpatient care of alcoholics in Finland," Paper presented at the 23rd International Institute on the Prevention and Treatment of Alcoholism, Dresden, GDR, 1977.

Saila, S-L. Huoltola - paihdyttavien aineiden vaarinkayttajien hoitolaitos. (An Institute for the Treatment of Alcoholics.) Helsinki: Reports from the Social Research Institute of the State Alcohol Monopoly, 28/1967.

Scott, R. W. "Professional employees in a bureaucratic structure - Social work," In A. Etzioni (ed.), The Semi-Professions and Their Organization (Teachers, Nurses, Social Workers). New York: The Free Press, 1969.

Shain, M. The Small Tesidential Care Center as an Organization. Toronto: Addiction Research Foundation, Substudy 1-42, 1971.

Siren, P. "Tavoitteet ja todellisuus," (Objectives and the Reality) Sosiologia, 1:101-112, 1965.

Siren, P. Nuorisoasema poihteita kayttavien nuorten hoitoyhteisona. (The Youth Clinic as a Therapeutic Society for Young People) Helsinki: Reports from the Social Research Institute of the State Alcohol Monopoly, 111/1977.

SOU Stockholm: Nykterhetsvardens lage, del 1. (Official Studies of the State. The Situations of Alcohol Care, part 1), 36, 1967

Trice, H. M. "The outsider's roles in field study," In W. J. Filstead (ed.), Qualitative Methofology - Firsthand Involvement with the Social World. Chicago: Markham, 1971.

Wilmot, R., & Ogborne, A. C. "Conflicts and contradictions in a halfway house for alcoholics: Inside the black box," Journal of Drug Issues, 7:151-162, 1977.

Wiseman, J. P. Stations of the Lost: The Treatment of Skid Row Alcoholics. Englewodd Cliffs, NJ: Prentice-Hall, 1970.

GENESA: AN ADJUNCT TO ART THERAPY IN THE

TREATMENT OF DRUG AND ALCOHOL ABUSE CLIENTS

Robert B. Kent

Department of Art and
Program in Expressive Therapies and Creative Education
University of Georgia
Athens, Georgia 30602

The population of drug/alcohol users and misusers is a hetrogeneous one (age, mental abilities, ethnicity, race, and social background, etc.). Under such conditions, no one treatment method or ideology can apparently claim a consistent record of success.

The main emphasis of this paper is to examine a number of alternative's and somewhat radical approaches that may be employed therapeutically. These include movement, relaxation, and imagery, and are employed, particularly as they relate to art therapy.

MOVEMENT THERAPY

In recent years a wide variety of therapeutic methods involving integration of mind and body have become popular in the treatment of mental illness and physical disabilities. For example, eastern philosophies, religions, and methods such as Zen, Yoga, and the martial arts, such as T'ai Chi'uan have been increasing in popularity. T'ai Chi'uan the most majestic of the martial arts demands discipline with an ability to synthesize mental and physical processes (Mogul, 1980). It is the study of preventive health in the most fundamental way. Lao-Tse expresses this aptly in the Tao Te Ching:

>   Trouble is easily overcome before it starts...
>   Deal with it before it happens.
>   Set things in order before there is confusion.

New age healing, at times called wholistic health, embraces a multitude of methods and techniques which are employed to achieve integration of centering (Hendricks and Roberts, 1977). Some methods involve spiritual and paranormal healing, the generating of personal super energy, and biofeedback (Brown, 1974).

In a rather optimistic statement the editor of The New Age of Healing (1979) proclaims that:

The world is entering a time when astonishing events and amazing possibilities are becoming to be experienced by ordinary persons.

And the authority over our individual lives no longer belongs to people and institutions out there. Rather, it belongs to us. Especially in regard to our bodies; to our sickness or health, the responsibility and the choices are ours. (p. VIII).

There are many reasons why movement therapists believe that positive benefits accrue through its practice. The relevant factors will be discussed shortly.

Movement therapists believe that its practice can be helpful in increasing energy levels through fortifying psychic and physical processes in the treatment of drug use and addiction (Feder and Feder, 1981). Opposite of this is the conservation of energy which, when needed, is necessary to enhance one's personal environment which is so necessary for learning to relax.

Movement helps develop more positive mechanisms which are needed for problem-solving, thus facilitating the well-being of the individual (Riordan, 1975). A somewhat more controversial aspect is that movement can be a positive force in developing the capacity to image. The welling-up or shedding of deep physical muscular trauma increases the ability to more sensitively employ the senses, and consequently, frees the mind to image more richly (O'Donnell, 1969).

It is thought that imaging can help heal internal physiological damage. That is, the individual will be able to "see" the damage drugs and alcohol cause, and aid in the healing process. It is also thought that the practitioner of movement can focus-in on positive behavioral changes through imaging (Bruch and Kent, 1979). That is, one can picture in the mind's eye, a better person; consequently raising one's self-concept.

In recent years there has been a spate of literature extolling the merits of movement as an enhancer of mental and physical health, and as a stimulus of creative behavior. Szekeley (1979) states that movement, dance, and breathing are conducive to increasing creative abilities.

Body movement is thought to activate right-hemisphere processes which in turn enhances creative thinking (Rubenzer, 1979; Restak, 1979). It also helps induce a feeling of mental and physical integration, popularly called centering. This effect leads to relaxed awareness, or relaxed attention as described in body therapies, such as Sufi dancing (Riordan, 1975). Such relaxed but focused attention is suggested by Rubenzer (1979) to evoke primarily an Alpha state of brain-wave awareness which is predominantly a right-brained process, although both hemispheres may be involved. Bruch, Langham, and Torrance (1979) indicate that relaxed awareness is facilitative of creative incubation.

Morena (in White, 1976), a pioneer in the development of psychodrama, insists that physical warm-ups are essential for maximizing the actor's craft. This theory advocates that creative behavior is a direct concomitant of positive movement. Simply stated, directed, intelligent, physical activity can be the gateway to the reduction and possibly the elimination, of mental and physical tension. The body is then prepared to be able to make greater use of the senses which is a necessary precursor to basic learning and enhanced creative behavior. It is interesting to note that when muscles are relaxed memories come more easily to the surface (Feder and Feder, 1981, p. 180).

Related to the above is the utilization of movement therapies in athletics. Physical activity is often used as a form of meditation. Many joggers, weight lifters, etc. believe that physical activity can provide an outlet for reduction of tension and that these activities prepare one for greater physical and mental concentration.

A brief mention of body therapies and its implications to drug and alcohol abuse may be appropriate to this discussion. Body therapies have generated a large body of research, especially in its relationship to movement and brain patterns. While all therapies are built on the same principle--re-enducating the motor cortex--each has its own spokesman and special techniques.

The oldest of the therapies, the Alexander Technique, serves as the theoretical base for many of the other body therapies--particularly Feldenkrais (1977) and Houston and Masters. Alexander's method is relatively simple; one becomes aware of movement patterns by inhibiting ineffective patterns and substituting new ones (Myers, 1980b). Like many of the body therapies, Alexander's discoveries came out of a personal need. Alexander was an Australian actor who lost his voice. Intense self-study led him to the conclusion that inappropriate use of muscles was responsible.

Imgard Bartenieff's Effort/Shape Program is based on Laban's movement studies and work on body notation. The Bartenieff method encourages awareness of sensation and feelings, particularly through the use of sensory imagery. Since visual images and bodily sensations contribute to one's total image, the self is diminished by an inhibited range of physical activity and dulled kinesthetic perceptions. As one ages, without intervention, movement is likely to become less and less pleasurable and less cathartic (Myers, 1980a).

Jeri Salkin's Body Ego Technique developed as a result of her dance therapy in institutions. Her instruction deliberately used fantasy to stimulate explorations of a wide range of experiences and helps clarify the difference between reality and fantasy in those experiencing emotional difficulties. Salkin's work is predicated on the fact that confused body images and body boundries lead to exaggerated emotional problems; she suggests that her method leads to a reintegration of body image and identity (Salkin, 1973).

Alexander Lowen's "Bioenergetic Analysis," as many of the body therapies, can be either a group or individual therapy. Based heavily on Reich, the therapy aims to unlock traumas or archaic conflicts which lead to movement restrictions or locking due to tension (Lowen, 1975). Since long-term unresolved conflicts may lead to chronic attitudes and postures, working with those areas may work in reverse. When muscles relax, memories come to the surface, energy is unfrozen and utilized creatively. At this point, Lowen indicated, the subject has abandoned his or her Reichian "armor" which he or she used defensively therefore, restructuring therapy is an important next step.

Houston and Masters (1978) acknowledge a heavy indebtedness to the therapies of Alexander and Feldenkrais. Like the other body therapists, these theorists indicate that psychophysical re-education is of great therapeutic value in promoting good use of the body and awareness. When viewed as a preventive discipline, it enables one to avoid tensions, unconscious actions and sensory impairments which damage health, block potential, and produce emotional and mental problems.

Houston and Masters focus on the body image. Defining the concept as "body as we sense it to be" (p. 12), the concept includes kinesthetic awareness, appearance notions of how we look to others, and is a subset of the large concept of self-image. The body image may not coincide with the physical self as it ideally should. Since we tend to act in accordance with

Table 1

Originator	Year	Type	Primary focus
F. M. Alexander	1932	Body therapy	Becoming aware of movement patterns, inhibiting ineffective patterns, substitute new ones
I. Bartenieff	1965	Body therapy	Developing accurate visual images to bring about desired kinesthetic response
J. Salkin	1973	Body therapy	Reintegration of body image and identity
M. Todd	1962	Body therapy	Psychophysical body reeducation through visualization
L. Sweigard	1963	Body therapy	Ideokinesis as visualized movement without conscious voluntary direction
C. Selver	1938	Body therapy	A method of sensory awareness based on the wholeness of the nervous system
L. Sheleen	1954	Body therapy	Structured exercises based on symbolic spatial configurations and mask therapy as a nonverbal psychodrama technique
A. Lowen	1963	Body therapy	Bioenergetic analysis based on Reich which unlocks affect traumas or archaic conflicts
Houston & Masters	1978	Body therapy	Psychophysical reeducation as therapeutic in promoting positive use of body and greater self-awareness
M. Feldenkrais	1949	Movement therapy	To establish a more complete body image and more sensitive kinesthetic responses
Pesso	1969	Movement therapy	The freeing of unexpressed emotions through psychomotor training
D. Langham	1967	Movement therapy	A system of movement bringing about physical balance which enhances relaxed awareness

that image, it should be as objective as possible. Pain, disuse, and mental difficulties are responsible for the gap. (Table 1).

THE GENESA SYSTEM AS THERAPY

A review of various movement and body therapies indicates possible treatment methods. One that appears to be promising is the Genesa System. The Genesa crystal or model is a life-sized structure constructed of 1" PVC tubing (plastic pipe), which is an analog of the geometry of a biological cell. There are approximately four variations of the model. Each have specialized functions: the simple cube, the four-planed sphere and through the combination cube/sphere to more complex representations of the form of a crystal, such as model with 13 planes outlined.

Dr. Derald G. Langham (1978a), the originator of the Genesa System describes it as:

The coded matrix of your own life force and is directly related to the living energy moving in spirals in all forms of growth, development, and change. Everything is directly related at the core, and a knowledge of the relationships can serve as a key of understanding. (p. XI)

The term Genesa relates to Langham's long association with the fields of biology and genetics. In Sanskrit Genesa means the removal of obstacles to spiritual unfoldment and growth. Those desiring more information on Genesa can refer to Langham's publications (1967, 1969, 1974, 1975, 1977, 1978a, 1979b), and other publications derived from Genesa studies (The

Search for Satori and Creativity by E. Paul Torrance, 1979b; Bleedorn, 1979; Bruch, 1979a,b,c,; Bruch and Kent, 1979; Bruch, Kent, and Wechsler, 1980; Bruch and Langham, 1978; Bruch, Langham, and Smith, 1978; Bruch and Lewis, 1978; and thesis by Burchett, 1977 and Tobias, 1977). Langham has developed four main concepts which comprise the basic substance of Genesa.

1.   The first concept is Form, that is, the physical structure and all concepts related to its structure.

2.   The second concept is Flow. The Genesa is a structure which one stands in the center of, both physically and mentally. Within this structure the person participates in a series of relaxed body movements guided by arms and hands; swinging or pendulum movements; pumping or piston-like movements; and spiralling or circling movements. Movements are made to all areas of the three-dimensional crystal space, in relatively equal proportions of activity to the different spatial areas and movement variations may be effected after the person becomes well centered and comfortably relaxed with the movement processes. In Langham's view:

> Genesa is a uniform framework that intensifies experience of all
> levels so as to generate new ideas and thinking patterns. Genesa
> is thus an orderly yet free and open way of setting many things at
> once in their actual proportions and relations to each other.
> (Langham, 1978, p. XII)

3.   The third concept is Focus. This is based on Langham's system in plant genetics (1967, 1969). The mode may also be used as a systems organizer for either concrete component part designations to clarify logic or a gestalt set of relationships. For example, people working with a Genesa therapist, when centered, can usually develop greater visualization abilities. This enables them to be more successful in problem-solving situations. This process also can facilitate the development of wholistic thinking (Bruch, Langham, and Torrance, 1979). Systematic coding of the terms and relationships are shown as a two-dimensional diagram derived from positions in the life-sized Genesa model (see Figure 1). Positions are designated according to the points in the framework of the Genesa, with polarities or complementary relationships assigned to opposite poles of the same axis in accordance with principles derived primarily from Langham's initial designations in the system in plant genetics (Langham, 1967, 1969). The Center of the Genesa, or the Genesa zero, represents an integrated composite (mid-point relationships) of the complementaries, or the bipolar designates, of 13 rays or vectors.

4.   Function is the fourth concept. It occurs when effective centering is accomplished by the person moving in the Genesa or through the ideas conceptualized in the system. Further stages of growth, representing the next levels of development of the biological cell, or of the concepts or ideas evolving from the initial Genesa system, may be generated in the same fashion through elaboration of the system in each area of growth.

One of the direct outcomes of Genesa movement is centering. Centeredness appears to lead to relaxation for the person who becomes physically balanced with the crystal. This effect may be assumed to reflect relaxed awareness, or relaxed attention as described in body therapies, such as in Sufi dancing (Riordan, 1975). Such relaxed but focused attention is suggested by Rubenzer (1979) to evoke primarily a right-brained process, although both hemispheres may be involved.

Other effects of the Genesa centering movements appear to be the increase of a person's sensitivities to self and self-expressiveness to other persons and to aesthetic awareness (Bruch and Kent, 1979). The

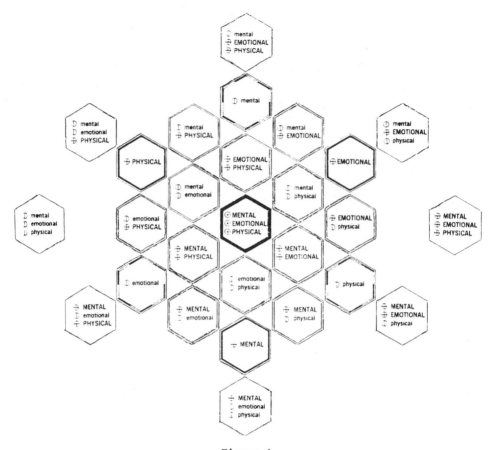

Figure 1.

movements seem to intensify feelings within self and for other people.
Emotions evoked through centering are generally "joy and inner peace". The
aesthetic sensing involves not only the usual senses (visual, auditory,
etc.) but especially enhances kinesthetic, or body-awareness sensing. It may
be that the patterning in the Genesa is similar to that described in Rolfing
or structural integration. Sobel (1974) suggests that the structure becomes
more receptive and sensitive to subtle energies (Bruch, Langham, and
Torrance, 1979).

Restak (1979, citing J. W. Prescott) emphasizes the importance of
movement for early brain stimulation, particularly for the cerebellum.
Without sufficient movement and physical closeness brain pathways that
mediate pleasure fail to develop. The limbic system is the area of the
brain most concerned with emotion. Restak (1979) reports that MacLean views
the limbic system as integrating internal and external experiences.

Recent experiments with college art students indicates that Genesa
movement can stimulate a variety of positive emotions and engender creative
behavior. Several papers are now being prepared which will describe the
various experiments. For example, in one particular experiment with college
art students significant gains were shown by the experimental group on four
of 13 "creative strengths" on Torrance Tests of Creative Thinking (Torrance
and Bale, 1978). Content analyses of open-ended responses also show

positive results for the experimental group with an increase of "relaxation," and other indicators ("calmness and security; playful enjoyment, and the like") (Bruch, Kent, and Wechsler, 1980).

THE NEW THERAPIES AS TREATMENT MODALITIES

During the past generation mental health practices have experienced a fundamental change. This is basically due to the tremendous growth of drug therapy in the treatment of drug and alcohol abuse and mental illness. This in turn has spawned the growth of a variety of helping professions and organizations, e.g., community mental health centers, community educational programs, community arts centers, church organization, etc.

Many of these groups disdain the traditional medical treatment model and view therapy and self-improvement as different sides of the same coin. For example, the services offered by dance/movement therapists and non-traditional-humanist art therapists (Jungian, Gestalt) are more apt to be received by nonmedical environments. With many clients being serviced by various local community centers and with a mobile client population, it appears that therapists will have to develop short-term therapies.

Treatment modalities and the time needed for both dance/movement and art therapy vary quite drastically. In more traditional settings, the length of treatment is longer; expressive therapies generally serve as adjuncts to the psychiatric method. In many nontraditional centers, the length of treatment is usually shorter with the expressive process (i.e., the dance/movement patterns and the art activities are the expressive modes) becoming the treatment process.

Unfortunately, reports of successful treatment for both therapies rests mainly on descriptive reports. Many stringent research designs that utilize both internal and external validity are needed. Both professions are attempting to build better training programs, develop substantive bodies of knowledge based on valid research applications, and maintain critical standards of practice. However, it must be said that these two professions are relatively new, but hold great promise, especially in the treatment of drug and alcohol abuse.

The reader should be aware that training for both dance/movement therapists and art therapists is rigorous. Both require the master's degree and extensive intern training programs. The American Art Therapy Association and the American Dance Therapy Association are the certifying and registration agencies.

GENESA AS AN ADJUNCT TO ART THERAPY

Art therapy as practiced today is a varied and diverse field. This can be seen in the definition of art therapy adopted by the American Art Therapy Association (AATA):

Within the field of art therapy there are two major approaches. The use of art therapy implies that the creative process can be a means both of reconciling conflicts and of fostering self-awareness and personal growth.

When using art as a vehicle for psychotherapy, both the product and the associate references may be used in an effort to help the individual find a more compatible relationship between his inner and other worlds (Directory, 1977-78).

The Genesa process is capable of encompassing both points of view. Through a series of rhythmically prescribed movements of body, arms, and hands, the client becomes physically and mentally centered which can help bring out feelings of well-being, induce physical calmness, develop moods of relaxation, and enhance levels of creativity. (Bruch and Kent, 1979).

The Genesa therapist observing and charting the clients movement is able to diagnose where possible tension and imbalance occurs. The therapist then develops an appropriate program of compensatory movements which incrementally attempt to diminish the points of tension and imbalance.

SUMMARY

This paper has discussed various movement therapies as possible primary and adjunct treatments for treating drug and alcohol abuse. Presently, there is little empirical evidence that these processes are viable treatment modalities for drug and alcohol abusers. However, some of the Genesa aspects are in the process of being tested clinically and experimentally. It is hoped that these experiments may generate bases from which hypotheses may be developed and tested, and thereby show linkages between biophysiological and psychological processes. This should consequently help to contribute an additional viable treatment modality to the current limited number of techniques used with this population.

REFERENCES

American Art Therapy Association, Directory, 1977-1979, p. 1.

Bleedorn, Bernice D. Strategies and resources for creative encounters with the future: Developing gifted thinking and leadership. A paper presented at the Second World Conference on Gifted Children, San Francisco, Calif., July 1977.

Brown, Barbara B. New Mind, New Body, Biofeedback: New Directions for the Mind. N.Y.: Harper & Row, 1974.

Bruch, C. B., & Langham, D. G. Genesa communications - a wholistic message. Paper presented at the Northeastern Regional Conference of the Association for Humanistic Psychology, Atlantic City, New Jersey, April 1, 1978.

Bruch, C. B., Langham, D. G., & Smith, T. L. Wholistic centering and communications through Genesa. Paper presented at the Midwest Regional Conference of the Association for Gifted Children, Houston, Texas, November 2, 1978.

Bruch, C. B., & Lewis, C. B. Expressive movement and creative drama using Genesa. Presentation at the Annual Conference at the National Assocition for Gifted Children, Houston, Texas, November 2, 1978.

Bruch, C. B. Educational applications of the Genesa model through movement learning. Paper presented at the Conference on the Relationship of Cerebral Lateralization to Education. Atlanta: Georgia State University, April 20, 1979(a).

Bruch, C. B. The arts and Genesa. Presentation at the Annual Creative Problem Solving Institute, State University College, Buffalo, New York, June 28, 1979 (b).

Bruch, C. B., & Kent, R. B. Applications of the Genesa model in mental health and education. Paper presented at the Wholistic Health Conference, Rock Eagle, Eaton, Georgia, March 11, 1979.

Bruch, C. B., Langham, D. G., & Torrance, E. P. "Genesa As An Aid to Incubation/Imagery and Brain/Mind Function." Gifted Child Quarterly, 23 (4), 1979.

Bruch, C. B., Kent, R. B., & Wechsler, S. Partial results of a Genesa experiment with college art students. Unpublished Paper, Department of Educational Psychology, University of Georgia, Athens, Ga., 1980.

Burchett, R. W. The arts and sciences as a unified system of knowledge. Unpublished Master of Arts Thesis, United States International University, San Diego, Calif., 1977.

Feder E. and Feder B. The Expressive Arts Therapies: Art, Music, and Dance as Psychotherapy. Englewood Cliffs, N. J.: Prentice-Hall, 1982.

Feldenkrais, M. Awareness Through Movement. New York: Harper & Row, 1977.

Hendricks, G., & Roberts, T. B. The Second Centering Book. Englewood Cliffs, N. J.: Prentice-Hall, 1977.

Houston, J., & Masters, R. Listening to the Body: The Psychological Way to Health and Awareness. New York: Delacorte Press, 1978.

Langham, D. G. 13-Dimensional Genetics. Fallbrook, Calif.: Genesa Publ., 1967.

Langham, D. G. Genesa: An Attempt to Develop a Conceptual Model to Synthesize, Synchronize, and Vitalize Man's Interpretations of Universal Phenomena. Fallbrook, Calif.: Aero Pub., 1969.

Langham, D. G. "Genesa: Tomorrow's thinking today." Journal of Creative Behavior, 8(4), 277-281, 1974.

Langham, D. G. Circle Gardening. Old Greenwich, Conn.: Devin-Adair, 1978a.

Lowen, A. Bioenergetics. New York: Penguin, 1975.

Mogul, Jerry, "T'ai Chi Ch'uan: A Taoist art of healing." Somatics, 4, part II, 43-49, 1980.

Myers, M. "Body Therapies and the Modern Dancer: Imgard Bartenieff's Fundamentals." Dance Magazine, March, 54, 88-92, 1980.

O'Donnell, P. A. Motor and Haptic Learning. San Rafael, Calif.: Dimensions Publishing, 1969.

Restak, R. M. The Brain: The Last Frontier. New York: Doubleday & Co., 1979.

Riordan, K. Gurdjieff. In C. T. Tart (ed.), Transpersonal Psychologies. New York: Harper & Row, 1975, 281-328.

Rubenzer, R. "The role of right hemisphere in learning and creativity; Implications for enhancing problem-solving ability." Gifted Child Quarterly, 23(1), 78-100, 1970.

Salkin, J. Body Ego Technique: An Educational and Therapeutic Approach to Body Image and Self Identity. Springfield, Ill.: Charles C. Thomas, 1973.

Sobel, D. S. Gravity and Structural Integration. In R. E. Ornstein (ed.), The Nature of Human Consciousness. New York: Viking Press, 1974.

Szekely, G. "Movement as a Basis for Teaching the Arts." Arts and Activities, Feb. 85(1) 22-25, 1979.

The New Age of Healing, Science of Mind Annual, Science of Mind Pub., Los Angeles, Calif., 1980.

Tobias, M. An Attempt to Simultaneously Analyze, Categorize and Synthesize Six Basic Approaches to Personality in the Genesa Model. Unpublished Master of Arts thesis, United States International University, San Diego, Calif., 1977.

Torrance, E. P., & Ball, O. E. Streamlined Scoring and Interpretation Guide and Norms for Manual for Figural Form B, Torrance Tests of Creative Thinking. Athens, Ga.: Georgia Studies of Creative Behavior, Sept. 1978.

Torrance, E. P. The Search for Satori and Creativity. Buffalo, New York: Creative Education Foundation/Creative Synergetic Associates, 1979.

White, A. W. The Effects of Movement, Drawing and Verbal Warm-Up upon the Performance of 4th Graders on a Figural Test of Creative Thinking. (Doctoral Dissertation, University of Georgia, 1976). Dissertation Abstracts International, 1976, 37A, 4248.

APPENDIX

The following photographs illustrate Genesa crystals.

HOLISTIC METHODS IN DRUG ABUSE TREATMENT

Ethan Nebelkopf

Walden House, Inc.
San Francisco, California 94117

INTRODUCTION

Recently many holistic treatment processes have begun to be applied in the treatment of substance abusers. These alternatives to the more conventional verbal/medicinal methods include herbs, acupuncture, nutritional therapy, megavitamins, bodywork, yoga, biofeedback, and a variety of other innovative methods. These approaches expose the user/addict and alcoholic to alternative approaches which emphasize the "whole person" rather than specific symptoms or substances abused.

The reader is asked to consider that these new approaches incorporate concepts of holistic health. Substance abuse is a very complex phenomenon. Interweaving emotional, physical, mental, social, economic, and spiritual problems in regard to the individual and society. Programs which take into account only a small portion of the healing spectrum (psychological counseling, methadone maintenance, peer pressure, therapeutic community) may be building failure into the long-run treatment outcome of substance abusers.

There is a growing awareness and concern to develop programs for substance abusers which utilize a holistic approach to deal with the mental, physical, emotional, and spiritual problems accompanying substance abuse. Holistic health relies upon a broad range of healing modalities in a balanced and natural alternative to drugs. Such factors as nutrition, relaxation, stress reduction, and physical exercises are perceived as important as peer pressure and psychotherapy in the treatment and rehabilitation of substance abusers. The purpose of this paper is to survey and evaluate some of the newer holistic methods in regard to the treatment of the drug addict/user and alcoholic.

HERBS

The use of herbs is a fascinating phenomenon in the treatment of substance abuse. Herbs are natural botanical substances which have noticeable effects on the human organism. Throughout history plants have been eaten for nutritional and healing purposes, as well as for getting "high" (Taylor, 1966).

The practice of healing with herbs is an ancient art. The earliest treatise on herbalism was written in China over five thousand years ago. In ancient Greece, Hippocrates, who is regarded as the father of modern medicine, wrote extensively about herbs and their applications to medical treatment, particularly in conjunction with diet, fresh air, and physical exercise. During the Elizabethan period in England, herbal medicine reached a peak from which much of our present-day herbal lore derives. With the discovery of the New World it was recognized that the Native Americans practiced an herbalism in which the relationship between people and plants was deeply rooted in spiritual concerns. In America during the early nineteenth century, European and Native American herbology combined into an eclectic naturopathy which was widely practiced by lay practitioners (Nebelkopf, 1981).

However, following the Civil War and the laboratory synthesis of morphine, professional medicine began its dependence upon advanced technology, especially laboratory produced chemical drugs. This practice is still regarded as standard operating procedure in the medical profession, although minor inroads are currently being made in light of the newly awakened interest in holistic health and "The New Herbalism" (Nebelkopf, 1980a).

There are herbs which can replicate the effects of most commonly used psychoative drugs. Herbal "uppers" include guarana seeds, cola nuts, and yerba mate, which, like coffee, contain high concentrations of caffeine and act as potent herbal stimulants. On the other hand, there are herbal stimulants which do not contain caffeine (e.g., ginseng, gotu kola, sassafras, sarsaparilla) but which do have definite stimulating effects on the nervous and circulatory systems without the negative effects of coffee (Heinerman, 1980).

In addition, there are herbal "downers" which include valerian root, hops, skullcap, and chamomile. Valerian root is a very powerful herbal tranquilizer and nerve tonic (Stary & Jirasek, 1976). Hops are a major ingredient in beer and are relaxing in their effects. Chamomile is an old-fashioned home remedy for cranky children when made into an herbal tea (Kloss, 1971).

Furthermore, herbs have proven effective in aiding detoxification from drugs like methadone and heroin. From a holistic point of view, the "getting sick" which an addict experiences during withdrawal is actually the first step in the process of healing. The symptoms of detoxification reflect the active changes the body is going through in order to find a healthier balance (M. Smith, 1979). Opiates depress the respiratory system and heroin addicts report that they don't normally get colds while they are strung out on opiates. However, once the addict begins to go through the process of detoxification, all of the symptoms which have been suppressed for so long by the opiates begin to become expressed with exaggerated effect until the bodily systems achieve a new balance. The emergence of such symptoms as colds, runny nose, and diarrhea dramatically point out that healing is taking place (M. Smith, 1979).

There has been moderate success in using herbal teas to combat the symptoms of detoxification and to promote healing. Detox Brew is an herbal tea blend consisting of comfrey, mullein, spearmint, rose hips, orange peel, and golden seal. It acts as a tonic for the respiratory system for addicts who are detoxifying from methadone or heroin. Comfrey and mullein are old folk remedies for respiratory disturbances. Golden seal is a tonic for the liver and kidneys in helping to eliminate toxins from the bloodstream. Rose hips and orange peel are high in Vitamin C, which is helpful in detoxification. Spearmint is mildly relaxing and has a pleasant, sweet taste (Nebelkopf, 1978a).

Detox Brew, because of its beneficial and healing effects on the lungs, is also an excellent aid to detoxification from tobacco. Many nonsmokers have used this herbal tea blend as an aid in their transition from smoker to nonsmoker. Detox Brew helps to expectorate phlegm and soothe the inflammed throat membranes caused by tobacco smoke (Nebelkopf, 1981).

Another popular herbal mixture used by recuperating addicts/drug users is Relaxo Brew. This blend consists of a mixture of valerian root, spearmint, and chamomile. Relaxo Brew acts as an alternative herbal tranquilizer and an aid to relaxation. Relaxo Brew can be used as a healthy substitute for people who drink to relax or who abuse such substances as valium, quaaludes, and barbituates. Next to opium, valerian root is nature's best non-narcotic and nonaddicting sedative (Nebelkopf, 1979). Valerian root is an excellent tonic for a hyperactive nervous system. It is used in Eastern Europe in pharmaceutical preparations to treat insomnia (Stary & Jirasek, 1976).

Nutra Tea is another herbal tea blend which has been used by alcoholics to regain health and substance. Nutra tea is a highly nutritious herbal blend, rich in vitamins and minerals. It has no psychoactive effects. It contains lemon grass, alfalfa, dandelion root, comfrey leaf, and red clover. Alfalfa is called "the father of all foods" and is extremely high in nutriments, including vitamins A, D, E, K, Niacin, sodium, potassium, phosphorus, calcium, magnesium, sulfur, and iron. Dandelion is an effective hepatic stimulant and red clover is high in iron (Nebelkopf, 1980b).

Energy Tea is a mild herbal "upper" which does not contain caffeine. Coffee is a much abused substance, particularly among human service workers. Energy Tea contains peppermint, gotu kola, red clover, sarsaparilla, damiana, and red raspberry leaf. Gotu kola, high in magnesium in an easily assimilated form, is used to enhance alertness by natives of Sri Lanka. Sarsaparilla was one of the most popular invigorating and stimulating herbs during the patent medicine era, primarily used as a blood tonic. Energy Tea is a very mild herbal stimulant, energizer, and tonic for the nervous and circulatory system. Since it contains no caffeine, it is a healthy substitute for coffee, tea, amphetamines, and cocaine (Nebelkopf, 1981).

The ways in which humans relate to plants and herbs is just beginning to be understood. Drug addicts/users may be said to assign "magical powers" to their favorite herbs on an unconscious level. The heroin addict may "worship" the opium poppy. The cocaine addict is enamored of the seductive coca queen and the wino is hooked on sour grapes. It appears that most substance abusers are hooked into a belief system regarding the powerful effects of plants on the human organism. These beliefs can be transmuted into a positive force if they are expanded to include plants than can facilitate health (Nebelkopf, 1978b). By choosing such plants as comfrey, golden seal, mullein, ginseng, and valerian root, the substance abuser can begin to explore a healthier and more natural life style. Herbal treatment, as well as life style, permits the person to actively take responsibility for the substances put into the body, rather than being irresponsibly and passively dependent upon a chemical substance (Table 1).

ACUPUNCTURE

The use of acupuncture as a healing tool originated over four thousand years ago in China. It is based on the ancient Chinese Philosophy of Taoism, in which the person's life energy, chi, is manifested in two polar

Table 1

Herb	Botanical name	Chemical constituents	Therapeutic uses
golden seal	Hydrastis canadensis	berberine, hydrasticine	natural antibiotic, strengthens liver
comfrey	Symphytum officinale	allantoin	soothes stomach, enhances respiration
rose hips	Rose canina	vitamin C, iron	strengthens immune system
cascara sagrada	Rhamnus purshiana	anthraquinone glycosides	used for constipation
blackberry rt.	Rubus fructicosus	tannins	used for diarrhea
mullen	Verbascum thapsus	mucilage	respiratory problems
dandelion Rt.	Taraxacum dens-leonis	taraxacin	strengthens liver
cyenne	Capsicum frutescens	capsaicin, vitamin C	blood purifier
garlic	Allium sativum	scordinin	lowers blood pressure, natural antibiotic
ginseng	Panax ginseng	adaptogens glycosides	helps with stress, regulates blood pressure
valerian rt.	Valeriana oficinale	chatinine, velerine	natural tranquilizer
gotu kola	Hydrocotle asiatica	hydrocoltylin	promotes mental alertness
ephedra	Ephedra nevadensis	ephedrine	used for allergies, stimulant
guarana	Paullinia cupana	caffeine	nerve stimulant
cola nuts	Cola nitida	caffeine	nerve stimulant
parsley	Petroselinum crispum	vitamin A	natural diuretic
hawthorne	Crataegus oxyacantha	Amygdalin	heart problems
aloe vera	aloe vera barbedensis	aloin	burns
red raspberry	Rubus stringosus	iron	pregnancy
black cohosh	Cimicimfuga racemosa	female hormones	menstrual regularity
sassafrass	Sassafrass oficinale	safrole	circulatory stimulant
spearmint	Mentha spicata	essential oil	settles the stomach, mild relaxing effects
chamomile	Anthemis nobilis	chamazulene bisabolol	anti-inflammatory, settles the stomach
alfalfa	Medicago sativa	vitamins/minerals	nutritive
hops	Humulus lupulus	lupuline	sedative

opposites, yin and yang, which are continuously interacting. Yin represents the feminine receptive pole of this energy, while yang signifies the masculine creative aspects of chi (Veith, 1972). The normal functioning of the body depends on a balance of yin and yang energy. This flow of energy takes place along 12 lines of the body called meridians. Stimulation of these points by needles, heat, or electrical currents can correct the imbalance of energy flows, release blockages of energy, and enhance the body's natural capacity to heal itself (Sharps, 1977). See Figure 1 and Figure 1A.

The application of acupuncture to the treatment of drug addiction was pioneered in Hong Kong by H. L. Wen, who developed a method of suppressing withdrawal symptoms by applying a five to six volt electric current to needles inserted in acupuncture points in the client's ear. A normal treatment lasted about 30 minutes a day for seven days. In most cases it was reported that within 15 minutes cramps, nausea, runny nose, and other symptoms were ameliorated (Wen and Cheung, 1973).

David Smith of the Haight-Ashbury Free Clinic in San Francisco writes that "Acupuncture when used for the treatment of heroin withdrawal on an outpatient basis is as successful as non-narcotic medication on an outpatient basis, but costs about one-third less and doesn't have the associated problem of drug diversion" (D. Smith, 1978).

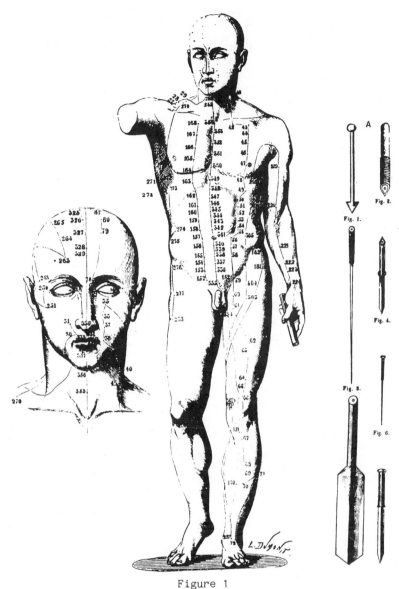

Figure 1

The Lincoln Detox Program in the South Bronx, New York, has been utilizing acupuncture to treat drug abuse problems. According to Michael O. Smith, "Acupuncture has been found to be the most dramatic and effective natural technique for the detoxification of drug addicts. Most of these treatments are simple and inexpensive to apply, and can be adapted to drug-free programs and medication programs alike" (M. Smith, 1979).

At the Lincoln Detox Program ear acupuncture is primarily used, with one to four thin needles inserted in the outside portion of each ear. Many clients are also treated at acupuncture points on other parts of the body. Careful selection of these other points is very important. Each day different points need to be stimulated according to what symptoms are currently occurring, the rhythm of the various pulses, differences in sensitivity at different points, and blockages of chi. People detoxifying from methadone usually being acupunture treatments when they are down to 20 mg. of methadone (Shakur and Smith, 1977).

Acupuncture is one of the most promising tools in the substance abuse treatment field. Although traditional western medicine has difficulty understanding the principles of acupuncture, recent research indicates that the pain-suppressing effects of acupuncture may be due to the activation of endorphins, the body's natural pain killers (Wen, 1977).

NUTRITION

In a holistic approach to substance detoxification and rehabilitation nutrition plays a vital role. Drug addicts/users and alcoholics are notorious for their poor diets. Drugs such as heroin, tobacco, coffee, and alcohol deplete the body's store of essential vitamins and minerals (Airola, 1974). Drugs may produce dietary deficiencies by destroying nutrients, preventing their absorption, and increasing their excretion (Kirschmann, 1975).

In working with drug addicts/users on improving their diet, it is important to be flexible, nonmoralistic and to emphasize a wide variety of wholesome, unrefined, and unprocessed foods. White sugar, white flour, excess salt, and greasy fried foods should be slowly and gradually eliminated from the diet (M. Smith, 1979).

Unfortunately, sugar products are the staple ingredients in the diet of many users/addicts. Michael Smith (1979) suggests that the abnormally high rate of sugar consumption by narcotics addicts is due to the following factors: (1) intestinal spasm caused by narcotics makes digestion and absorption of more complex foods difficult; (2) the strain on the liver caused by injected toxins leads to reduced food assimilation and glycogen storage function; (3) an addictive component of hypoglycemia, which stems from the "sugar-overreactive-insulin-more sugar" cycle; (4) lack of concern for self-help common to most addicts; (5) intense marketing of sugar products, especially in poor communities.

The condition of hypoglycemia is very common among substance abusers. In order to combat this condition Airola (1974) suggests six to eight small meals a day with whole grains, seeds, nuts, and fresh vegetables forming the basis of a diet. All refined and processed foods, including white sugar, white flour, pastries, cookies, ice cream, candy, coffee, doughnuts, salt, and soft drinks should be given up. A small amount of fresh fruit can be eaten, along with lots of cottage cheese, yogurt, and kefir. Also suggested are large amounts of vitamin C, vitamin B complex, vitamin E, potassium, and magnesium. Vitamin C aids the detoxification and elimination of foreign substances from the body. Vitamin B complex helps to deal with stress and aids in building up the adrenal glands which are often exhausted and depleted in persons with substance abuse problems and/or hypoglycemia. Vitamin E improves the glycogen storage and potassium helps to normalize the mineral balance.

Nutrition is an area conductive to many trends and fads. Probably the most important aspect in helping to change the diet of substance abusers is to avoid regidity. Simplicity is also important. Small changes in the direction toward a more natural and wholesome diet should be reinforced. In this way substance abusers can learn to listen to their own bodies for their own nutritive needs. (Table 2).

Table 2*

Element	Natural sources	Function	Ailments relieved
Vitamin A	Leafy green and yellow vegetables	Builds resistance to infection, especially of of the respiratory tract; improves liver function	Dry, itchy skin, loss of appetite, infections
Vitamin B	Brewer's yeast, meat, egg yolk, whole grains and cereals, nuts and seeds, yogurt	Together the B vitamins promote growth, aid in assimilation, and are essential for normal functioning of nerves, muscles, heart	Low energy, skin problems, anxiety, emotional upsets, headaches
Vitamin C	Citrus fruits, berries, greens, cabbages, peppers, sprouts	Strengthens connective tissue, promotes wound healing, builds resistance to infection	Infections, sore teeth and gums, low energy
Vitamin D	Cod liver oil, eggs, milk, butter, sunshine	Regulates use of calcium and phosphorus in the body, aids in healing of nerves, bones, and muscles	Soreness in bones, muscles, nerves, tooth decay
Vitamin E	Wheat germ, sesame seeds, green leaves, meat, eggs, whole grains and cereals	Prevents destruction of vital fats in the body, improves liver function, improves circulation, helps dissolve scars	Excellent treatment for "tracks" from prolonged self-injection of drugs; also other skin disorders
Calcium and Magnesium	Dolomite or bone meal, milk and other dairy products (calcium); green vegetables (magnesium)	Builds and maintains bones and teeth, helps blood to clot, aids vitality and endurance, essential for normal functioning of nerves and muscles	Bone pains, nervous energy, anxiety, insomnia
Zinc	Wheat germ, pumpkin seeds, liver, whole grains	Helps normal tissue function, aids in absorption of vitamins	Stress, inflammation, wounds and burns

*Developed by the Haight-Ashbury Free Medical Clinic.

During the first few weeks of drug detoxification your body is working overtime to get rid of the toxins, or poisons, in your system. Most of the discomfort you feel (runny nose/eyes, sweats, chills, fever, diarrhea, pain in the joints) during this time comes from your body's processes of eliminating these toxins. Now is a good time to begin to help your body by taking vitamins and minerals which aid the elimination process. This table presents the ones that are the most important to you during the detox phase. They should be added to your diet even after detoxing.

MEGAVITAMIN THERAPY

Another relatively new approach in the field of substance abuse treatment is megavitamin therapy. The treatment of physical and mental conditions with large doses of vitamins has been an important contribution from the field of orthomolecular medicine.

Pawlak (1974) describes the following symptoms as indications of the "drug wipe-out syndrome": general discontent, a sense of impending disaster, difficulty in verbal communication, extreme emotional lability, feeling unfulfilled, some loss of physical coordination, and a loss of former drives, beliefs, and goals. The drug wipe-out syndrome can be brought about by taking psychedelic drugs, by smoking too much marijuana, taking pills, being addicted to opiates, or by any sort of biochemical imbalance. In many instances, the drug wipe-out syndrome reflects cases of undifferentiated borderline schizophrenia.

Pawlak (1974) reports that high doses of niacin (1000 mg. three times a day) has had positive effects with people suffering from the drug wipe-out syndrome, as well as from borderline schizophrenia. Niacin helps to clean out impurities from the bloodstream, aids circulation, and inhibits the production of adrenalin. Usually the first effects after taking megadoses of niacin are less fatigue and less depression. An interesting side-effect of niacin is that the user's skin complexion turns red and tingles with a mild heat sensation. Niacin has also been used to ease the comedown period after a long amphetamine run, for psychedelic flashbacks and bummers, and as an aid in withdrawal from alcohol, barbituates, and heroin because of its anti-convulsant effects.

Another interesting orthomolecular approach has been the use of large doses of Vitamin C in the treatment of drug addicts. Libby and Stone (1977) report that massive doses of vitamin C (25-85 grams of sodium ascorbate a day) in a four to ten day treatment program can ameliorate even the most sever cases of drug addiction. According to Libby and Stone, drug addicts suffer from chronic subclinical scurvy and the Hypoascorbemia-Kwashkiorkor syndrome. They speculate that sodium ascorbate (Vitamin C) competes for and displaces the narcotic from the opiate receptor sites in the brain. Using their megavitamin treatment, Libby and Stone claim that "no withdrawal symptoms are encountered, and there is a great improvement in well-being and mental alertness. In a few days appetite returns, the client eats well, sleeps restfully, and the methadone-constipation is relieved."

However, there are certain methodological issued raised by the Libby and Stone study on the effectiveness of utilizing massive doses of Vitamin C as a treatment for drug addiction. First of all, it is relatively easy to detoxify addicts from opiates. The problem is going back to using heroin. Any treatment in the field of drug abuse which claims 100% success without any follow-up research, and which also ignores the psychosocial aspects of drug addiction is questionable. In addiction, neither is subclinical scurvy, nor the Hypoascorbemia-Kwashkiorkor syndrome adequately defined. Finally, there is no evidence that sodium ascorbate is actually absorbed in the opiate receptor sites of the brain (Goldstein, 1978).

In a more recent study, researchers at the San Francisco Drug Treatment Program, using a tightly controlled double-blind experimental design, found that heroin addicts undergoing detoxification experienced significantly fewer withdrawal symptoms when given megadoses (25 to 48 grams a day) of sodium ascorbate (Free and Sanders, 1979).

BODYWORK

The goals of bodywork are to alleviate chronic muscular tensions, to release suppressed emotions, to enhance relaxation, and to integrate the body and mind (Kogan, 1980). Bodywork with drug addicts/users takes several forms, including yoga, mediation, dance/movement therapy, breathing exercises and massage.

Yoga and various forms of meditation have been practiced in India and China for thousands of years. The purpose of yoga is both spiritual and physical: to come in contact with the divine forces within oneself as well as to increase the body's flexibility and strength. There are yoga exercises to improve the functioning of all the organs of the body as well as meditation exercises to facilitate relaxation and ultimately to produce peace of mind.

Many addicts just do not know how to relax and this is exactly where yoga and meditation are most helpful. One effective meditative device is the mantra. The repetition of the sound "OM" can help in dissipating the continuous mental chatter and inner dialogue which impedes concentration and peace of mind. Many clients report that they get "high" and achieve a consciousness akin to a drug-induced state but without chemicals after a session of chanting "OM" (Meltzer, 1977).

In order to practice the mantra "OM", it is necessary to sit on the floor, or in a chair, but with the back held straight as possible. The client should inhale deeply and utter the mantra "OM" on the out-breath. This sound can be sung softly in the manner of a lullaby or fiercely as if a primal scream, or continuously so as to create a feeling of unity. Meltzer (1977) reports that repetition of this mantra enables the client to release blocked emotion, and that many addicts appreciate the deep relaxation they experience with yoga.

Usually very simple yoga postures are effective with drug users/addicts, with an emphasis on relaxation and meditation interspersed between the various postures. One of the basic principles of yoga is not to strain the body--to stretch it to its limit, but to stop at a point just short of where pain is felt. Thus the movement is not to be forced, but gentle, slow, and gradual.

Yoga is very helpful in promoting relaxation for substance abusers who are moving away from pills and who have difficulty sleeping. The most effective yoga exercise to facilitate sleep is called savasana or the "corpse" pose. This yoga position is very similar to the behavioral technique of progressive relaxation. In the "corpse," the client lies on his or her back with eyes closed, legs straight out, and arms relaxed along the sides of the body. Breathing should be slow and regular and utilize the abdomen and diaphragm. The breath should be taken in through the nose rather than the mouth, and the client should concentrate his or her attention on the breathing process, especially if the mind begins to wander. The "corpse" exercise begins with focusing the attention on the feet. The client is instructed to "Pay attention to your feet. Become aware of whether your feet are relaxed or tensed. Tense up your feet and then relax them. Begin to feel a wave of relaxation flowing upward through your body beginning at your feet." See Figure 2.)

This process of relaxation is continued upward through the ankles, calves, knees, thighs, etc., to the top of the head. Addicts easily learn this process and can do it themselves without verbal instruction after one or two trials. By practicing this exercise, the client conditions himself to relax by merely paying attention to the tensions in the muscles which are tight. For someone who is jittery and cannot go to sleep, a cup of Relaxo Brew followed by the "corpse" meditation is effective in alleviating anxiety and treating insomnia.

The 3HO Drug & Alcohol Program in Tuscon, Arizona, offers a drug-free residential detoxification and rehabilitation program based on principles of kundalini yoga. Clients receive messages and hot baths to promote relaxation, eat a well-balanced vegetarian diet, and participate in a daily schedule of yoga activities. The average length of stay of clients in the yoga ashram is five to six months. Meditation exercises help the client to control the mind and to look within as a way to solve one's problems. Yoga exercises help to rebuild the physical body, the vegetarian diet is designed

SAVASANA

Relaxation — on mat.

Sun
Energy

WATER — SAVASANA

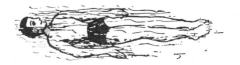

FLOATING ON THE WATER

Concentrate on **Relaxing** every external part of the body.
Start with the toes until you reach the top of head.
Continue by relaxing all internal organs.

Benefits:
Brings tranquility and calmness.
Restores vitality to muscles and internal organs.
Equalizes blood circulation.
Rejuvenates the entire body and mind.

Figure 2

to cleanse the body of toxins, and the life style involves a family situation with a constructive and positive "vibration" (Khalsa, 1977).

Dance/Movement therapy involves the therapeutic use of body movement and dance to enhance emotional awareness, self-expression, and interpersonal communication. It has been used to help chemically-dependent women to release tension, build self-esteem, and improve the body image (Myer, unpublished).

Another innovative technique related to bodywork and the facilitation of relaxation for drug addicts/users is the use of the flotation tank or "Samadhi Tank" developed by John Lilly in his experiments with sensory deprivation (Lilly and Lilly, 1981). In the Samadhi Tank, the individual floats in a shallow pool of warm water supersaturated with epsom salts. Walden House, a therapeutic community in San Francisco, utilizes the Samadhi tank to alleviate the symptoms of withdrawal for residents who are detoxifying from methadone.

BIOFEEDBACK

Another technique being utilized to facilitate relaxation and detoxification from drugs, particularly methadone, is biofeedback. In an experiment by Kuna, Salkin, and Weinberger (1976), nine addicts were trained to lower anxiety as measured by the Taylor Manifest Anxiety Scale. This occurred in a four-week period as a result of biofeedback relaxation training.

Biofeedback techniques employ instrumentation to provide a person with immediate and continuous signals regarding changes in somatic functioning of which the person is not usually aware. Such functions as skin temperature, sweat, fluctuation in blood pressure muscle tone, and brainwave activity are worked with in biofeedback sessions. Thus the information received from the biofeedback machine enables the person to learn to control what are typically "involuntary" functions. Biofeedback training has been effective in controlling tension headaches, heart rate, blood pressure, and migraines (Brown, 1977).

Biofeedback training is another promising avenue in the holistic treatment of substance abusers. Addicts have the ability to learn relaxation and they respond well to the biofeedback situation (Kuna, Salkin and Weinberger, 1976).

CONCLUSIONS

There is a need for systematic investigation and development of programs utilizing holistic techniques in the treatment of drug users/addicts and alcoholics.

One of the strongest obstacles in implementing this approach is the categorical funding policies of governmental agencies. Thus heroin addicts, alcoholics, PCP users, etc., must participate in separate programs funded separately, rather than in holistic programs geared toward the treatment of substance abuse (Pawlak, 1978). In addition, abusers of tobacco and coffee are placed in separate categories and must relate to different government agencies and departments.

Another constraint against the development of holistic programs are the competition between, and the vested interests in, maintaining the dominant forms of drug abuse treatment, such as methadone maintenance, and therapeutic communities. Methadone maintenance provides a substitute addiction in a medically supervised setting. Theoretically, many of the social, economic, and criminal consequences of drug addiction are ameliorated (Nebelkopf, 1981). There is a modicum of psychological counseling available in the methadone maintenance model, although methadone counselors spend an unrealistic part of their time monitoring the urine. The basic premise in methadone maintenance is that since opiate addiction is due to a biochemical imbalance (akin to diabetes) the client will remain on methadone for the duration of his life.

The other major sanctioned use of methadone is in detoxification programs which may be inpatient or outpatient and vary in duration as well as quality. The arbitrary time limits and rigidity of detoxification schedules (usually 21 days) severely limit the effectiveness of this model. In most short-term detoxification programs using methadone, there are little ancillary services (psychological counseling, vocational evaluation, etc.) and nothing which comes close to a holistic approach. Twenty one days are not enough to eradicate years' worth of substance abuse, but with proper psychological and nutritional counseling, and using herbal tea blends as an adjunct to counseling, the effectiveness of methadone detoxification programs are bound to improve. There also is a need for educational programs connected to methadone detoxification so that clients can understand the process of healing.

In contrast to methadone, the therapeutic community emphasizes a drug-free lifestyle. The focus is on developing self-responsibility, discipline, emotional honesty, and the appropriate expression of feelings. Unfortunately, most therapeutic communities ignore nutritional and orthomolecular factors in their approach. Although there is a strong negative emphasis on drugs such as heroin, alcohol, and marijuana, there is considerable abuse of "softer" drugs like coffee, sugar, and tobacco.

Since residents of therapeutic communities eat three meals a day in a residential setting, it would seem relatively easy to incorporate principles of good nutrition in these programs. However, food service systems in residential programs usually provide "institutional" foods which are high in starches and sugar. A move toward utilizing whole unprocessed foods, herbs, and vitamins in therapeutic settings would definitely improve the services rendered.

Changing the dietary patterns in therapeutic communities is not easy. This is due to a lack of awareness and inertia. Many of these programs are so crisis-oriented that prevention through proper nutrition is a very distant concept. In an experiment at Walden House, natural foods were prepared and eaten for several meals and the residents' responses were tape-recorded. It is interesting to note that many contacted residents had positive feelings concerning institutional foods. These were the residents who had never eaten three regular meals a day. Such natural foods a tofu and seaweed were received with very negative reactions by the majority of residents. There was a small minority who expressed extremely positive feelings toward natural foods. By and large the residents felt that their diet could be improved and that this could enhance their health. What was learned from this experiment was that a move toward natural foods in therapeutic communities should not be forced, and that the natural foods selected should be as common and simple as possible. There was universal interest in having fresh salads and fruit available, and that brown rice should be substituted for white rice if it was cooked properly.

For the alcoholic, there is a need toward establishing halfway house programs based at least in part on orthomolecular principles. There has been much experience treating alcoholics with niacin, B-complex vitamins, trace minerals, and well-balanced diets (Pawalk, 1978). Many alcoholics are also hypoglycemics, and this condition can be corrected or controlled through megavitamin therapy and orthomolecular medicine (Cheraskin, Ringsdorf, & Brecher, 1974).

The weak point in most substance abuse programs is the utilization of monodimensional approaches (Stern, unpublished). Peer counseling, methadone psychotherapy, acupuncture, extended family networks, herbs, vitamins, nutritional seminars, biofeedback, bodywork, yoga, and meditation are all being used with some degree of effectiveness when applied separately in the treatment of substance abusers. However, if all of these techniques can be seen as tools in an integrated holistic approach, which is being attempted at Walden House in San Francisco, then we can look forward to a major breakthrough in the treatment of substance abusers. What is needed is a truly holistic approach emphasizing the development of individual treatment strategies in case conference format that includes the client and is tailored to the specific treatment needs of that client.

The training of substance abuse counselors in the principles and practices of nutrition and bodywork can help them to be more effective with their clients, and can diminish the sense of hopelessness and burn-out in the field. The use of herbs, nutrition, bodywork, yoga, and biofeedback are available and are relevant alternatives in the field of substance abuse treatment. The incorporation of these methods in holistic programs that are geared toward acutalizing the whole person, mentally, physically, emotionally, and spiritually, regardless of the substance abused, should realistically be the direction of the future.

REFERENCES

Acompora, Alfonso "Methadone to abstinence, paper presented at the
        International Conference on Therapeutic Communities, Manilla, The
        Phillipines, November 1981.
Airola, Paavo How to Get Well Phoenix: Health Plus, 1974.
Brown, Barbara New Mind, New Body New York: Harper and Row, 1977.
Cheraskin, E., Ringsdorf, W., & Brecher, A. Psychodietetics New York:
        Bantam, 1974.
Free, Valentine, & Sanders, Pat "Megascorbate: A Detoxification
        Alternative," Journal of Psychedelic Drugs, II, 1979.
Goldstein, Avram Public Letter to the Director of the Department of
        Substance Abuse San Diego County, January 4, 1978.
Heinerman, John Science of Herbal Medicine Orem, Utah: BiWorld, 1980.
Khalsa, Mukhia Singh Sahib Sat Nam Singh "Ascent from Addiction," New
        Directions, 25, 1977.
Kirschmann, J. Nutrition Almanac. New York: McGraw-Hill, 1975.
Kloss, Jethro Back to Eden New York: Beneficial Books, 1971.
Kogan, Gerald Your Body Works Berkeley, California: Transformations Press,
        1980.
Kuna, D., Salkin W., & Weinberger K. "Biofeedback, Relaxation Training, and
        Methadone Clients," Contemporary Drug Problems, New York: Winger, 1976.
Libby, Alfred, & Stone, Irwin "The Hypoascorbemia-Kwashkiorkor Approach to
        Drug Addiction Therapy: A Pilot Study," Journal of Orthomolecular
        Psychiatry, 1977.
Lilly, John, & Lilly, Toni "Exploring the Limits of Consciousness," Journal
        of Holistic Health, 1981.
Meltzer, Gloria "Yoga as Treatment for Drug Addiction," Yoga Journal, March,
        1977.
Nebelkopf, Ethan The Herbal Connection: Herbs, Drug Abuse & Holistic Health
        Orem, Utah: BiWorld, 1981.
Nebelkopf, Ethan The New Herbalism Orem, Utah: BiWorld 1980a.
Nebelkopf, Ethan "Nutra Tea: Dandelion & the Alcoholic," The Herbalist,
        June 1980b.
Nebelkopf, Ethan "Detox Brew," The Herbalist, 3(2), 1978a.
Nebelkopf, Ethan "A Holistic Approach Using Herbs," U.S. Journal of Alcohol
        & Drug Dependence, 2(11), 1978b.
Nebelkopf, Ethan "Valerian: Help for the Drug Addict," The Herbalist,
        April, 1979.
Pawlak, Victor Megavitamin Therapy & the Drug Wipe-Out Syndrome Phoenix: Do
        It Now, 1974.
Pawlak, Victor "Orthomolecular medicine: Its application to drug abuse and
        alcoholism efforts" paper presented to American Society of Criminology,
        Dallas, 1978.
Sharps, Holly "Acupuncture & the Treatment of Drug Withdrawal Symptoms," The
        PharmChem Newsletter, 6(7), 1977.
Smith, David "Acupuncture an Alternative in Treating Pain," U.S. Journal of
        Alcohol & Drug Dependence, April, 1978.
Smith, Michael O. "Acupuncture & Natural Healing in Drug Detoxification,"
        American Journal of Acupuncture, 7(2), 1979.
Shakur, Mutulu, & Smith, Michael O. The use of acupuncture to treat drug
        addiction paper presented to National Drug Abuse Conference, San
        Francisco, May 1977.
Stary, F., & Jirasek, V. Herbs: A Concise Guide in Color, London: Hamlyn,
        1976.
Taylor, Norman. Narcotics: Nature's Dangerous Gifts, New York: Dell, 1966.

Veith, Ilza. *The Yellow Emperor's Guide to Internal Medicine*. Berkeley: University of California Press.

Wen, H. L. "Fast detoxification of drug abuse by acupuncture and electrical stimulation in combination with naloxone," *Mod. Med. Asia, 13*(5), May 1977.

Wen, H. L., & Cheung, S. Y. C. "Treatment of drug addiction by acupuncture and electrical stimulation," *Asian J. Medicine, (9)*, 1973.

MARATHON GROUP THERAPY WITH ILLICIT DRUG USERS

Richard C. Page

Department of Counseling and Human Development Services
University of Georgia
Athens, Georgia 30602

Marathon group therapy offers the skilled therapist a means of creating a therapeutic environment in which illicit drug users can work to resolve their personal problems and to improve the quality of their interpersonal relationships. It is the purpose of this paper to describe how marathons operate to help illicit drug users and to summarize research conducted on the group process and the resulting attitude changes occurring among marathon group participants.

THEORY OF MARATHONS

From 1974 to 1980 the author conducted fifteen 16-hour-long marathon groups with either male or female illicit drug users. These groups were all conducted by two or more co-therapists because of the intensity and length of marathon groups. The facilitators attempted to create a permissive group environment in which the members felt accepted and safe with one another and with the facilitators. Under no circumstances was anything which the group members revealed during the marathon group discussed with persons outside of the group. The members of these groups all had long histories of drug use and drug-related problems; the primary drugs misused by the members were heroin, amphetamines, phencylidine (PCP), and barbituates. The members were all serving prison terms for drug-related offenses and were either in prison or in residential drug treatment centers when they participated in these groups.

This writer has primarily used marathon group therapy as a supplement to regular group therapy with illicit drug users. This enables therapists to provide illicit drug users already participating in group therapy, either in residential drug treatment centers or in institutional settings, the opportunity for an intensive group experience. This writer, however, has also facilitated marathon groups successfully with female illicit drug users on a one time basis in institutional settings such as prisons. The members have always been interviewed before being accepted into a marathon. Most illicit drug users who are not psychotic and who express the motivation to participate have been accepted as members in marathons. The marathon group facilitators need to be skilled group therapists with previous group therapy experience with illicit drug users. These groups probably should not be conducted more than once every two months with the same clients or facilitators because of the intensity of these groups.

The marathon groups conducted with these illicit drug users had certain common characteristics. The group processes of long term, two hour counseling groups (Page, 1979) and of 16-hour-long marathon groups (Page, 1980) have already been described. Both marathon groups and shorter groups conducted over a longer time span have the group stages described by Page, 1979, which are as follows: (1) support of current life styles; (2) anger at authority; (3) self-revelation; (4) working through problems; (5) new ways of relating; and (6) ending stages. Marathon groups, however, are more intense then more classical therapy groups because of the effects of the length of these groups. The members of marathon groups spend more time than the members of regular therapy groups in confronting one another and in attempting to form meaningful relationships with each other within the group (Rogers, 1970; Stroller, 1968). Less time is spent in marathons discussing the personal problems members have outside the group. Marathon groups with illicit drug users are grounded in the "here and now" and the members strive primarily to improve the ways they relate to one another and to the facilitators in the group.

Marathon groups, therefore, enable illicit drug users to learn about the effects of their behavior on others more than most types of group therapy because of the length and intensity of these groups. Many illicit drug users are adept at hurting other people and avoiding the consequences of their behavior. The manipulative behavior of most illicit drug users enables them to cope in the drug culture but also often alienates drug users from people in society (Canadian Government Commission of Inquiry, 1970). Illicit drug users cannot avoid being confronted with the games they play with others in a marathon because of the length of the group. Unless they leave the group, which the members can do but never have done in any of the groups facilitated by this writer, the members have to live with one another for the 16 hour duration of the marathon. The members receive positive feedback in addition to negative feedback and by the end of the group almost always feel as if they have shared meaningful feelings with other members and that the other group members have reciprocated.

GROUP PROCESS RESEARCH

The content of one of the 16-hour-long marathon groups conducted with male illicit drug users (Page & Wills, 1983), and descriptions of the specific types of interactions that occurred during this group (Page, 1982) have been reported elsewhere. The Hill Interaction Matrix--Form-G (HIM-G) was used in each of these studies to assess different aspects of the group process of this marathon. Three different raters, who were masters degree students in the counseling department of a large university in the United States, used the HIM-G to answer questions about taped segments of this group. Statistical procedures were used to compile these ratings to a meaningful pattern (Page, 1982; Page & Wills, 1983). The marathon group conducted with the illicit drug users was compared on several dimensions to 50 therapy groups used by Hill to develop the norms of the HIM-G (Hill, 1961; Hill, 1965). The actual operations of this marathon with illicit drug users were compared to what occurs in more traditional therapy groups. One of the most important findings of this research is that the marathon was ranked in the 80th percentile in the Relationship area when compared to the 50 norm groups described by Hill (Page & Wills, 1983). The types of interactions occuring in the Relationship category emphasize the establishment and analysis of relationships among the group members and facilitators. This research shows that marathons with illicit drug users emphasize the development of positive interpersonal relationships among the group members more than most therapy groups. Other research by Page (1982) additionally ranked this group in the 79th percentile in the Quadrant 4 area which is characterized by Hill (1965) as being high both in interpersonal threat (topics are discussed which involve interpersonal risk-taking) and work (a member takes the role of patient and actively seeks self-understanding). Hill (1965) indicated that Quadrant 4 groups were the most therapeutic of the four classes of groups described in the HIM manual.

The marathon was also rated according to the amount of time spent in 16 group cells which represent distinct types of therapy groups (Page, 1982). It was determined that this marathon ranked in the 95th percentile in the Personal, Confrontive cell, and in the 99th percentile in the Topic Confrontive and Relationship cells when compared to the norm groups. The primary types of interactions occuring in these cells involve the efforts of members to clear away non-essentials from a discussion or to resolve group paradoxes; to confront one another on the effects of the behavior of one member on another member; and to discuss the actions of a member to illustrate the actual problem of this member. The types of relating that occur in these cells involve the efforts of the members to improve the ways in which they interact with one another in their group. This group was very low in interactions characterized by abstract discussions or discussions of the personal problems members had outside the group.

The members of this marathon also took a more active part in sponsoring the types of interactions occuring in this group than occurs in most therapy groups (Page & Wills, 1983). The facilitators of the marathon did not assume traditional leadership roles in which they encouraged the members to discuss personal problems, but instead interacted more as equals with the members. This group was therefore characterized by highly therapeutic interactions and the members were as active as the facilitators in sponsoring these types of interactions. The question, however, might still be posed about the types of outcomes which result from this type of marathon group faciltated with illicit drug users.

OUTCOMES OF MARATHONS

Four outcome studies have been conducted on the marathons with illicit drug users facilitated by this writer (Page & Kuciak, 1978; Page & Mannion, 1980; Page, Mannion, & Wattenbarger, 1980; Page & Miehl, 1982). These outcome studies provide information about the types of attitude changes which can be expected to occur among illicit drug users participating in unstructured marathon group therapy. The research design of these studies specified the subjects be randomly selected into marathon and control groups in all except one study (Page & Kubiak, 1979; Page & Mannion, 1980; Page & Miehl, 1982). One study used an analysis of covariance statistical procedure to control for the initial differences between the control and marathon group members when the subjects were not randomly selected into control and experimental groups (Page et al., 1980). Certain consistent patterns of change appear among the attitudes of the participants of these groups when the results of all of these research studies are examined.

The evaluative and potency scales of a semantic differential, (Osgood, Suci, & Tannenbaum, 1957) were used to assess the outcome of all of these research studies. The most consistent finding of these research studies is that the attitudes of the subjects toward counseling become more positive after participation in a marathon. The marathon participants evaluated the following concepts higher than the control group members:

Counseling $(t[20] = 2.84, <p\ .02)$ (Page & Kubiak, 1978);

Group Counseling $(t[26] = -2.07, <p\ .05)$ (Page & Mannion, 1980);

Group Counseling $(t[19] = 2.55, <p\ .02)$ (Page & Miehl, 1982)

Counselors (t[19] = 2.17, <p .05) (Page & Miehl, 1982).

The members of the marathons rated the potency of two concepts higher after participating in the marathon than the control group members: Counselors (F[1,22] = 8.9, p <.01); Page et al., 1980); Drug Treatment Programs (t [19] = 2.32, p <.05; Page & Miehl, 1982). The marathon group participants perceived counseling more positively than the control group members on all of these scales of the semantic differential after participating in 16-hour-long marathon groups.

The attitudes of the marathon group participants also changed toward various other concepts as measured by the administrative or semantic differentials. Ths marathon group participants evaluated the Future higher (t[20] = 2.88, p <.01; Page & Kubiak, 1978), Reality higher (F[1.22] = 4.69, p <.04); Page et al., 1980) and Guilt lower (t[26] = 2.35, p <.05; Page & Mannion, 1980) than the members of the control groups. The members of the marathon group additionally rated the potency of the following concepts higher than the members of control groups: My Real Self (t[20] = 2.33, p <.05; Page & Kubiak, 1978); Reality (F[1.22] = 11.47, p <.01; Page et al., 1980); and Guilt (F[1.22] = 7.25, p <.01; Page et al., 1980). Most of these research results are consistent with what may be expected to occur in this type of group.

The outcome studies of the four marathons described in this paper support the therapeutic value of marathons. The particular research studies cited in this paper may be consulted for a more detailed interpretation of these research results.

SUMMARY

Marathon groups therapy can be an effective method of treating illicit drug users. The groups facilitated by this writer were equalitarian groups in which the members and facilitators assumed similar roles. The facilitators and members each were free to express their feelings about whatever topics they wished. Both the facilitators and members received feedback about the ways their behavior affected others in the group. These marathons were conducted for 16 continuous hours and the amount of time the members were together in their group contributed to discussions emphasizing the "here and now" interactions of the members. Statistical analysis of assessed process and outcomes of these groups supports the notion that the process of unstructured marathon groups conducted with illicit drug users is therapeutic and that these groups can change the attitudes of illicit drug users in positive directions. More research is needed on the process and outcomes of different types of marathon group therapy, i.e. family marathon group therapy, conducted for illicit drug users.

REFERENCES

Canadian Government Commission of Inquiry. The Non-Medical Use of Drugs: Interim Report of the Canadian Government Commission of Inquiry. New York, Penguin, 1970.

Hill, W. F. HIM Interaction Matrix Scoring Manual. University of California: Youth Study Center, 1961.

Hill, W. F. HIM: Interaction Matrix. University of Southern California: Youth Study Center, 1965.

Osgood, C. E., Suci, G. T., & Tannebaum, P. H. The Measurement of Meaning. Urbana: University of Illinois Press, 1857.

Page, R. C. "Developmental stages of unstructured counseling groups with prisoners." Small Group Behavior, 10, 271-278, 1979.

Page, R. C. "Marathon groups: Counseling the imprisoned drug abuser." The International Journal of the Addictions, 15, 765-770, 1980.

Page, R. C. "Marathon group therapy with users of illicit drugs: Dimensions of social learning." The International Journal of the Addictions, 17, 1107-1115, 1982.

Page, R. C., & Kubiak, L. "Marathon groups: Facilitating the personal growth of imprisoned, black female heroin abusers." Small Group Behavior, 9, 409-416, 1978.

Page, R. C., & Mannion, J. "Marathon group therapy with former drug users." Journal of Employment Counseling, 17, 307-314, 1980.

Page, R. C., Mannion, J., & Wattenbarger, W. "Marathon group counseling: A Study with imprisoned male former drug users." Small Group Behavior, 11, 399-410, 1980

Page, R. C., & Miehl, H. "Marathon groups: Facilitating the personal growth of male illicit drug users." The International Journal of the Addictions, 17, 393-397, 1982.

Page, R. C., & Wills, J. "Marathon group counseling with illicit drug users: Analysis of content." Journal of Specialists in Group Work, 8, 67-76, 1983.

Rogers, C. On Encounter Groups. New York: Harper and Row, 1970.

Stoller, F. H. "Accelerated interaction: A time-limited approach based on a brief, intensive group." International Journal of Group Psychotherapy, 18, 220-235, 1968.

DEALING WITH THE RELAPSE:

A RELAPSE INOCULATION TRAINING PROGRAM

Meir Teichman

Institute of Criminology, Tel-Aviv University
69978 Tel-Aviv, Israel

Most alcohol treatment programs -- outpatient as well as inpatient --
are plagued by many patients dropping out of treatment (e.g., Backel and
Lundwall, 1977; Galanter and Panepinto, 1980). Often, the dropping out of
treatment is associated with a relapse or a "slip." The relationship
between the "slip" or the relapse and dropping out of treatment is complex.
Some patients have a relapse and therefore leave the treatment; others
terminate their participation in a theraputic process and following such an
act they relapse. One may argue that the latter case represents a behavior
that is aimed at justifying the drink. When a person "slips," or relapses,
the dropping out of treatment can be explained by several reasons. First,
in most programs a prevailing norm is "don't touch a drink." The person who
has violated such a law and who has to face it in front of the community,
may experience an intense feeling of shame and failure. The relapse also
presents an internal conflict. We may speculate than an internal dialogue
occurs between the person's "super ego," "ideal self," or the moral
principles and his addictive patterns of functionings. Such a conflicting
internal dialogue may result in emotional reactions like anxiety,
depression, or anger on the one hand, and may intensify drinking behavior,
on the other hand.

The phenomenon of relapse is related to the concept of _craving_.
Craving can be defined as a subjective experience of desiring or needing the
euphoric and tension-reducing properties of the alcohol. Ludwig and Wikler
(1974), Ludwig, Wikler, and Stark (1974), and Nace (1978) viewed the concept
of craving as a cognitive representation of a withdrawal syndrome which is
induced by internal as well as environmental cues. All of us are familiar
with patients who were granted brief excursions and have relapsed during it.
Often we learn that two basic processes occurred while the person was out in
the "cold." First, the person experienced strong craving for alcohol (or
drugs) when exposed to his neighborhood environment. Applying the AA
terminology, the exposure to people, places, and things have triggered the
sense of craving for the psychoactive substance. This process could be
understood within the framework of conditioning theories (e.g., Wikler,
1980). The craving, according to this explanation, is a result of strong
association (conditioning) between various external, as well as internal,
stimuli and the intake of alcohol in spite of the person's desire and
motivation to extinguish this behavior.

Second, when the intense feeling of craving is experienced, the
alcoholic too often tells himself that "one drink will not harm" him and

"slips." Following the relapse the person experiences frustration, guilty feelings of lack of self-worth, and self-loathing. These feelings of failure and inadequacy result in the continuous substance abuse.

Two cognitive-behavioral theories may explain the relapse process and have significant implication for treatment of substance abusers. Bandura's (1977) self-efficacy theory assumes that an individual's belief that he or she can successfully control the required behavior (being sober) needed to produce the desired outcomes (sobriety). It is suggested that positive outcomes (abstinence and sobriety) can be achieved through the enhancement of expectations of personal efficacy.

The second cognitive-behavioral explanation was proposed by Marlatt (1979). Marlatt suggested that when a relapse occurred, regardless of its sources and reasons, a complex cognitive process, which he has labeled the Abstinence Violation Effect (AVE), can be identified. The Abstinence Violation Effect itself is characterized by two cognitive factors: (a) a cognitive dissonance state, (Festinger, 1964), which is triggered by the violation of the only relevant rule for a recovering alcoholic, and which results in conflict and guilt feelings; (b) a personal attribution effect (Jones et al., 1972), in which the person attributes the relapse to personal weakness rather than to external factors. Marlett (1979) suggested that the following elements should exist for a "slip" to become a full-bloom relapse:

(1) The abstinent person is persuaded that he is "in control" until he encounters high risk-loaded cues;

(2) The person is lacking appropriate and effective methods to correctly identify those cues and thereafter to cope effectively with the situation;

(3) The person experiences positive expectancies about the properties and outsomes of the previously abused substance;

(4) The person "slips" and experiences the AVE;

(5) The probability of another "slip" is markedly increased.

Following this line of reasoning an experimental treatment program, which we labeled "the relapse inoculation" approach, was planned (Meichenbaum, 1977). This intervention approach is intended to help the person correctly identify the external as well as the internal cures that have previously triggered the drinking behavior; to develop cognitive and behavioral coping mechanisms that will minimize the craving and avoid relapses; and to enhance personal efficacy and self-control. Conversely, the experimental program was designed to provide the alcoholic with a variety of appropriate responses to the feelings of craving rather than the single, maladaptive response he has previously mastered -- the drinking.

THE RELAPSE INOCULATION TRAINING

The Participants

Ten patients who were admitted to an alcoholism unit during the winter of 1979 participated in the program. Their mean age was 31 years old, and they reported drinking heavily for an average of 10 years. All stayed in the unit for 25 days as was required.

The Setting

The alcoholism unit in which the experimental training program took place is an inpatient service within the Department of Psychiatry of a general hospital serving an urban community in southeast Philadelphia. The program offers emergency treatment detoxification; the first stage of the rehavilitation process. The unit works closely with Alcoholics Anonymous and other rehabilitation facilities and social service agencies. It also offers continuing outpatient care. The length of hospitalization is 28 to 30 days. Patients are referred to the unit by the emergency room staff, various community organizations and services, employers, family members or friends. Few are self-referrals. Following discharge, patients are referred to the unit's own outpatient service or to AA and other inpatient or outpatient programs in the Philadelphia area.

The Procedures of the Program

The intervention approach consists of several phases. The first stage was the diagnostic one in which the preceding cues and stimuli of the relapses were identified and categorized. The diagnostic stage focused upon two components. The person's recollections of his previous relapses were scrutinized and analysed for unique stimuli, common places, events, people, and situations that have been associated with those relapses. Second, each patient presented during group sessions his personal "good reasons" and common "excuses" to which each alcoholic has attributed the relapses. During the group sessions those "good reasons" and "excuses" were challenged and dealt with. Following the identification of the high-risk "people, place, thing. and situations," and after learning to monitor their occurance and meaning, the training stage began.

The relapse immunization training followed the general procedures developed by Meichenbaum (1977) and Novaco (1978). The training program involves three phases. The first phase was an educational stage. The patients were provided with conceptual framework that explained the important roles of the previously mentioned high-risk components in the craving relapses and drinking. Furthermore, it was explained to the patients that being able to monitor, deal, and cope with these risky components may significantly reduce the chances for another relapse.

The second phase was designed to provide the patients with a variety of coping techniques that will be employed whenever the person encounters the risky components. During five group sessions the alcoholic patients acquired coping skills and rehearsed them with their fellow patients. The patients, for example, staged their encounter with drinking buddies adjacent to the neighborhood bar, and rehearsed their inner talk that will be used to balance and counteract the pressure to join those people for a relaxing drink. Furthermore, the patients developed some unique behavioral responses that were aimed at reducing the stress. One of the patients who was an artist working with leather prepared a coin-holder necklace. During the training group session, whenever a risky and stressful situation was staged, he touched the coin and a dime and silently told himself: "I can call my (AA) sponsor..."

The procedures of Meichenbaum's stress-inoculation-training involved three phases. The third phase included the in vivo practicing and testing of the coping mechanisms. Unfortunately, alcoholic patients who are admitted to an inpatient program were unable to put to test their acquired coping skills under actual and real risky conditions. The patients, who constituted the training group, were granted permission to leave the alcoholism unit premises. They always returned to the unit in time with no recorded relapses. This may indicate that the training was effective, at least for a short period of time. These short excursions are the only periods during which the patients may and could test their coping skills. Thus, they were experimenting with their newly acquired skills while being alone in a generally hostile and unfavorable environment. Creating and staging stimulations and limitations of real-life elements may only partially help the patients to test their newly learned coping mechanisms.

Ten patients participated in the experimental program. Following their discharge they were referred to the unit's outpatient clinic and to one of the AA groups in Philadelphia. Five out of the 10 patients were still in treatment six months after their discharge with no recognized and recorded relapses. The other five patients have dropped out of their outpatient treatment programs and have not been traced.

DISCUSSION AND CONCLUSION

The treatment of addictive behaviors emphasizes two aspects. The first one is the development of effective "cessation: therapeutic strategies and the second one is the developing of procedures to assure long-term, robust behavior change. Past research has clearly indicated that most addicts (alcoholic, drug, as well as smokers) relapse shortly after treatment termination (e.g., Hunt, Barnett, and Branch, 1971; Hunt and Bespalec, 1974; Condiotte and Lichtenstein, 1981). Thus, all treatment programs should recognize that the production of the initial detoxification and cessation is only the beginning and onset of treatment. The maintenance of non-drinking patterns of behavior and life-style should become an integral part of all treatment modalities. Paradoxically, the occurance of a relapse or a "slip" often hinders the maintenance of abstinence programs rather than enabling us to block the "tide" before it is too late. Our experimental program was aimed at adding an additional and needed therapeutic effort stratagy -- this will provide the recovering alcoholic with coping skills rather than substituting the patient's involvement in undergoing follow-up treatment and/or participation in AA.

One of the major limitations of the relapse inoculation training program carried out in this study was the lack of the third phase, during which the alcoholic was able to test in vivo acquired coping skills.

The short excursions are risky opportunities for the patients to test themselves in real-life situations. The lack of emotional, social, and professional support during these crucial periods interfere, and in addition drug misuse hinders, the effectiveness of such a training program. Nevertheless, our initial data supports the applicability of "inoculation training" as an additional modality that may increase the probability of reducing the relapse.

REFERENCES

Backeland, F., & Lundwall, L. K. "Engaging the alcoholic in treatment and keeping him there." In B. Kissin and h. Begleiter (Eds.), Treatment and Rehabilitation of the Chronic Alcoholic. N. Y.: Plenum, 1977.
Bandura, A. "Self-efficacy: Toward a unifying theory of behavioral change" Psychological Review, 1977, 84, 191-215.
Condiotte, M. M., & Lichtenstein, E. "Self-efficacy and relapse in smoking cessation programs." J. of Consulting and Clinical Psychology, 1981, 49, 648-658.

Festinger, L. Conflict, Decision and Dissonance. Stanford, Calif.: Stanford University Press, 1964.

Galanter, M., & Panepinto, W. "Entering the alcoholic outpatient service. Application of a systems approach to patient drop-out" In M. Galanter (Ed.), Currents in Alcoholism, Vol. VII N.Y.: Grune and Stratton, 1980.

Hunt, W. A., Barnett, L. W., & Branch, L. G. "Relapse rates in addition programs." J. of Clinical Psychology, 1971, 27, 455-456.

Hunt, W. A., & Bespalec, D. A. "An evaluation of current methods of modifying smoking behaviour." J. of Clinical Psychology, 1974, 30, 431-438.

Jones, E. E., Kanouse, D. E., Kelley, H. H., Nisbett, R. E., Valins, S., & Weiner, B. (Eds.), Attribution: Perceiving the Causes of Behaviors. Morristown, N.J.: General Learning, 1972.

Ludwig, A. M., & Wikler, A. "Craving and relapse to drink." Quart. J. Studies on Alcohol, 1974, 35, 108-130.

Ludwig, A. M., Wikler, A., & Stark, L. H. "The first drink: Psychobiological aspects of craving" Arch. Gen. Psychiatry, 1974, 30, 535-547.

Marlatt, G. A. A cognitive-behavioral model of the relapse process In N. A. Krasnegor (Ed.), Behavioral Analysis and Treatment of Substance Abuse. Research Monograph 25, Rockville, Md.: National Institute of Drug Abuse, 1979.

Meichenbaum, D. Cognitive-Behavior Modifications: An Intergrative Approach. N. Y.: Plenum, 1977.

Meichengaum, D. & Novace, R. Stress inoculation: A Preventative approach. In C. D. Spielberger and I. G. Sarason (Eds.) Stress and Anxiety, Vol. 5, Washington, D.C.: Hemisphere, 1978.

Nace, E. P. The Use of Craving in the Treatment of Alcoholism. Paper presented at the 32nd International Congress of Alcoholism, Warsaw, Poland, 1978.

Wikler, A. Opiate Dependence: Mechanism and Treatment. N. Y.: Plenum, 1980.

ALCOHOLISM -- A RESPONSE OF MINORITY COMMUNITIES TO REJECTION

OF THEIR CULTURAL IDENTITY BY THE DOMINANT CULTURE

Saul E. Joel

Atkinson College
York University
North York, Ontario
Canada

The reader should be aware that I am a newcomer to the study of alcohol and drug addiction and that my readings and research in the field are limited. I believe that the causes of alcoholism are polygenic. It is in this spirit that my interest was aroused in developing an understanding of the presence of alcohol among my people, the Jews of India now settled in Israel. This led me to search for meaningful intervention in order to redirect the community's energies from alcohol misuse/addiction and its social self-destruction to joining other Israelis for a place of their own in the future state.

"I am that I am" (The Holy Scriptures, 1917). A statement! A pronouncement of existence and identity by a strong God who knows who He is, what He is, to whom, where, and when. In other passages of our history, we are told that God created man in His image. But once cast out of the Garden of Eden, man began his long search for self-knowledge, an identity, a reason for existence, and a rationale to continue the process of evolution in pain, suffering, love, and joy. This universal search of self-knowledge embedded in the establishment of an identity returns man partially to that Garden.

The need for a strong self-identity as one of (if not the main) the variables to accomplish and integrated, productive, and contented life, is well documented in numerous case studies conducted over the past several decades by practitioners and researchers in the helping professions. A successful identity aids normal human functioning, which is seen as functioning effectively with some degree of happiness and the achievement of something worthwhile for onself within the rules of the society in which one lives (Glasser, 1960). Those familiar with William Glasser's work will follow my extension of his concept of individual identity to group identity.

The establishment of self-identity is facilitated mainly by the individual belonging to a group or community which takes its place among other groups in the macrosociety. Here again several human developmental studies, both specific life phase and life-cycle studies, have documented that subgroups with well-established, positive identities create and aid the establishment of successful individuals who may be described as functioning normally within the group's boundaries and in the macrosociety. It will be helpful here to clarify what is meant by identity. Bell and Evans' (1980) descriptive definition can be operationalized in a holistic manner from the individual to the group level. This definition sees identity as consisting of three parts: (1) How I see myself; (2) How I see myself in relation to others; (3) How others see me. For intergroup relations within a macro-society these are: (1) How the group sees itself in society; (2) how the group sees itself in relation to external groups in society; (3) How external groups in society see the group.

In multiethnic and multicultural countries, each subgroup draws some of its identificational strength from belonging to its national state in reference to the international community. This is true for both majority and minority groups. Within a nation, the relative strength and effective functioning of a separate subgroup identity is directly dependent on its relative acceptance by and integration with other groups. The latter are either the major or the dominant group, inasmuch as they hold the political, economic, and social power of the nation.

In order that a subgroup maintain and develop a positive, strong group identity in relation to the wider community, the following needs must be satisfied:

1. Acceptance of its separate identity as being equal and unique by the dominant group.

2. Genuine expression and demonstration of being needed by the dominant group.

3. A unique history which gives the group a sense of pride in its contribution to the development of man, and, in particular, to the development of its country.

4. Equal opportunity in the marketplace, in the fields of education, employment, housing, government, social services, etc., together with an opportunity to produce its own elite, acceptable to its own members and also to the dominant group.

As an introduction and providing a theoretical background, I submit that subcultural or subethnic groups, like individuals, function more successfully in the environment when they manifest a strong, positive identity. This identity must be accepted and nurtured by the society at large in order to maintain its equal but unique status within the macrosociety.

The corollary of the above is that subcultural and/or subethnic groups whose identities have been rejected or damaged by the dominant group in society will manifest negative behaviors. These are operationalized as "fight or flight" responses described by Hans Selye (1978). Fight can be defined as acting out and violence; flight by self-destruction and withdrawal. In this respect, the excessive use of alcohol by a subgroup is a form of withdrawal from the real world and leads to the self-destruction of the group.

In the fall of 1978, the Department of Social Welfare of the City of Dimona* invited me to work with them to understand and attempt rehabilitation

---

*Dimona: A developing town in the Negev, established in 1946, and located 35 kilometers south of Beersheba. Population is 30,000, mainly Sephardic from Morocco, approximately 3,000 Bene Israelis from India, and 1,000 Black Hebrews from the United States. Politically, bureaucratically, and economically the town is controlled by the Sephardic leadership with Bene

of an identified number of Bene Israelis* who were excessive users of alcohol. The task began by the Department transferring to me a caseload of individuals and families for treatment and rehabilitation. Together with this I had hoped to develop an understanding of the factors influencing the high consumption of alcohol by the community. In a matter of weeks the reasons given by the community for excessive use of alcohol refocused my attention from microsituations to macrosituations. In addition, a macroassessment seemed most suitable after an informal survey undertaken by the Department, the local community leaders, and myself showed an alarming percentage of heavy users of alcohol.

Historically the community in India had used alcohol in limited quantities as a natural medicine, a relaxant, and as part of celebrations. Some families were total abstainers while some who were closely associated with the Anglo-Saxons used alcohol in what is termed "social drinking." Nevertheless, it has been confirmed by the majority of community elders that alcohol was never used to the excessive amounts that are currently experienced in this country.

It is true that the legal, easy, and cheap availability of alcohol in Israel, as compared to India, partially accounts for the increase in consumption. However, this factor alone does not explain the excessive and compulsive utilization of alcohol by men and women, and its use by young second- and third-generation members of the community born in this country. Further, a significant percentage of the members of the community have accepted alcohol as the only effective and immediate solution to life's problems.

A further examination of the community's experience in Israel and its quality of life identified several significant contributing stresses which damaged and rejected the maintenance and development of its identity among other groups. The following were identified as the main stresses.

1.   The directives of the chief rabbinate in Israel in the early sixties questioned the authenticity of the Jewishness of the community and forbade marriage with other Jews. To a Jew who believes he has maintained his Jewishness for some 2,000 years, isolated from the rest of Jewry for 1,700 years or more, these directives were the strongest systematic attack on the identity of this group and made them vulnerable to social ridicule by the rest of the population.

2.   A large number of this community had emigrated from a traditionally developing country into a modern developing country. They were ill-equipped with the necessary language skills. They found themselves eligible for only nontechnical, labor-oriented jobs with no opportunities for career development.

3.   The internal protaxia system** within developing towns excluded them from meaningful employment, career development, fair housing,

---

Israelis holding three seats in the local Municipal Council and the majority of the latter employed in the skilled and unskilled labor force. The main wave of immigration of the Bene Israelis from India to Dimona took place in the 1950s.
*Bene Israelis are Jews from India whose origin and past is shrouded in mystery. Some scholars are of the opinion that they have lived in India for approximately 2,400 years and were revived to modern Judaic practice and beliefs about 1,000 C.E. Total population in Israel is approximately 23,000 with 6,000 left in India.
**Family and social network influencing systems into the political, bureaucratic, and economic structures of the community.

Table 1

	Bene-Israel (minority group)	Macrosociety (dominant group)
Tertiary-- individuals in acute states of addiction	Meet basic needs of food, clothing, shelter Medical services; detoxification Balanced diet Crisis, individual and group counselling	Information about minority group and alcoholism to primary control agents, medical personnel and helping agents
Secondary-- individuals at high risk	Immediate 300% increases in cost of alcohol Decreases availability (reduce outlets, supply on advance orders only) Strict enforcement/penalties for intoxication and/or consumption at place of work Change to new environment migration and/or immigration for children of alcoholics Counselling support services	Information about minority groups and alcoholics to teachers and employers Improved quality of working life Equal rights legislation supported by social animators to act as agents for protaxia
Primary-- unaffected individuals	Alcohol education programs Total ban on alcohol advertising Assertiveness training	Strict control/penalties for drinking, driving information about minority groups to increase level of undertanding and tolerance

supportive and selective education programs, etc. These factors made the Bene Israeli community victims of selective discrimination. This selective discrimination was reinforced by the stigma attached to the label of "Hodin" (Hebrew word for one who comes from India).

4. Culturally different and striving to maintain its own traditions, the community isolated itself from the dominant Ashkenazai and Sephardic culture orientations and complemented the wider society by playing its labeled role. This factor is known to produce extraordinary stress in cases where a group has to learn new ways of existing in two worlds and to develop a double consciousness. To be successful in these dual worlds calls for the development of social behavioral skills described as "successful schizophrenia" (Bell and Evans, 1980), which causes further stress, pain, and negative self-perceptions.

In summary, the community's Israel experience can be described as a rejection of their Jewishness, stigmatization as "Hodim" (Indians), and discrimination by the power of protaxia. Its youth manifest a poor self-image, are lost, and lack the spirit to become. This is a good indicator of where the community is in relationship to the wider society.

The community responded to this rejection, insult, ridicule, and nonacceptance of their cultural identity by withdrawal and self-destruction rather than by violence or assertiveness for their equal and rightful place in society. This passive and withdrawal response was in keeping with their cultural background. Their use of alcohol as a solution to serve as identity-resolving forums led to further social, cultural, and economic isolation.

Table 2

Bene-Israel (minority group)	Macrosociety (dominant group)
Information about introduction of policy and programs	Information about introduction of policy and programs
Epidemiological study	Attitudinal study
Maintain cost increases of alcohol Employment restraining Human relations and assertiveness training Alcohol education programs	Extension of equal rights to minority cultural rights as human right with supportive programs
Support of minority groups cultures (music, dance, drama, drama, dress, food) Leadership training Alcohol education programs Supportive educational programs	

The response by the Bene Israelis in Israel is not unique in man's experience. The following are a few minority ethnic and cultural groups whose excessive use of alcohol has been related to their powerlessness and discrimination in the society they live in.

The famous work on the Irish by Robert F. Bales (1962) is a most useful example. In this study he presents the complex cultural relationship that exists between the Irish and their use of alcohol. As one of the variables he states "drinking is inextricably tied up with the expression of aggression in the Irish culture, not only covert aggression against the elders,...but most importantly, overt and active and persistent aggression against the English.... In brief it is quite clear that drinking and the activities closely related to it were and, to some extent still are, instruments of political aggression and resistance in Ireland. The activities were important because they provided an outlet for the expression of a fund of tensions centering around the political relation to England.

In addition to Bales' study of the Irish culture, if the reader considers the situation in Northern Ireland, it is clear that social, economic, and political exclusion has intensified the establishment of a separate Irish identity. This community manifests both withdrawal (alcohol) and acting out (violence behaviors as a means to cope with the rejection, ridicule, etc., of their identity by the dominant cultural group.

The Irish situation is extremely complex, multivariant, and has a long history. In contrast, let us look at a more recent and less complex situation as presented by the Australian Aborigines. J. Green (1976), in his report entitled 40 Gallons A Head, examines excessive drinking among Aborigines in northern Australia. An abstract of his findings reads:

Reasons indicated for the heavy drinking include the "bravado" that accompanies the Northern Territory "drinking myth," the lack of community feeling due to transient life-style, isolation,, cultural loss, and dehumanization due to racism, lack of adequate entertainment, and the ease of obtaining alcohol.

The above reasons parallel the stresses indicated earlier when considering the Bene Israeli situation. It demonstrates a similar rejection of the Aborigines' unique identity and their nonacceptance as equals by the European white-dominated macrosociety.

Almost similar conditions are cited in Peru where the Peruvian Indians use of alcohol as an escape from the continual deprivations experienced by them. Here "the Spanish conquest, which shattered one social organization but did not replace it with another, upset this (social and cultural) balance and from then on the Indians moved increasingly to the use of alcohol as a drug" (Saavedra and Mariatequi, 1970).

In considering damage to group identities, the most relevant situation is that of the Canadian Indians. Their excessive use of alcohol completely permeated the fabric of their social and cultural life. But it was the solution to live lived in a subgroup of the Canadian society described as the most impoverished in the country.

Compared with the total population, they have the poorest health conditions, shortest life expectancy, the lowest standard of living. Their rate of unemployment is the highest in Canada and very few are representedin the skilled labor force or professions. They are over-represented in jails and correctional institutions, but under-represented in secondary and post-secondary schools. Chronic unemployment, low income, and the lack of collateral places them at a disadvantage in dealing with private financial institutions, and as a consequence, they have been aliented from the forces of capital open to other Canadians. (Breton, Reitz, and Valentine, 1980)

Historically, the Government of Canada employed different strategies in their efforts to assimilate and control native peoples. These were accommodation, domination, and integration into the Canadian way of life. These strategies were directed to the eventual absorption of the Indian into the white society and took various forms. Between the State and the Church the Canadian Indian was raped of his pride, his beliefs, and his unique identity. The identity of the "free and brave" was replaced with dependence, disorientation, and drunkenness. The Indians today have begun to move from flight to fight as their awareness of their situation arises in their group consciousness. It was Chief Dan George who wrote that he could understand his white master's actions which robbed his people of their land, their homes, etc., but could never forgive the white man who robbed his people of their identity. Harold Cardinal, writing in 1969, said "the history of Canadian Indians is a shameful chronicle of the white man's disinterest, his deliberate tramplings of Indian rights, and his repeated betrayal of our trust." This is followed by a call to his people -- "Our identity of who we are, this is a basic question that must be settled if we are to progress.... The challenge of Indians today is to redefine that identity in contemporary terminology. The challenge to the non-Indian society is to accept such an updated definition.

Canadian Indians today have gathered together to reestablish and recreate their unique identity. It is clear that the strategy of planned and forced integration has failed. Even with the passing of generations, the goals of the government have not come any closer to reality. Clearly, an Indian identity must be developed in order to establish a working dialogue with the dominant groups in Canadian society.

In the above case presentations, I believe that among the polygenic factors for the excessive use of alcohol by minority groups are their political, economic, religious and social exclusion, and discrimination by the dominant cultural groups. As a means of reducing the consumption of alcohol, the repertoire of planned intervention by governments and citizen-based social agencies can be described as being of two types: those arising from the selective control of the distribution of alcohol and those therapeutic treatments directed at individuals and families of identified alcoholics. The latter have used a wide variety of environmental, medical, and psychological therapies.

In democratic and free countries, selective control of the availability of alcohol creates problems for the development of equal rights in the wider society. In addition, on a long range basis, such controls have proven ineffective as measures to reduce overall consumption. On the other hand, intervention, focused on individuals, and families, have as their goal the rehabilitation of alcoholics, their reentry into the world of work, and the marginal economic and social self-sufficiency in society. Following is a typical rehabilitated case: The alcoholic is identified as the deviant. He, and in a smaller number of cases his family, is the recipient of therapeutic intervention. His successful rehabilitation into society either terminates his treatment or he is seen on a follow-up basis. This temporary strengthening of his self-identify, at the high cost of treatment, slowly weakens as he re-enters his minority group culture which continues to maintain an unsuccessful group identity in relation to the wider society. Most therapists who work on an individual level are now aware of the dynamics, power, and influences that the minority group's cultural identity bears on the individuals adjustment to the wider society. Thus the identified alcoholic succumbs to this unsuccessful group identity, loses his resolve, and becomes a statistic on the revolving door of rehabilitation programs.

In considering the broad variety of available therapeutic intervention programs, it is Alcoholics Anonymous that has been able to present the most significant and successful statistics of short- and long-term rehabilitation (Maxwell, 1962). This program requires the individual to make a new formal commitment to a group which holds common goals, norms, expectations, and beliefs leading to the creation of a strong group identity.

The only other macrostrategies strongly supported by several countries and directed at their total populations are alcohol Education programs. Here the emphasis is on the use and abuse of alcohol and its social destructiveness. Understandably, these programs are more effective for groups maintaining successful identities than for those with damaged or destroyed identities.

To return to the examination of the excessive use of alcohol by special populations, the solution of the problem moves directly into the political arena. It is my belief that in order to reduce the overall comsumption of alcohol, comprehensive, interdependent strategies must be introduced to effect short- and long-term change within the minority group, and concurrently with the dominant group controlling society. Macroprograms designed and directed at this latter group must aim at raising their cultural tolerance level and increasing their knowledge of other subgroups. Legislation must be introduced to bring about fair and equal exchange in the marketplace. The program strategies which I propose reflect both orientations -- the social-cultural and the distribution of consumption models. Although these strategies are identified specifically with tertiary, secondary, and primary prevention, their impact will be progressive throughout the society. Incorporating the above, the accompanying chart presents macrointervention strategies. Although this plan was developed to address the situation of the Bene Israel in Israel, it can be adapted to meet the needs of other minority ethnic and cultural groups exhibiting maladaptive withdrawal through the use of alcohol. The onus is therefore on the dominant group in society to establish social policies stemming from their political commitments. Once cultural rights are accepted as a basic human right, macroprograms must be designed to support the development of subethnic and cultural groups so that they too can grow to have their equal place in the sun.

REFERENCES

Bales, Robert F. "Attitudes toward drinking in the Irish culture." In D. J. Pittman and C. R. Snyder (Eds.), Society, Culture and Drinking Patterns. New York: John Wiley and Sons, 1962, pp. 173-174.

Bell, Peter, and Evens, Jimmy. Alcohol Use and Abuse in Black America Unpublished paper, 1980, pp. 22, 17-23

Breton, R., Reitz, J. G., and Valentine, V. F. Cultural Boundaries and the Cohersion of Canada. Montreal: The Institute for Research on Public Policy, 1980, p. 118.

Cardinal, Harold. The Unjust Society. Edmonton: M. G. Hurtig Ltd., 1969, pp. 1, 25.

Glasser, William. Mental Health or Mental Illness. New York: Harper and Row, 1960, p. 1.

Green, J. "40 gallons a head -- Alcohol in Alice." Aust. J. Alcsm. Drug Depend. 3: 5-6, 1976, In Classified abstract prepared by Quarterly Journal of Studies on Alcohol. New Brunswick: Center of Alcohol Studies, Rutgers University.

Maxwell, Milton A. "Alcoholics Anonymous: An interpretation," In D. J. Pittman and C. R. Snyder, (Eds.), Society, Culture and Drinking Patterns, New York; John Wiley & Sons, 1962, pp. 577-585.

Saavedra, A., and Mariatequi, J. "The epidemiology of alcoholism in Latin America." In R. E. Popham, (Ed.), Alcohol and Alcoholism. Toronto; University of Toronto Press, 1970.

Selye, Hans. The Stress of Life. New York: McGraw-Hill Book Co., 1978, pp. 329-353.

The Holy Scriptures. Philadelphia: The Jewish Publication Society of America, 1917, Exodus Ch. 3, vs. 14.

PROBLEMS OF COMMUNITY DENIAL ON RECOGNITION AND TREATMENT OF

ALCOHOLISM AMONG SPECIAL ETHNIC POPULATIONS IN LOS ANGELES COUNTY

Marcia Cohn Spiegel

Los Angeles County Commission on Alcoholism
Special Target Group Committee and Alcoholism Advisory Board
Jewish Family Service of Los Angeles

INTRODUCTION

Every large city has its own unique alcoholism problems. In each urban area alcohol abuse will be an acknowledged problem in one or more population groups. A large portion of public health services and funds will be spent on meeting the needs of that target population. However, within the same community, other groups may try to hide overt signs of drunkenness and alcohol abuse, and the needs of that population will be overlooked. This paper describes Los Angeles County in the hope that the description of the problem and the recommendations for addressing the problems will be applicable to other communities.

DESCRIPTION OF THE PROBLEM

Los Angeles County, California is a large multiethnic community of 7,500,000 people divided racially as follows: 68% White, 12% Black, 6% Asian, .6% Native American, and 13% other. Each of these racial groups is further divided by differences in language and national cultures. For example, the 49,000 American Indians represent over 100 different tribes. While those listed as Asian come primarily from China, Japan, Korea, Indochina, and the Philippines, Asian immigrants from at least 10 other countries, each with its own language and distinct culture, are also included. Over 27% of the population is described as "Hispanic" which refers to Spanish-speaking individuals from Mexico, Central and South America, Puerto Rico, Cuba, and other Caribbean countries. The white community also represents a wide variety of national, religious, and ethnic groups (U. S. Census Report, 1981).

The Los Angeles County Office of Alcoholism and Alcohol Abuse estimates that there are 622,000 alcoholics in this diverse population. County treatment programs serve the 200,000 problem drinkers who are considered economically disadvantaged. When the Office of Alcohol Abuse and Alcoholism was founded in 1973, its primary target group was the public inebriate on "Skid Row." That population appeared to be mainly drawn from the impoverished white community. Enactment of special legislation (The Sundance Act) in 1978 focused attention on the acute problem in the American Indian Community where rates of alcoholism are disproportionately high: although they represent .6% of the population, they receive over 2% of the alcoholism services. At the other extreme, the Asians with 6% of the population receive .2% of the services (L. A. County Plan, 1981-82).

Attention has been drawn recently to the needs of other underserved ethnic minorities with high rates of alcoholism. The Black and Hispanic communities account for almost 40% of the Los Angeles population, yet they appear as only 27% of the client load at county facilities (L. A. Plan, 1981). As the advertising industry became aware of the consumer potential in these populations, massive campaigns have been mounted showing Blacks and Hispanics happy and successful implying that this success came from drinking certain alcoholic beverages.

In 1974 Anheuser-Busch tailored ads to the Spanish-speaking beer drinker. Within two years Budweiser Beer, an Anheuser-Busch product, was the number one beer sold in the state of California (Cal. Com. for Spanish Speaking, 1978). Jet Magazine (1981) recently portrayed successful, well-dressed upper-class Blacks in ads for Schlitz Malt Liquor, Michelob Beer, Smirnoff Vodka, Black Velvet, White Label, Canadian Mist, and Seagrams. Billboards in ethnic communities and ads in minority periodicals appeal to the customer and may add to the growing problem.

To combat the spread of alcohol abuse, the State of California and the County of Los Angeles established special commissions for the Native American, Pacific-Asian, Black, and Spanish speaking communities. These Special Target Commissions serve as technical advisors to provide information and assistance in preparing the county alcoholism plan, and to act as advocates in the preparation of contracts to insure that alcoholism programs reflect the special needs of each ethnic group. They monitor legislation to make sure that these needs are considered. In addition they serve as resources to the community to provide information on identification, prevention, and treatment of alcohol abuse through programs, training workshops, printed materials, and films.

As the county is becoming increasingly aware of problems of alcoholism among American Indians, Blacks, and Hispanics, there is popular agreement that two specific populations do not share these problems. The Asians and the Jews each account for approximately 400,000 people; together they are almost 15% of the county population. Very few people of either group are receiving county services for alcoholism. One obvious conclusion would be that there are no problems of alcoholism, therefore the people do not seek help. Recent observations indicate, to the contrary, that alcoholism does exist but several factors lead to its denial by community leaders, the community itself, and even alcoholics within the community. This denial delays or keeps these individuals out of treatment.

These two groups are each made up of a diversity of nationalities, languages, and religious beliefs. The Asians come from more than 15 countries and their religious beliefs include Christian, Buddhist, Shinto, Taoist, Confusian, and Moslem. Recent waves of immigration have brought an influx of newcomers from the Filipines, Vietnam, Korea, and Cambodia (U.P.A.C., 1978)*

Religious affiliation and observance among the Jews ranges from orthodox, conservative, and reform, to secular-cultural and non-observant Jews. Most Jews in Los Angeles are second or third generation Americans from European backgrounds, but there is a sizeable Sephardic community and many new immigrants from Russia, Iran, South America, Morocco, and Israel (Phillips, 1979).

---

*Special thanks to Dr. Herb Hatanka, U.C.L.A., for his overview of the Asian American Community and its drinking behaviors.

224

Despite the diversity of religious beliefs, languages, and national cultures, these divergent groups share some important values and behavior:

1. The family plays a central role in their lives.

2. There is cooperation within the ethnic group.

3. Self-control is considered very important. Extreme shame is connected with any loss of control.

4. Behaviors which could bring shame on the individual, family, or community are avoided.

5. Individuals will try to solve problems by themselves. When help is needed it will probably be sought within the community in order to preserve appearances and to protect confidentiality (U.P.A.C., 1978).

The importance of self-control and the avoidance of shame seem to have contributed to the avoidance of public drunkenness. Assimilation, acculturation, and alienation all appear to contribute to increases in alcohol consumption and a concommitant rise in alcoholism within these communities. Although individual social workers and alcohol treatment personnel have observed an increase in alcohol related problems among both Jews and Asians, informal surveys of religious leaders and directors of social service agencies evoked total denial of any such problems. Both Asian and Jewish doctors and psychiatrists denied seeing patients with alcoholism. The interviewers knew of alcoholic congregants, clients, and patients of each of those questioned (Spiegel, 1980; Casaclang, 1981).

One example of this denial occurred when the director of a service agency was asked if any clients had alcohol related problems. The director knew of none. At a training workshop attended by 75 social workers from that agency, a show of hands indicated that 80% of the social workers had seen or were seeing clients with problems related to alcohol abuse, including family tension, depression, health problems, suicide attempts, wife-battering, child abuse, unemployment, and driving violations. At another workshop it was learned that over 15% of the case load had alcoholism problems. This information was not requested by the agencies, therefore the information did not exist on agency records. It was also learned that clients concealed the relationship of alcohol to the problems which brought them into treatment.

Some difficulties are related to the length of time a family has been in the United States. While the adult new-immigrant is struggling with a totally alien culture and language, children and teens are thrust immediately into the public school system where they begin to learn English and adapt to the new environment, frequently turning their backs on the customs and values of their parents. Teens may band into ethnic gangs for emotional support and to protect themselves from other hostile groups. At one such high school in Los Angeles there were gangs of Blacks, Hispanics, Vietnamese, Russian Jews, Iranian Jews, and Israelis (Kaminski, 1977).

In Asian neighborhoods there have been documented reports of gangs of Taiwanese, Hong Kong, Chinese, and Koreans which fight with groups from adjoining neighborhoods. Gang violence has erupted between Filipinos and Koreans, as well as Samoans and Hispanics. In all of the gang activities beer and wine drinking play an increasingly important role. Second and third generation youth appear to be starting to drink at younger ages. Parents do not understand the behavior of their children and have diminishing power over the young people (Casaclang, 1981).

Earlier in the twentieth century, when this drama was acted out in the Jewish urban ghettoes, alcohol played a less important role among the youth. But even then it was a factor in their socialization. During the era of prohibition older teens associated drinking with the thrill of the illicit speakeasy with its aura of grown up glamor (Spiegel, 1977).

Because the appearance of alcohol-related problems is so recent and the community denial is so complete, there is, as yet, very little hard data on alcoholism in either the Asian or Jewish communities. Dr. Harry Kitano of the University of California at Los Angeles is completing a study of drinking practices among the Chinese, Japanese, and Koreans. Dr. Nabila Beshai of the Los Angeles County Office of Alcoholism and Alcohol Abuse is beginning a study of drinking patterns and alcoholism among the Jews. The findings of my own limited study of 16 recovered Jewish alcoholics parallels those of the Dropkin and Blume study of 100 recovered Jewish alcoholics in New York (Blume, Dropkin, and Sokolow, 1980).

ALCOHOLISM IN THE JEWISH COMMUNITY

In order to treat alcoholism in any ethnic group, an understanding of the beliefs, values, and behaviors of that group is necessary. The Jewish people have always stressed the importance of moderation and sobriety. Early religious codes defined the behavior of Jews as separate and different from that of the pagans among whom they lived and where orgies of drinking and sexual excess were frequently part of the religious ritual (Cohen, 1974). Rabbinic commentaries describe both the benefits of moderation and the dangers of excess. In the twelfth century the great teacher Maimonides stressed that moderation was good in all things. He pointed out that a little wine is good for a man, but he deplored any kind of sensual overindulgence. When Joseph Caro codified Jewish law in the sixteenth century Shulchan Aruch, he was very specific about why, where, how, and what a man or woman may drink, and how alcoholic beverages may be prepared, stored, bought, and sold. These codes became the standard for most Jews. Ritual drinking continues to be important in Jewish home observances: it precedes or accompanies a meal, and is a part of all life-cycle events. A boy tastes wine for the first time at his circumcision on the eighth day of his life.

Wine and alcohol have also been important in the economic development of the Jewish people. Moses learned of the fertility of the land of Israel when his scouts brought a single cluster of grapes so large that it took two men to carry it. Grapes were grown for food and wine first in Israel, then in Babylon and throughout the Mediterranean Basin and Eastern Europe as the Jews went into exile. Several rabbis, cited in the Talmud, earned their living as brewers or vintners; even the celebrated scholar, Rashi, earned his living in this way (Baron and Kahan, 1975). The Jews were given the rights of propination in Russia and Poland in the fourteenth century. This gave them a virtual monopoly on the manufacture and distribution of alcoholic beverages. It is estimated that 40% of the Jews in Eastern Europe villages earned their living in this way. Revocation of these rights in the 1890s caused more than 200,000 Jews to lose their source of livelihood, which, when added to other oppressive legislation and pogroms, caused large numbers of Jews to seek more favorable environments. Many came to America where they continued in family businesses which involved the manufacture or sale of beer, wine, and alcoholic beverages (Dubnow, 1916).

Public drunkenness was frowned upon except for a few rare occassions. The popular Jewish folk song "Shikkur is a Goy" (a non-Jew is a drunk) supports the belief that drunkenness was seen as non-Jewish behavior. The Babylonian sage, Rava, told the Jews that during the festival of Purim they should drink until they cannot distinguish Haman (the villain) from

Mordeccai (the hero). However, in the Orah Hayyim of the <u>Shulchan Aruch,</u>
individuals who are aware their drinking causes them problems are excused
from exercising this religious duty.

Glassner and Berg (1980) discovered that the understanding that
drunkenness was not a Jewish behavior was one of the reasons for the
continuing low rates of alcoholism among Jews. Other reasons which they
cite include the fact that wine is emphasized over other alcoholic beverages,
specific drinking practices learned in childhood, and a variety of avoidance
techniques which have been developed. My own interviews confirmed these
findings that recovering alcoholics perceived their drinking behavior as
shameful and non-Jewish. Several of them had difficulty defining themselves
as both Jewish and alcoholic.

The sample of 16 whom I interviewed included 10 men and 6 women whose
ages ranged from 21 to 79. While most were drawn from a middle-class
population, two received public assistance and three were upper-class. The
religious affiliation included one orthodox, two conservative, and eight
reform Jews. Two described themselves as observant but not affiliated with
a congregation, and three were not members of any Jewish organization but
maintained most of their social and business associations with other Jews.
One was foreign born, eight were first generation Americans, five were
second generation, and two from the third generation. One-half of the
sample held prominent positions in the Jewish community at the time they
were drinking most heavily (Spiegel, 1979). (See Figure 1)

Those interviewed did seek help for their problems. Between them they
saw 17 doctors and psychiatrists, four rabbis, and two social workers.
Three of the doctors consulted knew that their patients drank, but they did
not pursue the matter. Four of the subjects sought help from rabbis and
were reassured that "There are no Jewish alcoholics." Because these people
felt that they were not alcoholic, they continued to seek help for their
pain and malaise; but the help they sought was inappropriate to their
problems. Their drinking behavior was ignored and treatment dealt only with
the symptoms. All of these individuals were delayed from getting help for
the alcoholism problem which underlay all of their symptoms. Only one
psychiatrist actively intervened in identifying the alcoholism and insisting
that his patient attend Alcoholics Anonymous meetings in addition to
psychotherapy.

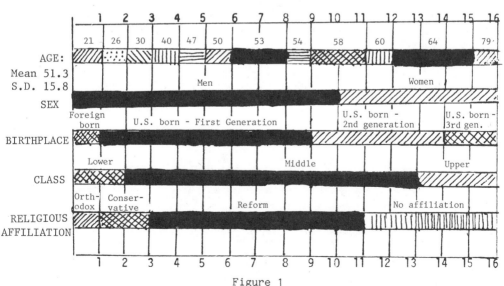

Figure 1

- Ike was visible to the whole community during his worst drinking bouts. As president of a large congregation he was designated the role of greeting the congregation at services on Friday evenings and making brief announcements -- a matter of a few minutes at the most. The brief announcements became lengthy monologues, take-offs on the rabbi's sermon, and comedy routines lasting up to 15 minutes. People began attending the synagogue services to hear what Ike would say next. When he finally admitted to the rabbi that he thought alcohol was beginning to interfere with his life, the rabbi's response was, "Don't be silly. You're Jewish."

- Frank was an important philanthropist whose doctor prescribed an evening martini for relief of stress and tension. The martinis became bigger and more frequent. The people close to Frank were unaware of his drinking problem, even when he led all of the board of a major Jewish organization in a frenzied drunken dance singing "Shikkur is a goy."

- Mary was president of a large women's organization. Although she suffered from frequent blackouts, she earned a reputation for her thoughtfulness because she called her hostess on the morning after any meeting or social event with gracious thanks, but really to find out what she had done the night before. No one suspected that she was alcoholic. No one ever saw her take a drink.

- Lola was a survivor of Auschwitz and an orthodox Jew. She developed hepatitis and diabetes. She was sent to an alcoholism recovery home for women when she finally sought help from a women's treatment program. She left the recovery home after only a few days because she could not eat the food which was not prepared according to the Jewish dietary laws. She was uncomfortable in the daily Alcoholics Anonymous meeting because she perceived them as being Christian in orientation.

Because they were aware of their inability to control their drinking, one-half of the subjects made an effort to avoid any social drinking. They drank before they went out and after they came home; they sneaked drinks during the social affair, and tried to avoid any appearance of drunkenness. They all used alochol to relieve stress or tension, to make themselves more comfortable in social situations, and primarily as a sleeping potion. They were unaware of the depressive, addictive nature of alcohol. They felt tremendous shame over their drinking and did their best to conceal the alcoholism from everyone, including their closest family members.

CONCLUSIONS

These four cases illustrate some of the problems which Jewish and Asian people may share in dealing with alcoholism. The following suggested steps can be taken to alleviate the problems of alcoholics and their families

- Gatekeepers must be educated to understand there is no predictable immunity to alcoholism. Jews and Asians, as well as members of other minority groups, may be alcoholic. It is the responsibility of the gatekeeper to explore the possibility that alcohol may be contributing to the problems which clients are presenting. Professionals should be able to recognize the symptoms of alcoholism specifically, chemical coping generally, their effects on the individuals as well as the various systems s/he is part of, aware of intervention techniques, and be able to make appropriate referrals.

- Members of a special ethnic group may avoid going outside of the community for help, so churches, synagogues, schools, and community centers can be used to house treatment programs and self-help groups which are exclusive to that group.

- Problem drinkers may be reached by working with spouses and/or family groups.

- Alcoholics Anonymous may be perceived as representing a Christian value system and thus be unacceptable to non-Christians. Alternative programs can be developed which would be relevant to each specific culture.

- Groups should be available to those who are monolingual in a foreign language.

- Workers should come from within the ethnic group being served. If this is not possible they should be bilingual and have a thorough grounding in the culture of the group they are serving.

- Personnel of existing treatment programs for the general community should be trained to recognize the needs of the different ethnic groups, especially as they relate to diet, cultural, social, and religious values (Kitano, 1980).

The County of Los Angeles has begun to implement some of these steps. There are now two Japanese language Alcoholics Anonymous groups for men. There are Korean language Alanon and Alcoholics Anonymous groups meeting in Korean churches. One Korean church has established a group for alcoholics and their families. One recovery home was set up to meet the needs of several Asian groups by providing volunteer translators to assist in treatment; this facility has been underused by the community it was meant to serve. A refuge for battered women is being established. The Pacific/Asian Alcoholism Commission has been established to advocate for the needs of the community, as well as to educate the community about alcoholism (Casaclang, 1981).

Jewish Family Service of Los Angeles has established an Alcoholism Advisory Board. In addition to introductory workshops for their professional and paraprofessional staff, a workshop was held to train social workers to recognize the possibility of alcoholism in individuals and/or families currently being treated. At this workshop alcoholism treatment personnel in non-Jewish settings were helped to understand the cultural issues involved in working with alcoholic clients who are Jewish. Several synagogues and one Jewish community center now hold Alcoholics Anonymous and Alanon meetings; however, none of these includes a Jewish component in the ritual, and all are open to non-Jews as well. The initial thrust of the advisory board is to educate Jewish professionals and the public about the disease of alcoholism through a speakers bureau and a column in the Jewish press. A roster of recovering alcoholics who are Jewish and willing to do outreach to alcoholics and their families is being set up, and a listing of AA and Alanon meetings at which a Jewish person would feel comfortable is being prepared.

These are only preliminary steps. The Los Angeles community is just beginning to recognize the problems and needs of special ethnic groups and to break through the denial to help alcoholics and their families to full recovery. It would be most useful if other communities in other countries could begin to communicate their concerns, their policies, and plans as well as their programs. An international clearinghouse, concerned with the needs and variable programs for special populations would be of immense help in facilitating effective intervention. Such a sharing may also go a long way in breaking down the geographical notion that "It can't happen here."

REFERENCES

Baron, Salo W., Kahan Arcadium, et al. Economic History of the Jews. Edited by Nahum Gross. Jerusalem: Keter Publishing House, 1975.

Blume, Sheila, Dropkin, Dee, & Sokolow, Lloyd. "The Jewish Alcoholic: A descriptive Study, "Alcohol Health and Research World, United States Dept. of Health & Human Services, Summer, 1980: 21-26.

California Commission Alcoholism for the Spanish Speaking, Fact Sheet, 1978.

Caro, Joseph ben. "Laws of Purim", Sections 686-697 of Orah Hayyim. Shulchan Aruch. Brooklyn, N.Y.: Kroizer Press, 1930.

Casaclang, Rene D. Project Director, Pacific/Asian Alcoholism Commission. Interview, September, 1981.

Cohen, H. Hirsch. The Drunkenness of Noah. University, Ala.: University of Alabama Press, 1974.

Dubnow, Simon. History of the Jews of Russia and Poland(translated by I. Friedlander, 3 volumes). Philadelphia: Jewish Publication Society, 1916.

Glassner, Barry, and Berg Bruce, "How Jews Avoid Alcohol Problems." American Sociological Review, 1980: 637-644.

Jet Magazine, November 5, 1981.

Kaminski, Rosalie. Report on the Involvement of Jewish Youth in Gangs at Fairfax High School, Los Angeles: Hebrew Union College, class report, 1977.

Kitano, Harry. Social Service Need of Asian Americans in West Los Angeles Area. Los Angeles: Assistance League Family Service and United Way, Inc., Western Region, 1980.

Los Angeles County Alcohol Plan, 1981-82. Office of Alcohol Abuse and Alcoholism, 1981.

Phillips, Bruce. Los Angeles Jewish Community Survey: Overview for Regional Planning. Los Angeles: Jewish Federation Council, 1979.

Spiegel, Marcia C. Interviews with Lottie, Hank, Martin, Ike, and Bill, 1977.

Spiegel, Marcia C. The Heritage of Noah: Alcoholism in the Jewish Community Today. Ann Arbor, Michigan: University Microfilms International, 1979.

Spiegel, Marcia C. Workshops and trainings, 1980-81.

United States Department of the Census Report, 1981.

U.P.A.C. Understanding the Pan Asian Client: A Handbook for Helping Professionals. San Diego, California: Union of Pan Asian Communities, 1978

WHEN ARE SUPPORT SYSTEMS SUPPORT SYSTEMS:

A STUDY OF SKID ROW*

Steven E. Hobfoll

Ben Gurion University of the Negev
Beersheva, Israel

Dennis Kelso

Altam Associates

W. Jack Peterson

University of Alaska
Anchorage, Alaska

It has been clearly demonstrated that health disorders, as well as social disorders are, in part, a function of stressful life events. (Dohrenwend and Dohrenwend, 1974). The magnitude of the resultant disorders, however, may well vary or be moderated by the social-environmental conditions in which the individual lives (Dohrenwend, 1973; Myers, Lindenthal, Pepper, 1974; Nuckolls, Cassel, and Kaplan, 1972) as well as by personal characteristics of the individual (Antonovsky, 1979; Murphy, 1974).

Social support has been one aspect of the individual's social-environmental condition which has received increased attention in community psychology. Social support may be defined as social sources which provide guidance, information, assistance with tasks, and comfort and sanctuary during times of stress (Caplan, 1974). In this light, researchers have viewed social support as a moderator of the effect of stressful situations (Brennan, 1977; Dean & Linn, 1977; Hirsch, 1979; Sandler, 1980).

However, if community psychologists are to build competency-enhancing interventions from resources in the natural community (Rappaport, 1977), it will be necessary to de-romanticize community psychology's views of support systems. Specifically, whereas the term "moderator" has been used to describe the role of social networks, it has consistently been used to imply a globally positive moderator effect (Cobb, 1976; Dean & Linn, 1977; Hirsch,

*This study was made possible by a grant from the Division of Social Services, Department of Health and Social Services, State of Alaska and was further supported by the Center for Alcohol and Addiction Studies, University of Alaska at Anchorage.

1979; Sandler, 1980). In other words, social networks are seen as good and helpful, and people that have them are presumed to be better off than those who do not. While this may be true of support systems, it is not true of all social networks, and it is here that the distinction between support systems and social networks becomes blurred.

For example, Hirsch (1980), in an otherwise careful study of social networks, defines "natural support systems" as "the set of presently significant others who are either members of one's social network or affiliated non-mental health professionals." Thus, the social network and the support systems are conceptualized as synonomous and this is a misconception. Certain social networks provide support and assist individuals in coping, others are neutral, and still others may be counterproductive. Only those social networks that assist the coping process should be viewed as supportive.

Support systems, when viewed from the perspective of professional models of intervention, are those which help people cope in terms of the values of the dominant cultures: work, health, and independence. What this disregards is the ecosystem. From the perspective of one ecosystem, some social networks are disruptive (i.e., result in poor coping), but the same social networks may be a support system (i.e., assist in coping), or provide an ecological niche in other ecosystems. In this regard, recent research has recognized that social networks vary in their quality (Alissi, 1969; Hirsch, 1980; Mitchell, 1980), but has not emphasized that quality can only be defined within a given ecosystem. A pond frog has very little of "quality" in its repertoire when placed in a desert; his marvelous adaptability is defined vis a vis his interaction with his environment.

Factors such as network size (Alissi, 1960; Mitchell, 1980) density (Brennan, 1977; Craven and Wellman, 1973; Hirsch, 1980) and multidimensionality of relationships (Hirsch, 1980; Kapferer, 1969) are important concepts in the search for the working parameters of social networks. Each of these critical factors will operate differently depending on the group and the ecosystem under study. In order to integrate an ecological paradigm (Catalono, 1979; Kelly, 1979) with the increasing knowledge about how social networks aid in coping, we would need to posit that these factors would operate under different rules depending on the characteristics of the organism and the nature of the ecosystem under study.

This interaction between organism (which we usually study in groups in the social sciences) and ecosystem may be called "ecological congruence"; the suitability/adaptability of the organism to the requirements of the environment as experienced by the organism. Not only do different ecosystems allow different niches and life cycles but all too often we overlook that organisms possess inherent potentials and needs. In some ecosystems large size or greater density may be facilitative, whereas in others the opposite may be the case. Hirsch (1980), for example, found density of a widow's social network to be negatively related to coping, contrary to most prior research, as it limited individual's support. It makes little sense to speak about social support factors out of the context of a specific ecosystem, and without attention to the needs and potentials of the group under study, at least at this early point of theory development. Hopefully more developed theory should/may clarify some general rules that apply across ecosystems.

The current study, carried out in 1978, focused on members of Skid Row in Anchorage, Alaska. While to the passer-by or general public the Skid Row person may appear to be a totally isolated individual, in-depth knowledge of Skid Row reveals it to be a complex subculture with many working interrelationships.

The very fact that these individuals remain alive and can usually obtain liquor refutes the naive first perception of dysfunctional isolation. The fierce cold, lack of hygienic facilities, presence of physical threat, and very high prices of food and other commodities pose a formidable challange to the most "connected" individual in the Alaskan environment. Historically, the modern form of American Skid Rows emerged from post-Civil War hobohemia (Anderson, 1923) and expanded and progressed to the Skid Row area as we know it today (Lovald, 1963; Bahr, 1969). Most recent changes in Skid Row have been attributed to urban renewal, expanded social security coverage, increasing difficulty in finding day-labor, and advances in technology (Sigal, Peterson, and Chambers, 1975). Rubington (1971) has described the contemporary Skid Row population as consisting of older white men and younger men from minority groups; the entry of younger minority men (and women) being attributed to the shorter "social distance" they need to travel to Skid Row from their already precarious position in society (Littman, 1970; Sigal et al., 1975). Skid Row members have typically been found to have impaired styles of social affiliation (Blumberg, Shipley and Moor, 1971; Spradley, 1973), although minority individuals on Skid Row have been found to have less impaired life-styles than nonminority individuals (greater presence of family and male-female relations) (Brody, 1971; Kuttner and Lorinez, 1967).

In the present study an attempt was made to only study the parameter of contact with the elements of the social network. It was felt that by limiting the investigation in this way that the relationship between contact and support of the social network could be more carefully evaluated in terms of its ecological congruence to the Skid Row environment. This limitation was also in no small part, due to the budget and time considerations, as well as the reality that many other areas of concern were being evaluated in the greater project of which this study was but one of a number of components.

For each of the four stated hypotheses which follow, predictions were based on preexisting knowledge of the urban, Alaskan Skid Row ecosystem:

1.   Greater family contact was predicted to be related to more healthful coping. It was thought that for persons on Skid Row family contact would: moderate feelings of social isolation; increase the chance of existence of shelter; and place demand characteristics for greater employment in order to provide for family needs.

2.   Greater contact with employment was predicted to be related to more successful coping. Aside from the reality that the income producing aspect is in iteself supportive, employment was further seen as creating demand characteristics of responsibility which are inconsistent with other Skid Row behaviors.

3.   Greater contact with professional providers was predicted to be related to less successful coping. Prior research on Skid Row exposed utilization of more than two or three agencies as tending to be manipulative and nonproductive and local agency services were seen as tending to be overlapping and not well intercoordinated.

4.   Greater social contact (friends) was predicted to be negatively related to successful coping. Personal contact on Skid Row tends to be focused on sustaining Skid Row living and as such would not tend to facilitate more successful coping.

Coping was operationalized two ways, both of which related to independent functioning (the authors being aware that independent functioning as a measure of coping is a value judgement). It is possible and perhaps even preferable that coping could be defined in other manners (e.g., survival, ability to get liquor, or even successful manipulation of welfare agencies). Independent functioning was measured by employment status and extent of agency utilization during the prior month. Health-related measures reported elsewhere were associated with coping defined in this way (Hobfoll, Kelso, and Peterson, 1980, 1981). In some cases dependent and independent variables overlapped. This to some extent was the "nature of the beast" and the handling of this is discussed in the results and discussion sections.

METHOD

The Sample

Subjects were 206 members of the Skid Row community sampled from the resident population, the "street," and from agencies serving Skid Row in Anchorage, Alaska during March 1978. Of the study sample, 57.3% were Alaskan Native (Eskimo, Indian, and Aleut), 39.5% were Caucasion, 2.6% were Black, and 5% Asian. Modal marital status was "never married" (47.4%), 39.4% were separated, divorced, or widowed, and 13.2% were currently married or cohabitating. Religious affiliations included protestants (44.6%), atheist or claiming no religion (25.5%), Catholic (19.1%), and Russian Orthodox (9.9%), while 16.9% studied past 12th grade, and 23% did not complete 8th grade. Veterans constituted 44.2% of the sample and there was an usually high percentage of women (18.9) and a younger population than previously reported in studies of Skid Row (modal age was 26-35 years, 36.7%).

Interviewers

Ten paid interviewers were selected from among applicants who possessed considerable professional experience (e.g., nurses, paraprofessional therapists) with Skid Row populations. Interviewers were further trained using role-playing exercises designed to familiarize them with the survey instrument and to ensure more uniform administration. Interviewers were periodically monitored, on-site, by supervisors.

Survey Instrument

A comprehensive survey instrument was verbally administered to all respondents. The schedule is described elsewhere (Hobfoll et al., 1980). Briefly, it consisted of questions relating to demographic, social, family, work, drinking, and agency usage characteristics. Aspects of interviewees' social networks were investigated using one portion of the survey. This section—the Support System Survey—contained 16 items, keyed true-false, that were designed to be relevant and coherent to the sample population. Two of these questions, relating to alcohol consumption, were not analyzed as it was later decided that they did not relate to support systems.

The questions related to behavior during the past week are presented in Table 1. They were subdivided into four subnetworks: family, social, work, and professional help. Most questions clearly fit in these categories. Sexual contact (for men, with women; for women, with men) was placed with family as it involved male-female pairing which was thought to fulfill one function of family. Sleeping on the street was placed in the social sphere as only the most disconnected individuals are forced to sleep on the street.

Coping, as was previously mentioned, was measured in terms of employment status and extent of agency utilization. Three groups, employed, employable, and unemployed, were distinguished from the interview responses. Employed were persons with current, regular work. Employable Skid Row inhabitants had no regular work, but received unemployment compensation.

This required their having worked on a regular basis in the recent past. Unemployed subjects had no regular work and did not receive unemployed benefits.

Agency utilization was measured by interviewees responses concerning their agency utilization over the past month and was divided into four groups (zero, one, two to four, and five or more agencies used). Social services, Indian welfare, alcohol treatment, medical and convalescent care, and Evangelical support agencies were all available to individuals, with many agencies operating competing programs.

RESULTS

The percent of affirmative endorsement of items in the Support System Survey are presented in Table 1. As may be noted, "traditional" family support (spouse [14.8%] and children [9.5%]) was reported in a very low percentage of cases. Sexual interaction (36.5%) was a more common type of "family" support. Sleeping at "home" was quite common (77.4%), but it appears that "home" and "family" are not empirically related among this sample, as was initially assumed. This, in itself, is a sign of the isolation of these individuals.

Professional contact was a more common area of contact. A separate analysis indicated that 28.3% of the sample used zero agencies and 20.25% used only a single agency, indicating that the sample was fairly evenly divided between multiple and non-multiple (zero and one) agency utilization.

Table 1. Percentage Affirmative Responses
to Support System Survey Question Format

During the past week have you	
Family	Yes
1. Been with your spouse?	14.8
2. Been with your children?	9.5
3. Slept with a woman (man)?	36.5
4. Slept at home?	77.4
Social	
5. Hung out on street with friends?	44.2
6. Been to a friends place?	59.5
7. Slept in the street?	10.5
8. Slept at a friends place?	28.4
Professional	
9. Contacted police?	7.4
10. Contacted social service?	30.0
11. Contacted social workers?	23.7
12. Been to a hospital?	26.8
13. Been to a church?	36.3
Employment	
14. Been on job?	27.4

Table 2. Percentage Endorsement in Direction of Predicted Association with Coping for Analysis by Employment Status with Corresponding Chi-square Values for Statistically Significant Items[+]

Social	Employed	Employable	Unemployed	x^2
Hung out on street with friends? (no)	75	54.3	50.4	8.02*
Slept on street? (no)	97.7	100	84.3	11.18**
Professional				
Contact with social services? (no)	84.1	68.6	61.2	7.76*
Contact with social worker? (no)	96.4	80.0	66.9	7.17*
Been to hospital? (no)	81.8	82.9	65.3	6.8*

*p<.05. **p<.01.

[+]Higher percentage suggests more healthful behavior according to the hypotheses stated.

Table 3. Percentage Endorsement in Direction of Predicted Association with Coping for Analysis by Extent of Agency Utilization with Corresponding Chi-square Values for Statistically Significant Items

	Agencies Utilized				
Family	0	1	2-4	5+	x^2
Slept with women? (yes) (for woman, with man)	55	40.5	32.5	20	10.82*
Social					
Slept on street? (no)	100	91.2	89.6	77.1	10.48*
Professional					
Contact social service? (no)	98.7	94.6	54.5	51.4	35.21**
Contact social worker (no)	97.6	94.6	64.9	57.1	29.71**
Been to hospital? (no)	97.6	83.6	64.9	51.4	25.63**
Employment					
Been on job? (no)	46.3	27.0	20.8	20.0	10.06*

*p<.05. **p<.01.

Social contact among Skid Row members was found to be considerable. While Skid Row inhabitants tend to be isolated from general society, peer interaction was still relatively frequent. Many individuals hung out on the street with friends (44.2%) or had been to (54.5%), or slept at (28.4%), a friend's place.

Employment, on the other hand, was rather low among the study sample. Only 27.4% of the sample had worked even once during the past week. This not only affects income, but also utilization of time. Without work there are simply many more hours in the day to occupy with activity.

The Support System Survey was further analyzed using a chi-square technique. Two separate analyses were undertaken regarding the stated hypotheses. In the first analysis, the sample was divided on information gleaned from the interview schedule into three groups based on employment status: (1) those with steady work; (2) those receiving unemployment compensation; and (3) those reporting no steady employment and not receiving unemployment compensation. The three groups were labeled: (1) employed; (2) employable; and (3) unemployed, respectively.

In the second analysis, the sample was divided into four groups with regard to their extent of agency utilization as derived from the interview schedule: (1) zero agencies used; (2) one agency used; (3) two to four agencies used; and (4) five or more agencies used, during the period of the previous month.

Items whose chi-square value reached statistical significance ($p < .05$) in the first analysis are presented in Table 2, scored in the direction of predicted association with coping. Thus, for example, sleeping in the street is coded "no," while being with family is coded "yes." A higher percentage suggests more healthful behavior to the extent the hypotheses are correct. Question 14, regarding employment was not included in the analysis due to its redundancy with the dependent variable of employment status. As may be noted in Table 2 there was a step like relationship between employment status and professional and social contact. Greater evidence of professional and social contact were associated with poorer employment status (coping). Employed persons tended to have less professional and social contact than employables and employables less than unemployables.

Extent of agency usage was negatively associated with having slept with someone of the opposite sex and having been regularly employed, and positively related to having slept on the street ($p < .05$) (3 of 9 significant). It was considered redundant to analyze extent of agency usage during the past month as an independent variable with agency contact over the past week as a dependent variable. However, without implying any causality, it was thought that this analysis might be descriptive of the process of professional contact for Skid Row persons. As may be noted in Table 3, the greater the extent of agency usage the more likely the individual contacted social service agencies, social workers and hospitals ($p < .01$). Again the relationship is steplike, such that for each increase in extent of agency utilization the association with "family" and employment contact was lesser and street/social contact greater.

DISCUSSION

The hypotheses were generally supported. Social network characteristics of the Skid Row population were related to coping as predicted within the model of ecological congruence. Unemployed persons had greater contact with professional services and were more integrated within the street-related, social subnetwork than employed persons. Family life, however, did not seem to differentiate employed vs. unemployed persons. It is possible that the very low overall contact with family may have limited the relationship, i.e., negative ceiling effect. Multiple agency users were less likely to have employment contact or have family contact. They had greater professional contact and were more involved in street life (social contact) than nonagency or single-agency users. It should be mentioned that nonsignificant items were consistently in the predicted direction.

The relationship between multiple agency use and professional contact is somewhat tautalogical. However, a negative relationship would have been found if size of the professional subnetwork would have been positively related to coping. That is, had prior services been effective, multiple agency users may have been still involved in treatment, but should have already been in some stage of recovery and not sleeping on the street or needing acute hospitalization. The relationship is, in any case, descriptive of the social network of high vs. low agency utilizers, and the results clarify the step-like relationship between extent of agency utilization and coping.

Two alternative styles of sustainable social networks appear to be available to this Skid Row population: income creating and agency dependent. These styles, of course, are related to and even derivative of the direction taken by the current study. Other styles may well be found to be operable and equally descriptive. Both styles can support heavy drinking and a sustained presence on Skid Row (that is, continued survival within the ecosystem). Other reports of this study suggest that the income creating style is related to better health and a higher level of functioning. The agency dependent style is more closely related to what we more typically (or stereotypically) consider the public inebriate (Hobfoll et al., 1980, 1981). Both styles may be achieving the goal of the individual, to remain on Skid Row.

While not necessarily typical goals, the reader is reminded that Skid Row life has the attributes of anonymity, cheap rents, permeable boundaries (few membership criteria), nonjudgmental personal relationships, and alcohol provision (those with money often purchase for those who do not and liquor can be shared among "bottle gangs"). In this regard, Brody (1971) has argued that there are special advantages to Skid Row life for the Native American, who were highly represented in the current study. Among these are the ability to drink without negative sanction, enhanced status by coexistence with whites, provision for shelter and food, and an available labor market in which s/he is accepted by employers. Such attributes may well be desirable to certain individuals and must be understood if the rules and dynamics and parameters of the ecosystem are to be accurately described and the Skid Row person's behavior predicted.

The findings of the current study support an ecological approach and suggest that "ecological congruence" may be a valuable explanatory concept within such a perspective. Various groups within each ecosystem will tend to have and make demands related to their needs in different ways. The successful coping of alcoholics, widows, persons with high level of prior success, or psychiatric patients will be related to (ecologically congruent with) different types of support systems and will vary further between environments. Coping strategies will also change in their ecological congruence over time as needs change or properties of the ecosystem shift.

It is reasonable to ask why the four support subnetworks differ in their ecological congruence (behave differently) between ecosystems. Comparing the current study sample to previous samples (see Cobb, 1976; Hirsch, 1980; Nucholls et al., 1972), the outstanding difference is one of deviance from cultural norms. Social support between members of deviant subcultures may tend to support the continuation of the deviant behavior. Likewise, professional agencies become entwined in perpetuating the system of the deviant subculture, both in order to perpetuate their own existence and as a result of manipulation. In the case of family and work there is a limit at which work is denied and the family degenerates in divorce and separation. Maintaining marital relationships, for example, is generally incompatible with most deviant life-styles (serious delinquency, Skid Row, or drug addiction). Steady employment requires (creates pressure toward) honesty, regularity, sobriety, reasonable health, and presentable appearance. All of these attributes are inconsistent with sustained membership in most deviant antisocial subcultures. (The deviant status is attributed by the professional model and it would be interesting to ask what happens if the subculture--the folk model--doesn't identify with the deviant valence.)

In addition to the subcultural deviance, the general functioning level (mental and physical) of the sample under study may effect ecological congruence (suitability of a given support system). One characteristic of psychologically disturbed individuals (not individuals in crisis) that has been noted previously is an impairment in their use of settings for routine social relations (Goffman, 1971). Individuals in different ecosystems, at different levels of functioning will have different hierarchies (i.e., food before shelter, shelter before sex) of needs. Beyond alcohol and drugs, needs for shelter, security, and comfort were seen a priori to be of primary concern to the Skid Row population. These needs must be met before individuals can be concerned with socializing, self-expression, and aesthetics (see also Cooper, 1979).

The tendency of certain aspects of the social network to behave as providers of support (supportive systems) within this ecosystem yields information on which to begin to build treatment recommendations, including the relevant training needs of intervention agents. Continuing to do more of the same (a greater number of agencies) does not appear to be legitimate or functional. In fact, "treatment" itself, as it has been generally defined and permitted to operate, may be less well suited to the level of hierarchy of needs of the sample. The Skid Row street people might gain more by the provision of shelter, hygienic facilities, and nutritious food, especially in the case of the agency-dependent group. Efforts to make nonexploitive occasional labor available might be the first priority of "treatment" after health/survival needs have been met. The income-creating group may not require "intervention" at all, as they do not seem to be choosing to be "treated." It is possible, however, that they may have evaluated the available treatment alternatives and judged them unsuitable. An overall intervention strategy might provide the health/survival related facilities, and then concentrate remaining resources on fewer, better-funded agencies to meet the treatment of persons suited and interested in recovery and rehabilitation. Indeed, a functional interface between intervention efforts which are based on and associated with professional and folk/street models might be the most efficient and effective choice.

The current study has a number of limitations. The Support System Survey, being used for the first time, is of unknown reliability and subnetwork representation among items is disproportionate. It is also questionable as to whether support from persons and from institutions are parallel [i.e., it is not known if one personal contact is equivalent to one agency contact (where a number of individuals might be involved)]. In addition, there was an overlapping of independent and dependent variables. However, this entanglement may be an aspect of the nature of the relationship of support systems to coping, as supportive contacts contribute to coping and are themselves a measure of coping. For example, employment contact is supportive and it is also a measure of the individuals' ability to successfully cope. This is not simply a chicken-and-egg dilemma, but rather a spiralling effect. By and large those who have more support cope better and, in turn, can seek more support and further improve successful coping. The nature of this spiral would be a fruitful area of future investigation.

FUTURE NEEDED RESEARCH

Future social network research should clearly define: (1) level of functioning; (2) needs of the sample being studied; and (3) special

properties of the ecosystem in order to make predictions regarding the
ecological congruence of the social network to the ecosystem. Whereas size,
density, multidimensionality, and subnetwork interaction of social networks
have been shown to be related to coping in past research, some of the
apparent inconsistencies across studies may be resolved by giving attention
to ecological congruence. These factors will behave differently between
ecosystems to the extent that at ecological model accurately applies to
support system theory.

ACKNOWLEDGMENTS

A special thanks to Professor Aaron Antonovsky who read and commented
on earlier drafts of this paper. Special thanks also to Dr. Bernie Segal
for his consultation during the study and write-up periods.

REFERENCES

Alissi, A. S., "Social work with families in group service agencies: an
     overview." Family Coordinator, 1969, 18: 391-401.
Anderson, N. The Hobo: The Sociology of the Homeless Man, Chicago:
     University of Chicago Press, 1923.
Antonovsky, A. Health Stress and Coping. San Francisco, Jossey-Bass, 1979.
Bahr, H. M. "Lifetime affiliation patterns of early and late onset of
     heavy drinking on Skid Row." Quarterly Journal of Studies on Alcohol,
     1969, 30: 645-656.
Blumberg, L. V., Shipley, T. E., and Moor, J. D., "The Skid Row man and the
     Skid Row status community (with perspectives on their future)."
     Quarterly Journal of Alcohol Studies, 1971, 32: 900-941.
Brennan, E. Social networks, support, and coping in older students.
     Unpublished senior honors thesis, University of Massachusetts at
     Amherst, 1977.
Brody, H. Indians on Skid Row, Ottawa, Canada: Department of Indian
     Affairs and Northern Development, 1971.
Caplan, G. Support Systems and Community Mental Health. New York:
     Behavioral Publications, 1974.
Catalono, R. Health, Behavior, and the Community: An Ecological
     Perspective. New York: Pergamon Press, 1979.
Cobb, J. "Social support as a moderator of life stress." Psychosomatic
     Medicine, 1976, 38: 300-314.
Cooper, C. Easter Hill Village. New York: Free Press, 1979.
Craven, P., and Wellman, B. "The network city." Sociological Inquiry,
     1973, 43: 57-88.
Dean, A., and Linn, N. "The stress buffering role of social support:
     Problems and prospects for future investigation." Journal of Nervous
     and Mental Disease, 1977, 1965: 403-417.
Dohrenwend, B. S. "Social status and stressful life events." Journal of
     Personality and Social Psychology, 1973, 28: 225-235.
Dohrenwend, B. S., and Dohrenwend, B. P. Stressful Life Events: Their
     Nature and Effects. New York: Wiley, 1974.
Goffman, L. Relations in Public. New York: Basic Books, 1971.
Gottlieb, B. H., and Todd, D. M. "Characterizing and promoting social
     support in natural settings." In R. F. Munoz, L. R. Snowden, and J. G.
     Kelly (Eds.), Social and Psychological Research in Community Settings.
     San Francisco: Jossey-Bass, 1979.
Hirsch, B. J. "Natural support systems and coping with major life changes."
     American Journal of Community Psychology, 1980, 8: 159-172.
Hobfoll, S. E., Kelso, D., and Peterson, W. J. "The Anchorage Skid Row."
     Journal of Studies on Alcohol, 1980, 41: 94-99.
Hobfoll, S. E., Kelso, D., and Peterson, W. J. "Human service strategy in a
     urban Skid Row." Journal of Alcohol and Drug Education, 1981, 26: 832-838

Kapferer, B. "Norms and the manipulation of relationships in a work context." In J. C. Mitchell (Ed.), Social Networks in Urban Situations. New York: Humanities Press, 1969.

Kelly, J. G. The High School: Student and Social Context in Two Midwestern Communities. Hillsdale, N. J.: Lawrence Erlbaum, 1979.

Kuttner, R. E., and Lorinez, A. B. "Alcoholism and addiction in urbanized Sioux Indians." Mental Hygiene, 1967, 51: 530-542.

Lavald, K. A. From Hobohemia to Skid Row: The Changing Community of the Homeless Man. Unpublished doctoral dissertation, University of Minnesota, 1963.

Littman, G. "Alcoholism, illness, and social pathology among American Indians in transition." American Journal of Public Health, 1970, 60: 1769-1787.

Maslow, A. H. Toward a Psychology of Being. Princeton: Van Nostrand, 1968.

Mitchell, R. E. Social networks and psychiatric clients: The personal and environmentalcontext. Unpublished manuscript, Stanford University, 1980.

Murphy, L. B. "Coping, vulnerability, and resilience in childhood." In G. V. Coelho, D. A. Hamburg, and J. E. Adams (Eds.), Coping and Adaption. New York: Basic Books, 1974.

Myers, J. K., Lindenthal, J. J., and Pepper, M. P. "Social class, life events, and psychiatric symptoms: A longitudinal study." In B. S. Dohrenwend and B. P. Dohrenwend (Eds.), Stressful Life Events: Their Nature and Effect. New York: Wiley, 1974.

Nucholls, K. B., Cassel, J., and Kaplan, B. H. "Psychosocial assets, life crisis, and the prognosis of pregnancy." American Journal of Epidemiology, 1972, 95: 431-441.

Rappaport, J. Community Psychology: Values, Research, and Action. New York: Holt, Rinehart and Winston, 1977.

Rubington, E. "The changing Skid Row scene." Quarterly Journal of Studies on Alcohol, 1971, 32: 123-125.

Sandler, I. N. "Social support resources, stress, and maladjustment of poor children." American Journal of Community Psychology, 1980, 8: 41-52.

Sigal, H. A., Peterson, D. M., and Chambers, C. D. "The emerging Skid Row: Ethnographic and social notes on a changing scene." Journal of Drug Issues, 1975, 5: 160-166.

Spivak, M. Archetypal Space. Cambridge, Mass.: Laboratory for Community Psychiatry, Harvard University, 1971.

Spradley, J. P. Public health services and the culture of skid row bums. Unpublished manuscript, Macalester College, St. Paul, Minnesota, 1973.

FAMILY TREATMENT OF DRUG ABUSERS

David K. Wellisch

Department of Psychiatry
UCLA School of Medicine
Los Angeles, California

TREATMENT OF ADOLESCENTS AND YOUNG ADULTS

To properly address the family treatment of the early adolescent versus that of the late adolescent/young adult drug abuser, a brief excursion into the developmental issues of these two stages becomes necessary. Although these two periods are separated by a small number of years, they are separated by important and pivotal developmental milestones. Of primary importance is the "second individuation process" (Blos, 1967) wherein the adolescent reworks earlier developmental individuation/separation issues. At this point, the adolescent is struggling with broad fluctuations in dependency needs and counter-dependency needs in relation to parents and is getting used to the concept of variable states of emotional dependence on a significant other. Needless to say, if parents cannot tolerate the side of the coin that involves counterdependence, real individuation and separation is hampered. If the parents' needs are pathological, the adolescent can remain faithful and loyal under the guise of separation by developing a real drug problem, which will endlessly repeat the dependence-counterdependence cycle, a process well described by Stanton (1978).

Important in the transition between early adolescence and young adulthood is the capacity to move from fantasized love relationships to attachments outside the family where intimacy and commitment are possible. Here again, the parent-adolescent-drug abuser triad blocks this process. The adolescent is woven into the structure of the marital relationship and does not have "permission" to develop emotional progress toward intimacy outside of the family.

Finally, the issue of realistic assessment by the adolescent of his/her capabilities to develop a role in the world outside the family, especially in regard to work, becomes very important.

Thus, for the early adolescent the primary task of family therapy is toward <u>functional integration</u> within the family, where the fluctuating dependency needs of the adolescent can be lived with by all concerned. Within this context the ego-centricity of the early adolescent is of special concern. For the older adolescent/young adult <u>functional separation</u> becomes the central issue. The primary task of the family therapy is the resolution of binds or loyalty conflicts that might impair or delay this crucial process. The adolescent, still mired in ego-centricity, has less ability to do this. Thus, family therapy with the younger adolescent will attend, by necessity, to how his/her behavior impacts on others in the family. Drug abuse can be one symptomatic face of extreme ego-centricity and failure to progress beyond the developmental status of early adolescence. The older adolescent is potentially also more able to see how others in the family impact on him/her. In addition s/he may be more artful in not being swept u and integrated into family and marital conflicts. For the younger adolescent, although this is often far less clear, it is nevertheless desirable to work on in family therapy.

For the older adolescent, a successful leaving of the home will require realistic arrangements which do not doom the leaving to failure, as a disguised way of continuing to balance a pathological family homeostasis. While success in leaving home is a theoretical "someday" issue for the younger adolescent, it is a very real "here-and-now" issue for the young adult filled with anxieties and fears. We have seen several young adults become so fearful that a pathological regression sets in, prompting a family response which symbolicaly turns back the clock and delays the separation. We have usually taken a firm stance in these circumstances that the notion of turnning the clock back is a dangerous illusion for all concerned. The older adolescent with a drug abuse pattern who fails to separate is at risk to beome more "hard core" over time without developmental progress. On the other side of the coin, we have seen several cases where older adolescents and their families dealt with the leaving home process by arranging a fight so as to fend off the pain of dealing head on and appropriately with the pain and meaning of the separation. Here again we are leery of this type of pseudo-separation, which can be a way of reuniting when feelings become less angry or negative. This reuniting provides the possibility that the pathological, drug abuse maintaining, family interaction will resume exactly as before.

A case example may serve to illustrate some of these points, expecially in regard to the difficulties of separation for the older adolescent. An 18-year-old polydrug abuser and his family had been in treatment for almost two years. Much had changed in the family, but more pivotal in the change process had been the parent's marital relationship. Initially the father had felt isolated from both his wife and his whole family. Through family therapy the father had come to feel more involved with both his sons, but especially with his wife. During the course of family therapy he had shown emotions that no one in the family had seen previously. He revealed portions of his life history that no one had heard, and he ceased to rage impotently against not only his former drug-using son but the others in the family as well. A crucial point in the last phase of the family work was a trip he and his wife took for six weeks back to his country of origin in Europe. This was a step that involved intimacy and sharing with his wife at many levels previously not experienced. The trip went well. The couple returned evidencing a more profound bond. However, the former drug-abusing son was now found to be dressing in his mother's clothes during home visits from his own treatment. Three steps were taken by the family therapists. First, this curious phenomenon was reframed as a sign that a loving but somewhat confused son was attempting to get close to his mother by wearing her clothes. Second, the mother was encouraged to help her son by remaining uninvolved in this unique but loving piece of behavior, thus resisting the need to recapture the symbiosis with her son, now reduced and directed, in a modified way, toward her husband. Third, this was not defined as prima facie evidence that the family work was not progressing, but defined as an interesting statement on the son's part in regard to the shifts and impending separation in the family. The son was given support to deal with his feelings of separation more directly than attempting to capture his mothers external skin. This process continues at the present time.

CRUCIAL ISSUES IN FAMILY THERAPY SUPERVISION

As with all other forms of psychotherapy, the therapist enters the arena of family therapy with his/her own internal life replete with conflicts, unfulfilled needs from the past, extensions of unfulfilled needs from the past reconstructed in the present, unresolved losses, and various islands of insecurity which can rapidly coalesce into continents of problems within the family therapy context. By and large, the family therapists at our agency are in their mid-20's, have had little or no therapy of their own, and enter the family therapy arena armed with degrees of certainty that the techniques they have learned, if properly applied, will carry the day. We have learned that it is well worth everyone's effort to go beyond the development of intervention strategies in supervision to explore what processes and issues are activated in the therapists by the families. We have found, for example, that no matter how well developed the particular intervention strategy might be, if the therapist has countertransference problems the strategy usually cannot be effectively applied in family therapy.

The countertransference problems invoked in the therapist working with families of drug abusers extend well beyond the notion of "He reminds me of my father" or "She reminds me of my mother." We are now talking about the summative effects of countertransference feelings in response to:

1. Individuals,

2. Relationships (i.e., between marital partners, between siblings, or between parents and children), or

3. Aspects of those relationships which may be quite subtle but very disturbing.

In terms of the third point, one of our family therapists reported "My buttons get pushed and I go out of whack" when a set of parents would shift their expectations toward their child in rapid order from the impossible to none at all. This invoked in her a blind panic which we traced to her own family and their style of dealing with her. Another trainee experienced a particular urgency to connect and "work things out" between two siblings which was something the trainee was in great need of but which the siblings could live without at that point. All these siblings seemed to be getting from the trainee was his feelings of urgency. In supervision, it became evident that this trainee had always been troubled by a sibling who promised a supportive, gratifying relationship but never delivered it. As a result, when this trainee found the opportunity to work this unresolved empty place out with this family, he rushed to do so, not in response to their needs but for his own needs. Another traditional problem for family therapists, heightened with families of drug abusers, is the tendency to move from the interactional field to more intensive interactions or interventions with one member. When this is not planned, it can often represent a successful resistance by the family which is mounted through the delegated member. Such a process is workable given the therapists proclivity to get seduced by this particular member. This can come from therapist needs to fight, punish, capture, or even retreat from a particular family member who invokes special feelings in that therapist.

One of the main foci of supervision is to enable the treater to "see" the family. As with a more powerful microscope lens, as represented by the supervisor, the picture becomes clearer and more rational and a reasonable intervention plan can be constructed in regard to the families needs and processes.

"Therapist-paralysis/stage fright" is a similar phenomenon to that described by Masters and Johnson (1966) in their description of the etiolog of male erectile dysfunction. They describe a performance anxiety stemming from the performer assuming a spectator type (internal) role and watching himself perform, with the spectator filing a negative report in regard to the proceedings. The family therapist, especially in working with the ofte emotionally volitile drug abuser's family, will focus on his or her fantasies, needs, and internal (imagined) criticisms to an extreme degree. As a result he/she cannot solve the problems at hand in the external field because he or she cannot really attend to them. Part of this stems from a lack of experience in how much weight to carry in the therapist/family interaction. We have found that most families come to therapy with the notion "All right, we're here, so tell us what's wrong and give us the answer on what to do." If the therapist accepts this message from the family, then he/she is carrying all the weight and the family none. In supervision, the therapist works out how to tell the family that they will generate answers by what they do in therapy, and the therapist works out his/her feelings that this is the most appropriate intervention strategy. The overwhelming tendency for the young or novice family therapists is to pick up this demand by the family and run with it, increasing the sense of paralyzing weight and responsibility. Families where parents assume an omnipotent or condemnatory stance toward their children can often invoke the same feelings in the therapists. They seem to say "You mean you don't know the answer, are you sure you're not too inexperienced to deal with us?"

Therapist's self-management of potentially paralyzing anxiety in the family arena can be managed in a number of ways:

1. We have had supervisors work in conjunction with new therapists for periods of time.

2. We have worked through one-way mirrors and,

3. We have worked out careful intervention plans to be put into place by trainees independently.

Trainees have often reported feeling like new jet pilots trying to lanc a jet at sea on the deck of an aircraft carrier. They have a plan, but things are moving so fast that they can't see and the damned target won't hold still so they can make the planned contact. We encourage trainees to slow down, expecially in terms of their own self-induced demands to be smooth and expert so they can begin to see and can establish contact. Family therapy, like a jet plane, can be the shortest, fastest way to get between two points but is a powerful vehicle that can get out of control if the pilot panics or freezes. Following the same analogy the relationship between the therapist and the supervisor has a dynamic life of its own and must be attended to. If, like the jet pilot, the family therapist misses the deck after the first pass, the supervisor needs to acquaint the therapist (pilot) that the deck (family) is still there, the same as before, for the second pass. Our families have been remarkably kind and patient in presenting us over and over with the same interactional patterns and problems for us to aim at. If the supervisor pressures the therapist-pilot, he or she will build up more internal pressure and not calm down enough to "see" the field of interest with more clarity. If the supervisor demands, covertly or overtly, that the therapist please him or her, this can obviously be counterproductive. The most desirable situation is to create a plan together that enables the supervisee to relax enough to follow through with the intervention after having seen and heard both the plan, on one hand, and the family on the other.

Close on the heels of therapist analysis comes the other side of the coin which is premature closure or "jumping the gun" with prescriptions. The same internal element in the therapist dictates this behavior as with previously mentioned paralysis, namely an attempt to bind anxiety. Neophytes often jump the gun and become prescriptive (as do many noteworthy, stellar family therapists) before they even know how the family structure works or what the real problem and issues are with and/or for a family. Thus, they are really prescribing to their own needs to reduce anxiety, to achieve a sense of mastery and closure, and to "know", rather than prescribing to the needs of the family. Again, the problems presented by the drug abuser can be laden with a sense of life and death urgency that accentuate this need in the therapist. For example, a trainee watched Braulio Montalvo move between a mother and her daughter in a videotape, achieving a particularly touching and heroic goal in a remarkably short time. When asked why he repeated the same maneuver as Montalvo, the trainee replied "This family is similar to the one Montalvo worked with, and what is good enough for Montalvo is good enough for me!" Here the trainee applies a good-looking maneuver with no real underlying reason to do so beyond the need to (1) Look like Montalvo; and (2) Quickly close the book on a complicated, emotionally charged interaction. Perhaps the most important messages here from the supervisor to the trainee are: (1) Link the intervention to the needs of the family in an appropriate time sequence; and (2) be a reasonable approximation of yourself vs. a lousy approximation of Salvador Minuchin or Braulio Montalvo.

Finally, the issue of emotionality generated by families of drug abusers is a very serious issue to be considered in supervision. Dennis Reilly in his classic chapter entitled "Drug Abusing Families: Intrafamilial Dynamics and Brief Triphasic Treatment" attends to several of these emotional issues in depth, (1979).

Two emotional themes of particular importance discussed by Reilly, which potentially reverberate off the therapist are: (1) Miscarried expressions of anger; and (2) Impaired mourning. To quote Reilly in regard to miscarried anger:

> Family members already feel deprived of love, affection, and attention, and this state of emotional deprivation creates an enormous sense of rage and frustration. However, the very rage is suppressed and repressed for fear that it will lead to further rejection and loss of love, or to complete loss of control over potentially murderous impulses. (1979, p. 118)

We have often seen the solution to this, albeit maladaptive, be a family/wide sense of chronic depression or of unbroken boredom which secures these angry feelings in place just out of family consciousness. This reverberates to the therapist who also develops depression and boredom after spending time in the family territory. If this becomes the dominant emotional theme of the interaction, needless to say, the therapy cannot be expected to prosper.

Of equal importance in Reilly's conception and in our own experience is the issue of impaired mourning in families of drug abusers. To quote from Reilly:

> The parents of youthful drug abusers have often suffered profound emotional losses within their own families or origin. They have a strong sense of having lost their own parents via death, divorce, rejection, or neglect. The conflicts over this loss have never been worked through; mourning is incomplete and the love/hate ambivalence so characteristic of such relationships is never resolved. The family of origin are never given up, and the

individual is never able to transfer affections fully to new love objects (such as the spouse and children of the family of procreation.) (1979, p. 119)

These are emotional dynamics that underlie the fierce collusive fusions in these families. Here again lies an extreme paradox for the therapist. When the work of the therapist has been skillful and decisive enough to move the family into a position where these pathological fusions begin to give way, this opens up the Pandora's box of repressed grief and pain buried under the protective barrier of these fusions. The reward, if it is one, is the release of a profound level of emotions which promise the possibility of change in the long run, but can be equally as frightening to the therapist as to the family in the short run. We have had the experience of more than one therapist in supervision saying "What have I done, is this really a help to this family?" The ability to remain objective, and not become caught up or mired in the evolution of the mourning process can become a major step in the therapist's intervention. This requires careful attention and support in supervision.

MARITAL INTERVENTION

Marital intervention in the drug abuse field can theoretically take place within two different marital situations which are:

1. Marital work with two drug abusers who are married, or with marital partners where one is drug involved and the other is not or,

2. Marital work with parents of a youthful drug abuser.

This section will focus on the steps toward marital intervention in the second circumstance, namely that of work with parents of a youthful drug abuser.

In detailing the steps in this complicated and lengthy intervention process, certain fundamental theoretical conceptual assumptions must be made. These form the bedrock of current family systems theory. They are:

1. The drug abuse of the offspring is symptomatic of disturbance in the wider family context;

2. Only with change in the wider context will this sympton remit with any real certainty; and

3. The marital subsystem has been shown by a vast number of researchers to usually reflect disharmony, and be woven into the fabric of many symptoms such as drug abuse exhibited by adolescents and young adults, with these symptoms protecting the fragile marital interaction.

The approach to a meaningful marital intervention for the parents of a youth drug abuser defies the geometric axiom that the shortest distance between two points is a straight line. In this case, the shortest distance to the marital partners is a triangle which first focuses on the child and his/her problems and involves joining at surface level with the parents.

Marital therapy begins with the "formation of a partnership" as described by Minuchin "to free the family symptom bearer of symptoms" (1981). This necessitates defining and building a common ground between family (or subsystem, in this case marital) and therapist. This, by definition, calls upon the therapist's use of self in innovative ways. At this point the therapist joins the marital partners in a common endeavor, but not the marital system, per se. To attempt to join and work directly with the marital issues prematurely in families of youthful drug abusers is the shortest route to nonsuccess and early termination of the family. Minuchin describes three positions of joining: (1) close position; (2) median position; and (3) disengaged position.

At this point, the "close position" is necessary, wherein the therapist will join with selected family members where the currency of affiliation is confirmation. This is a confirmation of their stresses and of their pains, expecially in terms of their difficulties with their offspring. No questions are asked at this point about other pains in their lives, past or present. The therapist, at this point, can also assume the median position of joining, which consists of active listening. The key ingredient here is "family tracking" where the therapist learns how the system works, including the marital subsystem. Minuchin makes clear the point that "joining is not a technique that can be separated from changing a family; the therapist's joining changes things."

Defining the Problem with the Child

In the median stage of joining, the therapist is an active listener "who helps people tell their story." This allows, then, their view of the child's problem, also yielding information on transcultural maintenance of the problem. At another level, the therapist now begins to become more involved with the couple. The therapist here learns their view and enters their reality.

Unifying Around the Problem with the Child

The therapist can now elect to unjoin, having previously worked at joining with the couple. This, be definition, forces the couple to continue to interact with each other around an issue (this being the child's problem). The therapist, at this point, can either block or facilitate the therapeutic exploration of the spouse's interaction. We are wary of exploration of interactional issues beyond the child at this point. However, at this point, we are very interested in how the couple interacts in relation to the child, especially in terms of aborted or effective sequences of communication about the child's problem. The issue here is unification around the problem with the child, and what it is going to take to change things. This will include now:

1. Work in effective communication;

2. Learning how each views the problem;

3. Spouses educating each other on their views; and

4. Spouses beginning to develop alternatives and possibilities toward remediation of the problem.

DEVELOPING AND EXECUTING CHILD FOCUSED INTERVENTIONS

All that has come before now culminates at this stage. Here the therapist will begin more active intervention with the couple as they interact around the child's problem. This is a chance for the therapist to actively intervene with the couple, which sets the tone for real marital work, but still does not focus on marital issues themselves. As the self-defeating interactional patterns emerge, it can be made clear to the couple that more purely spousal issues and feelings underlie their difficulties as parental figures. Interactions with the child are now profoundly reworked. In-depth planning of these intervention and pattern shifts can be worked out with parents at this point. For example, a frequent shift we often recommend is that when the school (or other civic authority) is displeased with the child, rather than the mother being called automatically, the father is notified.

Joining at a Deeper Level

At this point, the couple is engaged in an often frustrating process. They have been involved in developing and implementing strategies which can only work if they cooperate with each other. They can only, at this point, cooperate and interact within a narrow range, given their marital patterns and difficulties. Nevertheless, the development and implementation of a new strategy with their drug abusing child is of no small concern and can offer relief of no small consequence. However, in many cases we have seen, the couple at this stage clearly begins to see, with relatively little impetus from the therapist, that they have deeper conflicts which they each bring to the interaction. These conflicts, which culminate in the interactional difficulties, are highlighted by their difficulties in dealing with their drug/abusing child even with help. In many cases the couple is or may be more than satisfied with the symptomatic changes. They can begin to work together differently, especially if and when fathers are ready and willing to be more involved. The dividing line for this seems to be the early childhood experience of the couple. The more deprived they were in their own families of origin, the less they are able at this point: (1) to either effectively interact in implement a new strategy; or (2) to go into the needed depth of work on the relationship which will bolster their ability to help their child.

As such, in multiple family work, we have catagorized families into two groups, the low and high functioners.

This process will occur with some couples; with others it will not. It is the predecessor to real involvement in marital issues. The last stage sets the time for this to occur. The frustrations of not being able to effectively work together, seem to bring out the frustrations of things never having worked; in the marriage and in the respective spouses families of origin. At this point the stage is set for joining at a deeper level in which the therapist has access to the more profound feelings of each spouse and of the couple.

Recontracting

The time has come to make overt what has previously been covert. A new contract must and can be established. Marital issues are to be dealt with at this point. The overt understanding is that a child focus is no longer necessary, although the child will potentially benefit from this recontracting. For many couples this phase is not desirable or possible, for some it is.

PHASE II

Boundary Making

In spite of the fact that a new contract has been made, the pull to reinclude the buffer of the "identified patient" and his/her symptoms can be very strong. The pull from the direction of the "identified patient" by an upsurge of symptomatic behavior can be extreme. We have seen wrist slashing and other forms of suicidal power tactics designed to regressively recalibrate the triangular system. In one case, a 17-year-old bolted from residential treatment when his parents had made such a contract. The mother

Table 1. Phases and Issues in the Process of
Marital/Family Therapy with Parents of Youthful Drug Abusers

Defining the problem with the child
Unifying around the problem with the child
Developing and executing child focused intervention
Joining at a deeper level
Recontracting
Boundary making
Unbalancing
Intensity
Creation of reciprocal empathy

(a nurse) had changed her shift (after more than a decade) to be able to be with her husband. There is the therapeutic need to establish or delineate boundaries by the therapist. The therapist will generally want to create more mother/child distance. Minuchin terms this "handicapping the child's detouring strategy." Often at this point, the child should be excused not only psychologically, but also physically from the sessions.

Unbalancing

Boundary making aims at changing subsystem membership (i.e., disincluding the child from within the marital subsystem). The point of this is to provide individual members with expanded roles in relationships. Very often, in families of youthful drug abusers, the father's role is circumscribed and his is defined by the family as "out." Unbalancing would widen the father's "narrow" presentation of himself as defined by himself and the family. With a change in patterns of transaction between spouses there are changes in the perspective of the parents in relation to the child.

Intensity

In families of drug abusers, the drug abuser supplies, contains, and suffers from an overload of intensity. This function, if it is not to continue to be carried by the "identified patient," must be developed and tolerated by the marital subsystem. The therapist must repeat interactional sequences and be "immovable" in his/her stance that these sequences not be short circuited by the couple.

Creation of Reciprocal Empathy

After the couple has survived (emotionally) the increased intensity, they are prepared to survive the increase in feelings other than conflictual ones. They can then begin to learn and understand what each of them brought into the marriage and how this influences the other. This marks the end of a long course of therapy that began with the drug abuser but ends in the marital subsystem of the family

REFERENCES

Blos, P. "The second individuation process of adolescence." In R. S. Eissler et al. (Eds.), The Psychoanalytic Study of the Child, Vol. 22. New York: International Universities Press, 1967.
Masters, W. H., and Johnson, V. E. Human Sexual Response. Boston: Little Brown and Co., 1966.
Minuchin, S., and Fishman, H. C. Family Therapy Techniques. Cambridge: Harvard Press, 1981.

Reilly, D. M. "Drug abusing families: Intra-familial dynamics and brief
    triphasic treatment." In E. Kaufman and P. Kaufmann (Eds.), Family
    Therapy of Drug and Alcohol Abuse. New York: Gardiner Press, 1979.
Stanton, M. D., , Todd, T. C., Heard, D. B., Kirschner, S., Kleiman, J. I.,
    Mowatt, D. T., Riley, P., Scott, S. M., and Van Deusen, J. M. "Heroin
    addicts as a family phenomenon: A new conceptual model." American
    Journal of Drug and Alcohol Abuse, 5:125-150, 1978.

FAMILY THERAPY: A TREATMENT APPROACH WITH SUBSTANCE ABUSERS

IN INPATIENT AND RESIDENTIAL FACILITIES

Edward Kaufman

University of California, Irving Medical Center
Orange, California 92668

Family therapy is more easily done in residential settings for abusers than in outpatient settings because the family is a captive audience that can and should be required to participate in therapy if any contact is to occur. It is also essential to engage the family in treatment because if they are not a part of the solution they will frequently pull the patient out of the treatment setting. In addition the patient will frequently duplicate patterns in treatment which originate with the family and can only be recognized, understood, and changed if the patient is seen in the total family setting. Family therapy is presently being used by most substance abuse treatment agencies. A recent survey of 2012 agencies involved with treating substance abuse found that 93% provide some type of family therapy, and 75% of these agencies include the client and his or her entire family (Coleman, 1978). Additionally, 62% provide couple therapy and 36% group therapy for clients' parents (Basen, 1977). In drug abuse programs, the emphases have broadened so that both types of programs are viewing family therapy as involving three generational systems. Family treatment of substance abusers is a complicated process. In meeting the needs of the family as an entity, the spouse and parental subsystems, the sibling subsystem, and the individual needs of each person in the family must be considered. The therapy of each of these three systems must interlock and work in harmony. This can be done if the same therapist or therapeutic team treats all parts of the family. If not, there must be close communication between therapists.

It is not suggested that all the problems of the substance abuser can be solved through family therapy. In many cases, family therapy alone is not enough. In most cases, it is a valuable or essential adjunct to other modalities.

Although many substance abusers and their counselors may claim that there is no family or the family is not available, Stanton and Todd (1981) have demonstrated that in 77% of cases they were able to recruit both parents or parent surrogates of heroin addicts in a methadone program. This is generally a group where families are rarely available. Their success demonstrates that with concerted effort families can be brought into treatment. There is almost never a countraindication to a total family evaluation. The more family members the therapist sees, the better he/she is able to understand and change the family. After seeing an entire family the thrust of family work may be with the individual, his/her parents, spouse or children, but each unit may require additional contact and work from time to time. Families frequently must be brought together to help them separate. I have only once asked a nonaddicted family member to not participate in treatment. This was the mother of a heroin addict who disrupted a multiple family group by imposing her religiosity on the entire group. Four months later her son left residential treatment to return to her -- a consequence which probably could have been avoided if individual family therapy had been available. Substance-abusing family members other than the identified patient (IP) are a difficult problem in family therapy and will be discussed later.

GENERAL PRINCIPLES OF FAMILY TREATMENT

If the substance abuser is habituated to drugs or is unable to attend sessions without being under the influence of a chemical, then the first priority is getting him/her off the substance, at least temporarily. It is difficult to initiate family therapy unless the regular use of chemicals or a pattern of habituation is interrupted. Hospitalization for detoxification, long-term residential treatment in a therapeutic community for drug habituates or a 30-day hospital program for alcoholics provide a drug/alcohol free atmosphere in which individual and family treatment can be implemented. Thus initial family treatment with a chemically dependent IP is directed toward the patient's entering whatever setting is necessary to provide an inital drug-free environment. Frequently, intensive treatment of substance abuse cannot begin unless the family and significant social network, particularly the employer, are involved. Several authors, notably Bowen (1974) and Berenson (1976), have suggested approaches which permit work with the family while the IP is still drinking. However, treatment which takes place withing the "wet" or drug-abusing system runs the risk of perpetuating that system by offering a pretense of help while no actual change occurs.

Since spouses of alcoholics frequently present without the alcoholic, it is helpful to have a system to deal with the spouse. The spouse is offered three choices (Berenson, 1976): keep doing what you're doing, separate emotionally, or separate physically and is asked to join Al-Anon or a significant others group. The spouse is then helped to see that he/she is not helpless and can work at developing the choice between three alterntives with the help of a support group and family-oriented individual psychotherapy.

When another serious substance abuser is present in the family but not viewed by the family as a problem, it is helpful to relieve the symptom of the IP before another member's substance abuse is made an issue. At times the non-IP's substance abuse is focused on by relating it to the IP's symptoms; for example, parental alcoholism can be related to a teenagers's drug abuse as a means of separating the parents so that no limits are set on drug use. This must be done after the therapist has joined with the entire family and/or provided some focus on or relief of the IP's symptoms or the parent alcoholic will take the entire family out of treatment.

FAMILY SYSTEMS AND STRUCTURES OF SUBSTANCE ABUSERS

The first major step in the treatment of substance abusers is a diagnosis of the existing family systems and structures. This diagnosis is essential regardless of the therapist's approach. The family structures of drug abusers was the subject of a prior publication by this author (1981) and will be summarized briefly. In this study, 88% of mother-drug dependent child relationships were symbiotically enmeshed as were 41% of father-child relationships. The father is also frequently excluded and reacts with disengagement (43%), brutality, or increased consumption of alcohol. Siblings may either be fellow addicts whose drug abuse is fused with that of the identified patient or parental children, many of whom are quite successful. Male addicts frequently dominate their addicted or drug free spouse to assure themselves of being cared for. Mutually addicted spouse pairs are particulary difficult to treat even when both are involved in treatment, though the latter state is a prerequisite for any success. The author's experience with younger alcoholics, particularly those involved with the abuse of multiple drugs as younger alcoholics tend to be (Kaufman, 1976), is that their family dynamics are quite similar to those of primary drug abusers. Another arguement for the similarity in the family structures of alcoholics and drug abusers is the high incidence of parental alcoholism in both groups. In an Eagleville study, 58% of alcoholics and 46% of drug abusers had an alcoholic parent (Ziegler-Driscoll, 1979). There are also significant differences between the two groups which may depend much more on age, ethnic, social, and cultural factors than on primary substance of abuse. The literature is replete with studies of the dynamics of alcoholics and their spouses, and these studies will not be reviewed in detail (Whalen, 1953; Krimmel, 1973; Steinglass, 1976; Janzen, 1976). Bowen states that individuals seek spouses with equal lack of knowing who they are but opposite ways of dealing with stress (Bowen, 1974). Thus each person sees himself as giving in to the other. The one who gives in the most becomes "de-selfed" and is vulnerable to a drinking problem (Bowen, 1974). Drinking to relieve anxiety and increased family anxiety in response to drinking can spiral into a collapse or crisis or establish a chronic pattersn. These issues and many others are as germane to the spouse of the drug abuser as to the spouse of the alcoholic. So in both cases we should be attuned to triangulation (Bowen, 1976), hidden agendas, and rule-governed behavior (Haley, 1977).

A critical period in every relationship and every family is when one party gives up substance misuse. Then the nonusing family members must find an entirely different way of relating. There are totally new expectations and demands, and for the first time there is communication, an art which they may have never learned. Sex has been used for exploitation and as a means for asking total forgiveness until it is non-existant. Difficulties also arise because the recovering abuser has given up the most precious thing in his life (drugs or alcohol) and expects immediate rewards. The spouse has been hurt too many times and is not willing to give the substance abuser the rewards he feels he deserves. The ex-substance abuser may go through a period of mourning for months or years after giving up the precious substance-love object. This depression may be as crippling for the couple as substance abuse.

FAMILY THERAPY TECHNIQUES

This author's approach is basically an adaptation of Salvador Minuchin's structural theory (Minuchin, 1974) to the specific family problems of substance abusers. However, other frames of reference, including systems (Bowen, 1974), psychodynamic (Boszormenyi-Nagy, 1973; Zuk, 1967) communication (Haley, 1977; Satir, 1972; and Bateson, 1975) are incorporated. These varying systems are not so clearly delineated and have contributed much to each other. Minuchin (1974) divides therapeutic tactics into two major categories: joining, which consists of those tactics which are done to enhance the therapist's leverage within the family; and restructuring, which is composed of strategies designed to change dysfunctional relationships.

JOINING

The first joining maneuver takes place in the beginning of the interview as the therapist functions as a host and provides a comfortable social setting. The therapist should shake hands with each member, introduce him or herself individually, and remember everyone's name. They should be given a flexible choice of seating arrangements so that they can have a full range of possibilities for seating. Seating choices are a strong initial key to family structure. Social interaction should take place between the therapist and each family member before the problem is presented. The therapist must be capable of joining each subsystem including the siblings. He must enable each family member to feel his respect for the others as individuals as well as his firm commitment to healing. The therapist must make contact with each family member so that they are following him even when they sense he is unfair. Joining techniques include: Supporting the family structures and behaving according to the family's rules, supporting areas of family strength, rewarding affiliating with a family member, supporting a threatened member, and explaining a problem (Minuchin, 1974). In joining, the therapist adopts the content of family communications and uses the family's own special language to offer the therapist's ideas. The therapist also adopts the family's style and affect. If a family uses humor, so should the therapist, but without double binds. If a family communicates through touching, then the therapist should also touch (Kaufman, 1981). Joining with every member of the substance abuser's family is part of the art of therapy. The family therapist does not have to be tatooed and speak in four letter words in order to relate to the drug abuser, yet a thorough knowledge of the vernacular is essential. With older alcoholics, a certain respect for the pristine language of sobriety as well as a basic knowledge of the 12 steps of AA is likewise important.

It is important not to join the "healthy" members against the substance abuser. To do so will likely precipitate defensiveness and withdrawal by the IP and insidously imply that only the IP will have to change his/her behavior.

The therapist thus becomes the ally of all. During the initial stages, the therapist may briefly (for a few minutes at a time) join with one family member who is overwhelmed or victimized. The intent is to help that member rejoin the family interaction. Then the therapist returns to the position of relating to all the family.

Most families will feel quite defensive particularly about the feeling that their being asked to come in implies that they are the problem rather than the identified or index patient. At this early joining stage the therapist should gather information but not share it as this would increase the inevitable defensiveness of the family (Haley, 1977).

Before restructuring is attempted, a tentative diagnosis of the family dynamics of "family map" (Minuchin, 1974) must be made. In order to diagnose the family the author uses an integration of systems, psychodynamic and structural methods. A first step in diagnosis is the genogram. A genogram is a pictorial chart of the people in a three generational relationship system which marks marriages, divorces, births, geographical location, deaths, and illnesses (Guerin and Pendagast, 1976). A family chronology (Guerin and Pendagast, 1976) may also be used which provides a time map of the major family events and stresses.

RESTRUCTURING

Unlike joining, restructuring involves a challenge to the family's homeostasis and takes place through changing the family sets. In restructuring, the therapist uses expertise in social manipulation, with the word "manipulation" being used in a positive, rather than a perjorative sense (Minuchin, 1974). Techniques used for change production include the contract, probing, actualization, assigning tasks, the paradox, relabelling, interpretation, and marking boundaries. Joining is necessary as a prerequisite and facilitator for change production, and the therapist frequently alternates between joining and restructuring.

A. The Therapeutic Contract (Minuchin, 1974). A prerequisite for change is the therapeutic contract. The contract is an agreement to work on mutually agreed upon, workable issues. The contract should always promise help with the IP's problems before it is expanded to other issues. Goals should be mutual. If there is disagreement about them, then work on resolving disagreements should be made a part of the contract. Substance abuse and how it will be dealt with should be made a part of the contract including the commitments of other family members, such as Al-Anon. It is helpful if non-substance-abusing family members agree to remain in treatment even if the IP drops out. This is a particularly important with the spouses of alcohlics. The length of time that treatment will require should be included in the contract, but it can be extended at a later date.

B. Probing (Minuchin, 1974). The therapist affiliates with family systems and feels their pressures. Thus, the therapist's spontaneous responses will probably be syntonic with the family system. If not syntonic, the therapist's responses will challenge the system and thus be valuable as therapeutic probes. Probes may lead to learning by the family, activate alternatives, or unbalance homeostasis which results in long-term changes. They offer an opportunity to examine the flexibility of the family system and offer information which permits family diagnosis and shifts in "the map."

C. Actualizing Family Transactional Patterns (Minuchin, 1974). Patients usually direct their communications to the therapist. They should be required to talk to each other instead. They should be asked to enact transactional patterns rather than describe them and to deal with unresolved problems rather than use the therapist as a judge or arbitrator. When the therapist has made suggestions for changes in the family actualization permits observations of compliance with or resistance to these suggestions.

D. Task Assignment (Minuchin, 1974). The therapist may assign family tasks, assist members in choosing specific implementation, and reinforce action. A child can be asked to choose a peer-related activity they have always wanted to do, and then ensure that the family facilitates the child's action. A father who neglects his personal health because of worry about his wife's drinking can be asked to make an appointment with a doctor or dentist. If tasks are not successfully completed, then the difficulties can be the subject of family discussion. Compliance with tasks, such as homework, is enhanced by successful completion of a task within the session. Tasks should be prescribed near the end of the session and repeated until they are "heard." They are then always reviewed at the start of the following session. A wife who is overinvolved with the amounts of alcohol her husband is consuming, may be given the task of estimating how many drinks he has every day and writing it down without telling him. The husband can be asked to write down the actual amounts, so that they can be compared in a subsequent session. The discrepancies will demonstrate the futility of the wife's efforts and diminish her overinvolvement in his drinking. This task is also a paradoxical one, the nature of which will be described below.

E. The Paradox. The use of paradox is based on the observation that individuals do the opposite of what they feel they are being pushed to do (Haley, 1977). Paradoxical directives can be used to achieve change. Such tasks appear absurd because they require families to do what they have been doing rather than change because the latter is what everyone else has been demanding. If the family opposes the therapist, then they will reverse the symptom. If the family complies with the therapist, then they can acknowledge their power over the symptom and have the power to change it. The paradox uses the principle of the double bind to change the symptom. It is an overt message which urges the family to obey the opposite covert message. The best paradoxes are those where the prescription of a symptom changes the underlying maladaptive family system. A symptom may be paradoxically exaggerated in order to emphasize the family's need to change it. An example would be to encourage a family to continue the "glories" of overindulging and infantilizing the alcoholic. A symptom which occurs outside the family can be prescribed to be performed within the family so that the family can deal with it, e.g., adolescent stealing or secret drinking. Skillfully asking the family to not change or to not change too fast and identifying the reasons they are not ready to change is another paradoxical maneuver which can lead to change. When properly delivered, the paradox leaves the family chafing at the bit to make desired changes. Paradoxes are unnecessary with compliant families and work best with rigid, resistive families.

F. Relabelling or Reframing. These are methods of restating the meaning of symptoms or behavior. Simply, it offers alternative explanations that illustrate the multiple determinants and consequences of behavior (Minuchin, 1974). For example, a family with a drinking teenager may "frame" his behavior as simply bad or selfish. The therapist may "reframe" the behavior by pointing out that one motive is to gain the parents' attention as a couple and prevent an impending divorce. And in fact, the therapist may then point out that as a consequence of concern about the adolescent drinking, the parents are indeed working together. Reframing can be used to move the family from preoccupation with just the substance abuse to consider related motives, consequences, and family dysfunction.

G. Interpretation. Interpretations can be extremely helpful if they are utilized without blaming, guilt induction, or dwelling on the hopelessness of long standing, fixed patterns. Repetitive behavior patterns are pointed out to the alcoholic and family. The undesired consequences to themselves and their family are pointed out and they are now given tasks to help them change these patterns in the here and now. Interpretations of individual behavior patterns are often less useful than interpretations of family interaction.

H. Boundary Marking. This technique is used to define, establish, and reinforce the individual boundaries and responsibilities of each member of the family (Minuchin, 1974). This is part of the process of reducing the over-involvement and enmeshment which is so common in substance abuser's families.

Individuals should not answer for or feel for others, should be talked to and not about, should listen to and acknowledge the communication of others. Reacting to anticipated or assumed reactions (mind reading) is also discouraged. Nonverbal checking and blocking of communications should also be observed and, when appropriate, pointed out and halted. Boundaries may be established temporarily by the therapist's placing him/herself or furniture between members. In general, the boundary which surrounds the non-substance-abusing spouse and his/her children should be replaced by one which surrounds the parental subsystem, protecting it from intrusion by children as well as other adults in and outside the family. When members are deprived of a key role by a new boundary which does not generally exist in the family, they should be provided with a substitute role, such as relating to another family member or in their social network. If an adult or older sibling is placed behind a one-way mirror, that person may be given the role of expert observer and permitted to comment later. These artifical boundaries in the session may be so reinforcing to the family thay they will be continued or may be supported by tasks to be performed at home. Al-Anon and Al-Ateen are excellent reinforcers of individual boundaries (Davis, 1978).

GROUP TECHNIQUES

Multiple Family Therapy (MFT) is particularly useful with the families of substance abusers in residential settings. The techniques of MFT have been described elsewhere by the author (Kaufman and Kaufmann, 1977). However, many of the steps of MFT emphasize and compliment the tools of Structural Family Theraphy described above.

Generally, techniques which are most familiar to the treatment system are most readily used in MFT. Thus, in MFT's in Synanon based residences, confrontations and encounter are used in the service of structural changes. The therapist must join the treatment system as well as each family. New techniques can be gradually introduced to MFT and thence to the overall program or vice versa.

The evaluation of the effectiveness of family therapy techniques with substance abusers is far from complete. Reilly (1975), Hirsch and Imhoff (1975), and Stanton (1977) cite encouraging results. P. Kaufmann (1979) of Phoenix House has demonstrated that family therapy decreases the 12-18 month rate of recidivism to drugs from 50% without family therapy to 20% with it.

It must be emphasized that each therapist chooses those systems of family therapy that best suit his or her personality, utilizing those techniques which can be grafted onto one's own individual style and family background.

It is hoped that this article has presented a system and a group of techniques that most therapists can utilize in their family work with substance abusers in residential treatment.

REFERENCES

Basen, M. M. The Use of Family Therapy in Drug Abuse Treatment. H.E.W., ADAMA, Washington, D.C.: U.S. Government Printing Office, 1977.

Bateson, G., et al. "Towards a theory of schizophrenia." Behavioral Science, 1956, 1: 251-264.

Berenson, D. "Alcohol and the family system." In P. J. Guerin (ed.), Family Therapy. New York: Gardner Press, 1976.

Boszormenyi-Nagy, E., and Sprk, G. Invisible Loyalties, New York: Harper and Row, 1973.

Bowen, M. "Alcoholism as viewed through family systems therapy and family psychotherapy." Annals of New York Academy of Science, 1974, 233: 155-122.

Bowen, M. "Theory in the practice of psychotherapy." In P. J. Guerin (ed.), Family Therapy. New York: Gardner Press, 1976.

Coleman, S. B., and Davis, D. I. "Family therapy and drug abuse: A national survey." Family Process, 1978, 17: 21-30.

Davis, D. Alcoholics anonymous and family therapy. Paper presented at National Conference on Alcoholism. St. Louis, Mo. (April 30, 1978).

P. J. Guerin (ed.), In Family Therapy. New York: Gardner Press, 1976.

Haley, J. Problem Solving Therapy San Francisco, Josey-Bass, 1977.

Hirsch, R., and Imhoff, J. "A family therapy approach to the treatment of drug abuse and addiction." Journal of Psychedelic Drugs, 1975, 7: 181-195.

Janzen, C. "Families in the treatment of alcoholism." Journal of Studies on Alcohol. 1976, 38: 114-130.

Kaufman, E. "The abuse of multiple drugs: psychological hypothesis, treatment consideration." American Journal of Drug and Alcohol Abuse. 1976, 3: 293-304.

Kaufman, E. "Family structures of narcotic addicts." International Journal of Addicts, 1981, 16: 273-282.

Kaufman, E., and Kaufmann, P. "Multiple family therapy: A new direction in the treatment of drug abusers." American Journal of Drug and Alcohol Abuse 1977, 4: 464-476.

Kaufman, E., and Kaufmann, P. "Multiple family therapy with drug abusers," American Journal of Drug and Alcohol Abuse, 1977, 4: 464-476.

Kaufman, E., and Kaufmann, P. (eds.) In Family Therapy of Drug and Alcohol Abusers, New York: Gardner Press, 1978.

Kaufmann, P. "Family therapy with adolescent substance abusers." In E. Kaufman and P. Kaufmann (eds.), Family Therapy of Drug and Alcohol Abuse. New York: Gardner Press, 1979.

Krimmel, H. E. "The alcoholic and his family." In Alcoholism: Progress in Research and Treatment. P. G. Bourne and R. Fox (eds.), Academic Press, 1973.

Minuchin, S. Families and Family Therapy, Cambridge: Harvard University Press, 1974.

Reilly, D. M. "Family factors in the etiology and treatment of youthful drug abuse." Family Therapy New York: 1975, 2: 149-176.

Satir, V. People Making. Palo Alto: Science and Behavior Books, 1972.

Stanton, M. D. Some outcome results and aspects of structural family therapy with drug addicts Paper presented at National Drug Abuse Conference, San Francisco, Ca. (May 5-9, 1977).

Stanton, M. D., and Todd, T. "Structural family therapy with drug addicts." In E. Kaufman and P. Kaufmann (eds.), Family Therapy of Drug and Alcohol Abuse. New York: Gardner Press, 1979.

Steinglass, P. "Experimenting with family treatment approaches to alcoholism, 1950-1975: A review." Family Process, 1976, 15: 97-123.

Stuart, R. B. "Behavioral contracting within the families of delinquents." Journal of Behavior Therapy and Experimental Psychiatry, 1971, 2: 1-11.

Whalen, T. "Wives of alcoholics: Four types observed in a family service agency." Quarterly Journal of Studies on Alcohol, 1953, 14: 632-641.

Whitaker, C. "A family in a four-dimensional relationship." P. J. Guerin (ed.), In Family Therapy. New York, Gardner Press, 1976.

Wood, P., and Schwartz, B. How to Get Your Children to Do What You Want Them to Do. Englewood Cliffs: Prentice Hall, 1977.

Ziegler-Driscoll, G. "The similarities in families of drug dependents and alcoholics." In E. Kaufman and P. Kaufmann (eds.), Family Therapy of Drug and Alcohol Abuse. New York: Gardner Press, 1979.

Zuk, G. H., and Boszormenyi-Nagy, I. Family Therapy and Disturbed Families. Palo Alto: Science and Behavior Books, 1967.

ALCOHOL AND FAMILY VIOLENCE: THE TREATMENT OF ABUSING FAMILIES

Jerry P. Flanzer

Graduate School of Social Work
University of Arkansas at Little Rock
Little Rock, Arkansas 72204

All the world seems to know that alcohol abuse and family violence are often found together. Most cultures seem to have an array of jokes about the drunken husband who neglects his children or beats his wife. Yet few research studies have explored this alcohol-family violence association or causation. Maybe we have been afraid to find out.

The purposes of my paper are:

1. To review the known relationship between alcohol abuse and family violence;

2. To review their clinical-dynamic interrelationship(s); and

3. To begin to investigate the effectiveness of treating such families.

Data from a demonstration-treatment project focused primarily on adolescent abuse and alcohol connection will be used to illustrate the finding.

The relationship of alcohol to violence has long been researched. Over 20 years ago, Wolfgang (1958) noted the family association when looking at patterns of alcohol and criminal violence. However, the family systems approach only recently has tied child abuse, spouse abuse, and other interfamily aggressive behavior into the umbrella nomenclature -- family violence. Given the generic concept of family violence scientists and intervention agents could suddenly realize the endemic alcohol role and connection. The alcohol connection exists not only to one part, but to the whole family violence system. The research has begun to show us that alcohol and family violence are more closely related than alcohol and any other type of violence (Gelles, 1972). These realizations thus began to emerge in separate writings by Lystad (1975), Byles (1978), Hindman (1977), and others. Specific associations of alcohol misuse to family violence have been found in recent research. The resource table below outlines a few of these; several may be highlighted.

One study done by Behling (1977) illustrates the association of alcohol and family violence quite well. Behling reports on the incidence of alcohol-related child abuse at one large clinic:

Table 1. Resource Table: Alcohol Misuse/Family Violence

Author	Date	Subjects	Results
Behling	1977	59 alcohol & child abusers; Navy hospital records	High association of child abuse
Nau	1967	105 child abusers, 55 families, a forensic & psychiatric clinic and 55 apprehended by police	Significant number were perpetrators when not drinking
Flanzer	1981	65 child abusing families, 40 spouse abusing families, 20% overlap	Patterns of association
Spieker & Mousakitis	1976	42 child abusers, DWI-outpatient groups	Association of moderate drinkers and child abuse
Carder	1978	67 abused women in a shelter	Alcohol and spouse abuse association
Miketic	1972	64 children abused by parents	High incidence of alcoholism
Grislain	1968	32 cases of child abusing parents from hospital records	High incidence of alcoholism
Ramee & Michaux	1966	480 children in incestuous families	Alcoholic parents overrepresented
Virkkunen	1974	45 incest offenders from a psychiatric university hospital clinic	Alcoholic parents overrepresented

1. 57% of the abused/neglected children had at least one grandparent who was alcoholic or abused alcohol.

2. 65% of the suspected child/abusers/neglectors were alcoholics or abused alcohol.

3. 88% of the previously abusing parents were abused as a child by an alcoholic or alcohol-abusing parent.

4. 84% of the abused or neglected children had at least one parent who was alcoholic or abused alcohol.

Of extreme importance was Nau's (1967) finding that the alcoholic parent often committed the violent act when not intoxicated. This finding has led this researcher to note that four patterns may be seen.

In one pattern, the family seems to alternate between alcohol and aggression, drinking to avoid hitting, hitting perhaps to

avoid drinking. In another family pattern, violence appears <u>only</u>
<u>when alcohol is imbibed</u>. This style takes two forms: in form A,
the Abuser Form, the drinker has a few drinks and seems to feel
licensed to hit. In Form B, the Victim Form, the drinker inbibes
until he/she is in a helpless stupor and becomes the easy target
for abuse. In the third pattern, the heavy drinker becomes so
<u>preoccupied</u> with the habit that family members are severly
neglected and roles impaired. This pattern refers to an incipient
form of violence -- extreme passive aggression. (Flanzer, 1981)

A fourth pattern also exists, and this is where all parties are
<u>abstainers</u> -- they do not drink. This is mentioned here for two reasons.
The first reason is that alcoholism is not involved in all cases of family
violence. In fact, this is probably true in 20-50% of all cases, depending
on the type of family violence involved (e.g., alcohol seems not to be as
major a concern in sibling abuse cases.) The second reason is that a good
number of abusing spouses/parents who abstain turn out to be the children of
alcoholics!

Spieker and Mousakitis's (1976) findings were "eye openers." Most
alcohol connected studies have been done by following child abuse or spouse
abuse clinic populations, and then reporting alcohol incidence. Spieker and
Mousakitis, by following a population of an alcoholism treatment agency,
found child abuse prevalent among those with moderate drinking problems --
not alcoholics or severe drinkers. Subsequently, Carder (1978) found this
to be true following a spouse abuse sample. These data raise a most
important challenge for the researcher -- who and what should be the focus
of this work: A commonly accepted idea has been that the more one drinks,
the greater the level of abuse. Is that indeed true? Perhaps, we need to
worry more about the so-called moderate/episodic drinker than the heavy
drinker/alcoholic in regard to family violence?

The Questions

The current research is concerned about the following:

I. What level of drinking is associated with family violence?

II. Does it matter who is the drinker and who is the abstainer in the
family?

III. Does drinking level affect the severity of abuse and neglect?

IV. Does alcohol seem to affect any particular clinical manifestation
as related to family violence?

V. What is the order of occurrences between alcohol and family
violence?

VI. Can we treat these alcohol and family violence problem
associations?

The Sample

Delving into the data generated between 1978 and 1981 of a project
funded to specifically look at the alcohol relationships among a clinical
sample, one can deduce important trends and one can begin answering these
questions. The sample is limited to intact families (families whose
boundaries are clear and who consider themselves an operational family)
wherein at least one parent is abusing or severely neglecting his/her
adolescent child (aged 10-18). These client families have been referred to

this specialized university clinic through area child and adolescent protective services, youth services, and components of the juvenile and adult judicial systems. Some of these families are known to alcoholism treatment agencies, but due to laws of confidentiality, they have been first referred to the child protective service system. The families are described as abusive/neglectful in accordance with the relevant state statues -- though not all cases have been adjudicated. In nonadjudicated cases, ample evidence is available to admit the families to the project. (Families not meeting the project's abuse/neglect criteria either serve as teaching cases or are referred to more appropriate agencies.) Upon referral to the clinic, the families are engaged in treatment and screened for project appropriateness. Alcohol intake levels are reported, based on two self-report instruments and clinical observations at this time. Families are assigned social work therapists, and a family treatment approach is provided.

The Data

This part of the report is limited to the first 47 families who participated in the project and fully cooperated in completeing appropriate questionnaires during the course of treatment as one of June, 1981. Forty of these families had both parents in the home, seven were headed by a single mother. Not reported is fragmented, data gathered from clients (18 families) who either: (1) refused to participate in part or all of the evaluation/research prodedures; or (2) were not engaged in the "process" long enough to be considered part of the ongoing caseload. The modal length of treatment of these 65 (47 + 18) families was for 6-9 months, the range being 14-16 months.

The Instrumentation

The American Human Association standard survey instrument, as well as clinical impressions, was used to determine levels of physical abuse and neglect, emotional abuse and neglect, and sexual abuse. Selzer's Michigan Alcohol Screening Test, the MAST (1971), was the primary source for alcohol intake history. This test was supplemented by a modification of the Cahalan Quantity Frequency Index (1969). Results on Hudon's Short Form Clinical Likert type scales (1974): self-esteem, general contentment (depression), sexual and marital satisfaction, were used as the clinical measurements in association with drinking and abuse levels in this report.

The Findings

I. What level of drinking is associated with family violence?

When examining the alcohol habits of the population, the researcher must decide whether to focus on the client's biopsychosocial impact of drinking over time or to focus on the client's "here and now" actual alcohol intake.

The first focus led to utilizing the Michigan Alcohol Screening Test (MAST) which consists of 25 true-false questions which deal with medical, social-epidemiological impact on the client's alcohol use. The MAST scores were ranked as follows:

> 0   abstainers
>
> 1-5   no problem drinking
>
> 6-9   moderate problem drinking
>
> 10+ severe problem drinking

Table 2A. MAST and QFI Drinking Levels of the Parents

Drinking levels	MAST Parents		Modified QF1 Parents	
	Mothers	Fathers	Mothers	Fathers
Nondrinker	20	10	13	7
Low drinking	11	9	10	5
Moderate to high	5	5	15	14
	36	24	38	26
Missing	9	7	7	5
Total	45	31	45	31

Chi-square = mothers x fathers, MAST = 1.18, df = 2, sig .60, missing data deleted.
Chi-square = mothers x fathers, QF1 = 1.297, df = 2, sig .60, missing data deleted.
Chi-square = mothers' MAST x mothers QF1 – 6.4, df = 2, sig .05, missing data deleted.
Chi-square = fathers' MAST x fathers QF1 = 6.29, df = 2, sig. .05, missing data deleted.

Table 2B. Combined MAST and QFI Drinking Levels of the Parents

	M	F	T
Nondrinker	8	4	12
Low drinker	9	6	15
Moderate to high	13	11	24
Total	30	21	51

x^2 = .673, df = 2, sig .75.
Not everyone completed both instruments;
5 fathers and 8 mothers data will be deleted
if the results are paired.

Thus, as indicated in Table 2A, severe drinking was not heavily represented among the scores of the parents on the MAST. Contrary to popular belief, no significant difference was found between mothers and fathers in the MAST rankings.

The second focus, the here-and-now focus, led to utilizing a modified Quantity Frequency Index (QFI) (à la Cahalan and Jessor) which dealt with an absolute alcohol consumption over a 30-day period. Thus, as indicated in Table 2A, utilizing the same sample, different results are reported on the QFI than on the MAST. The QFI results indicate many more moderate and heavy drinkers, particularly among the fathers, although, once again, a statistically significant difference does not exist between the mothers and fathers as to the current alcohol intake.

Significant difference between the mothers' scores on the MAST and those on the QFI and the fathers' scores between those same instruments does emerge. The QFI shows that the mothers' and fathers' drinking patterns have certainly increased in recent times. Clearly, the two instruments are measuring different phenomena. And the truth is assumed to be in between these measures; thus a combined MAST and QFI ranking is ascribed. This gives a truer composite picture of drinking habits. Thus, subsequent comparisons of drinking as an independent variable takes the MAST, the QFI, and combined MAST-QFI (Table 2B) into consideration.

II. Does it matter who is the drinker and who is the abstainer in the family?

Who is the prominent drinker among the parents? As can be seen in Table 3A, the father is the prominent drinker in 44.6% (n = 21) of the families in this sample. That is, he is the identified heavy drinker. Note that as a close second, 42.6% (n = 20) of these families, the mothers and fathers share equal levels of drinking. That, of course, includes abstainers and light drinkers. Mothers are rarely viewed as the prominent drinker (12.8%, n = 6). Take away the seven single parents, and the mother is not in the picture at all. Thus, clearly, either the man is the drinker in the marriage or the husband and wife share the honors.

Discussion

There are alcohol-using families where the mothers are the prominent drinkers. These families do not appear to be child-abusing ones. Only when the fathers are at least equal to or greater than the mothers' drinking levels do these families appear among the child-abusing population. One would think that the fathers' drinking is the key factor throughout the alcohol-family violence connection. But, as elaborated herein, this is not the case.

Are the heavy drinkers significantly more often the perpetrators of abuse toward their children? And are the fathers then significantly more often the perpetrators of the abuse?

Contrary to expectations, no significant difference can be found between the perpetrator of abuse and drinking levels, nor can one be found between the perpetrator and sex of the parent. (See Tables 3B-E.)

Table 3A. Breakdown of Prominent Drinker between Partners

Prominent drinker	Number	%
Mother	6	12.8
Father	21	44.6
Equal	20	42.6
Total	47	100

Table 3B. Breakdown of Who Is the Perpetrator by Partners

Parent	Perpetrator	Nonperpetrator
Mother	16	7
Father	27	22
Total	43	29

x^2 = 1.43, df = 2, sig .49.

Table 3C. Combined Alcohol Levels by Perpetrator and Nonperpetrator

Combined alcohol	Perpetrator	Nonperpetrator	T_1
Non-drinker	10	2	12
Low	15	8	23
Moderate to high	14	7	21
T_2	39	17	56

x^2 = 1.365, df = 2, sig .50.

Table 3D. MAST Levels by Perpetrator and Nonperpetrator

MAST	Perpetrator	Nonperpetrator	T_1
Low	32	15	47
Moderate	4	1	5
High	3	1	4
Total	39	17	56

Collapsed low vs. mod-high; x^2 = .335, df = 1, sig .85.

Table 3E. QFI Levels by Perpetrator and Nonperpetrator

QF1	Perpetrator	Nonperpetrator	T_1
Low	34	16	50
Moderate	4	1	5
High	6	1	7
Total	44	18	62

Collapsed low vs. mod-high; x^2 = 1.10, df = 1, sig. 30.

Thus, alcohol is present through the sample. But who drinks does not seem to determine who the abuser will be. The nondrinking spouse is equally as likely to severely attack the teenage child in this sample.

Clearly, these findings may be biased by the particular sample, which is representative of the mid-southern United States. The strong influence of fundamental religions, a lower educational standard, and strong male-dominant marriages may have served to bias these findings toward a higher acceptable level of violence by both sexes. Comparable research with other regional samples is necessary.

III. Does drinking level affect the severity of abuse and neglect?

Neither correlations of MAST drinking levels or QFI drinking levels with severity abuse levels reveals a statistically significant difference at a .05 level (Table 4A), although it should be noted that the MAST physical neglect correlation yields a significance of .076 among the mothers. However, when combining the MAST and QFI, once again, into the new combined alcohol measurement both chi-square and Pearson-Product Moment correlational statistics lead to some outstanding significant differences (Table 4B). However, the physical neglect significant relationship disappears. All of the correlations are negative. Significant negative relationships are found between drinking levels and severity of physical abuse, emotional abuse, and perhaps sexual abuse. The more one drinks, apparently, the lesser the chance for severe physical and emotional abuse and the greater the change for a light level of abuse/neglect; or the less drinking, the greater the abuse. This is true even when controlling for the abstainers or for the married couples.

Discussion

The MAST purports to measure the effect over time, and thus physical neglect, which takes time, appears. Interestingly, mothers are the key. If the mother is drinking heavily, and few in our sample were, then physical neglect will appear.

The more intense, and believed to be more accurate, combined alcohol measurement does seem to ferret out the physical and emotional abuse. These two abuses go hand in hand and, indeed, consistently correlate highly in this study. (Note: the sample's clinical correlations are beyond the scope of this paper.)

The significance, however, is in a negative direction. Three factors contribute to this finding. One is that the sample includes only intact families at onset of treatment. The severe alcoholic has often lost his/her family and does not show up in this study. A second factor has to do with the relationship of discipline and drinking, which will be discussed further on. It appears that the parents began drinking more as they became frustrated with their burgeoning adolescents, and either are unable to or unwilling to abuse them further. A third factor is that adolescents are less likely to be severely physically abused than their younger counterparts -- they can fight back, they can run away, and they often do both.

One would conclude that we might be advised to encourage parents who abuse their children to drink more -- alcohol being used as a self-medication once again. However, the price would surely be in more physical neglect and broken homes.

In review: (1) alcohol is present in most of these abusing families; (2) alcohol does not account for which parent abuses the children; and

Table 4A. Correlation (Pearson) of Abuse Type
with MAST and QFI Levels by Partner

| | MAST | | QF1 | |
Abuse type	Mothers	Fathers	Mothers	Fathers
Physical abuse	1.000	.743	.650	.269
Emotional abuse	.676	.942	.341	.578
Sexual abuse	.433	.778	.442	.349
Physical neglect	.076	.554	.805	.480
Emotional neglect	.246	.224	.613	.287

Table 4B. Association and Correlation of Abuse Type
with Combined Alcohol Levels

	x^2	sig	pr^*	sig
Physical abuse	6.151	.05	-.2211	.052
Emotional abuse	7.79	.05	-.2917	.015
Sexual abuse	3.09	.54	-.1959	.076
Physical neglect	7.62	.26	-.0438	.375
Emotional neglect	11.16	+.08	-.0484	.363

x^2 is figured on collapsed 3 x 2 tables - alcohol-no, low,
mid-high; abuse - no-low; mod-severe.
*Direction is negative; more drinking, lighter abuse.

(3) alcohol may actually be partially accountable for a lower level of abuse
and for the interactional roles between the parents. This is contrary to
belief and certainly contrary to 1980 findings by this author (Flanzer, 1980),
where it was found that alcohol signaled the perpetrator of the abuse.

IV. Does alcohol intake seem to affect any particular clinical
manifestation as related to family violence?

The Hudson scales have been ranked so that a score of 25 or less equals
no problem in the clinical area, 26-35 equals expected level of problem
(under which two-thirds of the standardized norms lie), and 36 and above
equals open discontent.

Neither correlations of MAST drinking levels, nor QFI drinking levels
with the selected clinical measurements reveals statistically significant
difference at a .05 level (Tables 5A and 5B). Fathers do appear to have a
significant relationship difference (.086 level) between drinking levels on
the MAST and the General Contentment (depression scale). The more one
drinks, over time,the greater the likelihood for depression. Once again,
the combined MAST-QFI scale yields some significant results when compared to
the selected clinical measurements: An inverse relationship between
combined alcohol levels and sexual satisfaction. Apparently the greater one
drinks, the less the sexual satisfaction, i.e., for once, following the
known alcohol literature (Smith, 1974).

Table 5A. Correlation of Clinical Measurements
with MAST and QFI Levels of Partners

Clinical measurements	MAST		QF1	
	Mothers	Fathers	Mothers	Fathers
Generalized Content- ment Scale	.193	.086	.503	.572
Index of Self-Esteem	.166	.714	.219	.395
Index of Marital Satisfaction	.395	.627	.340	.755
Index of Sexual Satisfaction	.500	.453	.953	.212

Table 5B. Association of Clinical Measurements
with Combined Alcohol Levels

Clinical measurements	x^2	sig (.05)
Generalized Content- ment Scale	2.689	no
Index of Self-Esteem	5.805	no
Index of Marital Satisfaction	1.575	no
Index of Sexual Satisfaction	2.53	yes

Discussion

These findings clearly point out that alcohol level has little
relationship to these four clinical measurements (Tables 5A and 5B). At
least so it seems. Two factors are important here. The first is that
alcohol masks feelings and flattens clinical manifestations. Indeed, there
is clear indication at post-testing (Izard, 1980) that as individuals stop
drinking their self-esteem goes down and their expression of marital
difficulties goes up, i.e., their reality testing improves. A second factor
is that researchers have not been able to link abusive behavior to
particular clinical variables/portraits. Abuse seems to be primarily a
behavioral problem. It is most likely related to mastery and control over
one's environment and relieving stress. Other clinical measurements not
analyzed in this paper are being applied to test that theory. They include,
for example: The Welsh (1952) Anxiety-Stress Scales of the MMPI, the
Dominance-Submission Scales of LaForge and Suczek's (1955) Interpersonal
Checklist.

The clinical scores on the four Hudson scales are indicative of a
population under stress scoring a standard deviation above the expectancy,
but still, drinking alcohol has little effect on these scores. It would
seem that alcoholism defies clinical categorization, as does family
violence.

V. What is the order of occurrences between alcohol and family violence?

Fifty-one parents (41%) responded to the project's family discipline survey (25 or 33% chose to not complete this form). The responses to five of the questions relevant to this report are broken down in Table 6. This table helps delineate the comparison of each of these questions which look at the before, during, after, and "irrelevant" relationship of drinking to discipline.

Sample Findings

The overwhelming majority of respondents viewed their drinking as either (1) following their inappropriate/inadequate discipline of their child(ren); or (2) saw drinking and disciplining as two mutually exclusive phenomena. Of course, the question was irrelevant for those who did not admit to drinking, but note that many who drink moderately state here that they do not drink. Few parents have been told by others to cut down on their drinking because they were hurting their children.

Discussion

Implied in these findings is that half of the parents are afraid of abusing their children even more, so they turn to drink instead. It is not known if timing of alcohol use represents inadequate discipline techniques, fear of losing control, guilt, or other factors.

Relating these findings to the earlier ones, noting the relationship of the parents' drinking level to the battering and severe neglect of their adolescent children, leads one to believe that the severe drinkers are involved with the discipline of their children, or are "long gone." Many episodic drinkers, undoubtedly then, are drinking after a row with their teenagers. This implies timing, not causation.

The QFI scores imply that parents' increased drinking is recent. Coupled with the discipline findings, this points to the use of alcohol to handle new family stress, perhaps related to the adolescent's behavior or response to their parenting problems. These statements are true for half the parents in our sample. We do not see a relationship between parental drinking behavior and their discipline of their children. In the other half of the sample, part of this response may be masking the truth. This latter group is overrepresented by individuals who drink less than the former group. Thus, it is not surprising that those who drink less would not see alcohol as having a major impact on the disciplining of their children.

VI. Can we treat these alcohol and family violence abusers?

Family treatment has measured success with this clinical sample. Over half of the families have stopped abusive behavior, and drinking levels have diminished, particularly among the heavy drinkers. (See Sturkie-Flanzer, 1981, for a preliminary outcome report.) Measurable improvement in parental attitudes toward children has been achieved although significant clinical change on three of the measured instruments has not; excluding marital satisfaction (which, of course, is a family treatment issue).

It is a fact that alcoholics who still have intact families have a much greater chance of "making it." In fact, violent families often have a much greater chance of making it when problem drinking is around. This is partly due to superior alcoholism treatment options and partly due to the

Table 6. Family Discipline -- Drinking Timing Questions

1. Do you find you tend to discipline your children more often
    before you take a drink? 17
    while you are drinking? 1
    after you take a drink? 0
    irrevelant to drinking habits 16
    do not drink 7
    no answer 10
    Total 51

2. Do you find you tend to drink more
    before you have disciplined your children? 2
    while you are disciplining your children? 1
    after you have disciplined your children? 8
    irrelevant to drinking habits 18
    do not drink 10
    no answer 12
    Total 51

3. Do you find it easier to discipline your children____
    you have a drink?
    before 14
    while 1
    after 2
    irrevelant 17
    do not drink 8
    no answer 9
    Total 51

4. Do you lose control
    before dinking 2
    while drinking 3
    after drinking 1
    not appropriate 19
    no drinking 5
    no answer 21
    Total 51

5. Have you ever been told you disciplined your children
    while drinking and not remembered doing it?
    yes 1
    no 36
    irrelevant 8
    no answer 6
    Total 51

Summary by Question

Question number	1	2	3	4
Before	17	8	14	2
During	1	1	1	3
After	0	2	2	1
Irrevelant	16	18	17	19
Do not drink	7	10	8	5
No answer	10	12	9	21
Total	51	51	51	51

family's option of saving face by blaming drinking and allowing the
unmasking of denial as drinking decreases. Our least successful cases have
been at the two ends of the continuum, "rigid," often authoritarian
structured, and "loose," extremely unsocialized/sociopathic, heavy drinkers.
Fortunately there have been few.

A major question which is related to this research is whether or not
the alcohol consumption levels and discipline issues are any different than

those of the population at large. This researcher and others are currently investigating this issue. The fact that the same alcohol consumption ratios do not seem to exist among other clinical populations still weight the evidence in favor of a difference when alcohol is involved.

STUDY IMPLICATIONS

Several implications for research and treatment result from this study:

1. These results show the need for measuring alcohol intake, not only for current, but also past, behavior. A combination of the two appears to lead toward a more accurate picture of reality, and leads to new findings. These findings point to the advisability of measuring many behaviors besides drinking (e.g., discipline practice), and taking into account past and current histories as more reflective of the future probabilities.

2. The findings suggest that changing cultural, marital, and sexual role relationships may have great effect on who is the persecutor/aggressor. Increasingly, the male has been sharing the role of family disciplinarian, and the woman is becoming equally as likely as the man to exhibit her drinking. These cultural changes have shifted to give either sex an equal chance for aggressive behavior. Research and treatment must focus on these changing role relationships. Certainly, the bias against the man as the drunken aggressor needs to be re-examined.

3. "Shifting the blame to alcohol" apparently may be viewed as an effective treatment technique. But alcohol does not, in fact, appear to be the major problem among violent families. This research points to alcohol misuse as concurrent to or as a means of handling one's stress or guilt of acting violently. Focus on alcoholism first, then, would not and cannot be effective. The aggressive act must be dealt with first or simultaneously. An important therapeutic diagnostic, therefore, must be made before the intervention, i.e., the therapist must determine "the timing" of the incident(s) -- of drinking and family violence.

4. Many families do not drink at all. These are often the hardest to treat. Therapists must find new acceptable ways for these families to relieve stress.

Research in this area has just begun. The need for cross-cultural studies relating alcohol and family violence is clear. This study did not look at nonalcohol forms of drug misuse/addictions. Would the results between heroin addiction and family violence and alcohol and family violence be the same? Much work needs to be done.

ACKNOWLEDGMENTS

An interim report from the Arkansas Alcohol/Child Abuse Demonstration Project funded by the National Center on Child Abuse and Neglect, Grant Number 90-C-1757, Jerry P. Flanzer, principal investigator. The author wishes to thank the staff of the Family Center, Graduate School of Social Work, University of Arkansas at Little Rock, without whom these data could not have been generated or compiled.

REFERENCES

Behling, D. W., History of alcohol abuse in child abuse cases reported at naval medical center, Long Beach. Paper presented at the National Council on Alcoholism Forum, 1977.

Byles, J. A., "Violence, alcohol problems and other problems in disintegrating families." Journal of Studies on Alcohol, 1978, 39: 551-553.

Cahalan, D., Cisin, I. H., and Crossley, H. M. American Drinking Practices. New Brunswick, N. J.: Publication Division, Rutgers Center on Alcoholic Studies, 1969.

Carder, J. Families in trouble. Paper presented at the 24th International Institute on the Prevention and Treatment of Alcoholism, Zurich, Vol. 25, VI-VII, 1978.

Flanzer, J. P. Families that abuse: Adolescent and alcohol abuse. Paper presented at the American Association of Psychiatric Services for Children Annual Meeting, New Orleans, 1980.

Flanzer, J. P. "The vicious circle of alcoholism and family violence." Alcoholism Magazine. 1981, January/February, 30-32.

Gelles, R. J. The Violent Home. Beverly Hills: Sage Publications, 1972.

Grislain, J. R., Mainard, R., and DeBerranger, P. "Child abuse; social and legal problems." Annales Pediat, (Sem. Mop., Paris), 15: 440-448. 1968.

Hindman, M. "Child abuse and neglect: the alcohol connection." Alcohol, Health and Research World. 1977, Spring, 2-7.

Hindman, "Family violence." Alcohol, Health and Research World. 1979, Fall, 2-11.

Izard, C. Changes in the self-concept of alcoholic/adolescent abusers during family therapy. Master's Thesis, Graduate School of Social Work, University of Arkansas at Little Rock, 1980.

LaForge, R., and Suczek, R. F. "The interpersonal dimension of personality. III. An interpersonal checklist." Journal of Personality, 1955, 24: 94-112.

Lystad, M.H. "Violence at home: a review of literature." American Journal of Orthopsychiatry, 1975, 45: 328-345.

Miketic, B. "The influence of parental alcoholism in the development of mental disturbance in children. Alcoholism, 1972, 8: 135-139.

Nau, E., "Kindesmisshandling (Child abuse). Mschr. Kinderheilk, 1967, 115: 192-194.

Ramee, F., and Michaux, P. "De quelques aspects de la delinquance sexuelle dans un department de l'ouest de la France (Some aspects of sexual offences in a province in western France)." Acta Medical Legal Society, 1966, 19: 79-85.

Selzer, M. L. "The Michigan alcohol screening test: The quest for a new diagnostic instrument." American Journal of Psychiatry, 1971, 127: 1653-1658.

Smith, J. W. "Impotence in alcoholism." Addictions, 1974, 3 (5).

Spieker, G., and Mousakitis, C. Alcohol abuse and child abuse and neglect. Paper presented at Alcohol and Drug Problems Association of North America, 27th annual meeting, New Orleans, 1976.

Sturkie, D. K., and Flanzer, J. P. "An examination of two social work treatment models with abusive families." Social Work Papers, the University of Southern California, XVI: 1981, 53-62.

Virkkunen, M. "Incest offences in alcoholics." Medical Science and Law. 1974, 14: 124-128.

Welsh, G. S. "An anxiety index and an internalization ratio for the MMPI." Journal of Consulting Psychology. 1952, 16: 65-72.

Wolfgang, M. E. Patterns in Criminal Homicide. Philadelphia: University of Pennsylvania Press, 1958.

ALCOHOLISM TREATMENT: A FAMILY DISEASE MODEL WITH

PARTICULAR REFERENCE TO INPATIENT TREATMENT FOR FAMILY MEMBERS

Joyce M. Ditzler

The Falcon Foundation

James R. Ditzler

Broadway Lodge

Broadway Lodge is a treatment center in Avon, England, which has been open since 1974. It is a private, nonprofit foundation and has a bed capacity for 35 patients. It is, in fact, the only exclusively inpatient chemical dependence unit in a county with over 1.25 million people. Even so, our patients come from a wide geographical area in the United Kingdom. A small number come from Europe, the Middle East, and the United States. Referrals are from a variety of sources: physicians, psychiatrists, employers, the social services, ex-patients, Alcoholics Anonymous, and families of alcoholics.

The treatment is based on the Minnesota model, specifically on the type of treatment developed at the Hazelden Foundation (Kammlier and Laundergan, 1977). It is a milieu therapy concept, using a multidisciplinary team of doctors, psychologists, social workers, nurses,and counsellors, some of whom are recovering chemically dependent people. Some of the tools we use are techniques based on the ideas of Reality Therapy (Glasser, 1965) and Rational Emotive Therapy (Ellis, 1961), which stresses the need for ongoing change. We also utilize the terminology of the Alcoholics Anonymous program for concreteness and inorder to facilitate easy transition to it.

Increasingly, we have seen the need for more help for family members, whether they be spouse, parents, or children. We have developed, over the past five years, various means of assisting these families to recovery.

As Gacic says, "We believe that alcoholism is not an isolated individual problem, a personal problem, or a private matter of the individual. It is a mutual problem of all those surrounding the alcoholic." (Gacic, 1981.)

We find in our families, where the alcoholic has had multiple admissions, that the family does not expect to be listened to or included in the treatment and recovery process. The family systems concept (Ditzler, 1980; Williams, 1975) and the need for family involvement in treatment is new to them. They exhibit mixed feelings of relief and apprehension when asked to participate.

We see that the family member manifests many symptoms in common with the alcoholic. Denial is used often as a primary defense on many levels. Some even prefer for "their alcoholic" another diagnosis; schizophrenia, depression, emotional problems, or sexual problems. Feelings of shame may predominate as well as pride in the matter of spouse selection. At times there is an unwillingness to accept that there may have been an error in the choice of a marriage partner.

Probably the most widely heard initial denial is that the spouses themselves need help. They rationalize this by viewing themselves as copers. Their insistence that they are competent and necessary to the alcoholic is profound. They are unable to recognize that their continuing supportive behavior is enabling the alcoholic to avoid reality (Ward and Faillace, 1981). In short, some spouses refuse to accept that they are addicted to the addict -- a <u>coaddiction</u>.

Spouses experience frustration, resentment, a sense of inadequacy, lack of self-worth, and feel progressively more trapped and useless. They have neither the knowledge or the energy to change their own response to the alcoholic. Most seriously, they often fail to recognize the degree of their own pain. Their energies become centered on the dependent person -- <u>a dynamic codependence</u>.

Parents are often disregarded in the family of the drinker. They agonize on their apparent failure as parents. They blame each other and themselves. They persist in trying to control their drinking child. They fall into the trap of blaming the drinker's wife or husband -- the non-"alcoholic" spouse. This process is fostered by the alcoholic, who in turn is blaming the parents.

Alcoholics and their spouses are surprised when we express the wish that the children be involved. With Pilat (1981) we argue that all family members, including all children, need to learn about alcoholism. This permits having a placed to ask questions and to ventilate feelings safely. The parents, both drinker and spouse, often assure us that their children are ignorant of the drinking behavior. Conversely, we find that, given the opportunity, even the very young child of four to five years old is extremely aware. Leite (1979) describes how children in their simplicity often recognize the problem before the parents. The damage to the younger children is often missed by both parents. In many cases this is so because the drinker is drinking and their spouse is obsessed by the drinker and the drinking behavior.

The children learn many of the traits of the parent family member. They become frightened, feel guilty, responsible, and feel they are to blame. They learn manipulation, rebellion, or become copers. They become peace-makers in the home (Black, 1979) or surrogate parents. They hide many feelings, positive or negative. They learn that it doesn't pay to be open. As a result, a profound loneliness is created and experienced, and great difficulty in forming trusting relationships in early childhood occurs (Orford, 1977). This existential loneliness may continue during the parents' recovery if the child is not helped. As the parents' relationship improves, the child, who may have been used as an ally by the drinker, is excluded. The child may find manipulation less easy and jealousy can ensue.

In this type of changing family pattern, the sense of rejection becomes appalling.

New types and content(s) of communication in recovery is essential. The feelings of children need to be heard. As Cork (1969) said in her study of 115 children of alcoholics, "Many said they did not want other parents, they only wanted a chance to know and be understood by the parents they had."

From this brief outline of our experience, it can be recognized that all members of the family need specific and ongoing help if they are to achieve their own personal recovery. We attempt to intervene in the following ways.

METHODOLOGY

Conjoint Family Group Therapy

Once a week, all family members participate in a family group consisting of two 1.5-hour sessions. Each group consists of three to four families who, together with the counsellor, discuss alcoholism and drug dependence as an illness and explore the resultant damage to the whole family system. In the first group the concept of "enabling" in the family is explained, as well as ways in which the family members can change their behavior to become more detached emotionally from their dependent person (Maxwell, 1976). These groups are continuous ones. Participants who have been involved for some weeks progressively deal more openly with their feelings and assist newer members to do likewise. Later, in the second of the two groups, the drinking member of the family is present. This may be the first time s/he has soberly heard of their own behavior. Often this is the turning point in treatment for the whole family, as the drinker perceives the damage to the family and the family gains confidence to be open.

Children's Groups

We have found that young children tend to be dominated or intimidated by adults in the family groups. As a result we now have a separate group for ages 4 to 14. These weekly groups last .5 to 1 hour. In this setting the youngsters become open, express relief at being given permission to talk, and at finding they are not the only ones with a drinking parent. They begin to grasp that the drinking is not in any way related to them. They start to share their own destructive feelings. If appropriate, the children will tell their parents about these latter feelings. The alcoholic is often shocked and the spouse is appalled by the realization of the child's awareness about what has and is going on as well as their mutual insensitivity to the needs of their own child(ren).

These sessions can be very emotional. There is relief at risks taken and shared. It is usually the ending of the conspiracy of silence of years.

Individual Family Conferences

In most cases, at some point in treatment, in-depth discussion with the whole family is helpful. This occurs two to three times during the treatment process. Many painful feelings may be aired, the drinking behavior is addressed, issues of intimacy are explored, and the beginnings of change and trust can be attempted. These areas, as well as others, may not have been raised for years, let alone realistically discussed.

Aftercare Family Conferences

Many families wrongly assume that when the patient completes in patient treatment, most of their problems will be over. There are, however, many new ones. As Jackson (1954) states, "For the wife and husband facing a sober marriage after many years of an alcoholic marriage, the expectation of what marriage without alcoholism will be is unrealistically idealistic, and the reality of marriage almost inevitably brings a disillusionment." The former irresponsible child is now struggling to be an adult. The untrustworthy drinker wants to be trusted. There are often sexual problems (Mason, 1980). There is apprehension at the reemergence of tender and caring feelings.

Many patients return for aftercare treatment monthly for the first few months and are encouraged to bring their families. They can discuss at length, with a counsellor, the ongoing problems and look for solutions together.

Residential Program for Family Members

The criteria for admission to the in-patient program of Broadway Lodge were:

1. Initial assessment of problem

2. Association with Al-Anon for a period of time

3. May, or may not, have had other professional help

4. Referral to out-patient treatment has proved insufficient

5. Conjoint family group has been insufficient

6. Therefore a residential program has been recommended.

Experience has taught us that this part of our family program is unique and meaningful. Family members participate in the same groups, lectures, discussions, share bedrooms, live, and work with the alcoholic patients. They are admitted at a different time from their own dependent family member. Most family members are admitted for about four weeks. They attend two group therapy sessions daily and one on Saturdays. The lectures which they attend are on many aspects of recovery, including the physical, emotional, and psychological recoveries and are also on the tenets of AA and Al-Anon, with their practical application. Each family member is assigned their own counsellor who will see them two to three times a week, discuss their individual problems, and give them reading and written assignments. They also will write their life story as they see it with particular reference to their association with their alcoholic. Psychological testing (The Minnesota Multiphasic Personality Inventory and the Shipley) are used to assess all inpatient family members.

As with the alcoholic, the program is based on the ideas of Reality Therapy and Rational Emotive Therapy, which are adapted to the individual needs of the patient and stress also the need for ongoing change. The Al-Anon program is also utilized in order to make an easy transition after treatment. With the help of the counsellor and peer group, the family member begins to realize that they can only attain their own recovery by working on themselves. They begin to look at their own attitudes, personality problems, and feelings, rather than those of their drinking partner.

278

Table 1. The Broadway Lodge Program Schematically Presented

1. Assessment
2. Individual patient/plan interdisciplinary staff team
3. General program orientation
     (a) Small task oriented group meetings
     (b) Didactic "informational" lectures
4. Individualized program
     (a) Assignment of focal counsellor
     (b) Work assignment
     (c) Professional staff assistance
     (d) Clergy involvement
5. Phases (goals of treatment)
     (a) Help patient to admit and accept powerlessness over
         alcohol/the alcoholic and the unmanageability of life
     (b) Help patient to recognize there is a need to change;
         they have the ability to change;
         the program is the vehicle for change
     (c) Help patient to take action here and now to make a
         decision and commitment to change;
         to modify life style;
         practise doing it in treatment
     (d) Analysis of past character deficits related to
         behavior
     (e) Help patient summarize in-depth examination with
         counsellor
     (f) Positive plan for future growth
6. Discharge planning
     (a) Aftercare
     (b) Affiliation with Al-Anon

The family member begins to learn that, like the alcoholic, they cannot change and recover by themselves in isolation. They need others to help them see themselves and to suggest how they can change. These mixed groups of alcoholics and family members work remarkably well. Both parties can begin to see and feel how the other is affected by the disease. Repressed anger and self-pity come to the surface more quickly than in separate groups, and permit recognition of the blaming phenomenon (defense) on both sides. The alcoholics share their fear and anger at the family for trying to control their drinking and the resultant behaviors as well as their resentment at being treated like pathetic drunks. Family members are in their turn quick to spot the "victim" attitude of the alcoholic and how they are manipulated by the drinker to feel guilty and responsible. They see clearly, for the first time, by close association with other alcoholics, that they are not responsible for their own alcoholic's drinking.

The focus of treatment, after facing these difficult problems and feelings, is for the family member to perceive that change is essential and possible. They come to the fact that they themselves are responsible for this change regardless of whether they continue to live with their alcoholic. They come to realize, for perhaps the first time, that they are worthwhile people in their own right and not merely an appendage to an alcoholic partner or parent.

Since the inpatient family program commenced in 1978, we have admitted 41 people: 7 men and 34 women.

DISCUSSION

As would be expected of inpatient family members, their psychological state would tend toward pathology. In this study, pathology is defined as two standard deviations from normal on the Minnesota Multiphasic Personality Inventory (MMPI). Of the 31 patients remaining in the family program, 26 women and 5 men (those transferred to the chemically dependent program are not included in the MMPI data) only two scored within the normal limits on admission, one wife and one son.

In regard to the men in this study, Fox's Five Category schema seems to be present, at various stages in the relationship, as a way to cope with the stress of living with the alcoholic. ("The long suffering martyr, the unforgiving, self-righteous husband, and the dependent husband -- disappointed to find his previous self-confident wife now dependent on him.") In their own right they also correspond or have some features of Rimmer's (1974) findings of depression and other types of psychiatric disturbance.

Table 2. Treatment Status of Admissions to Inpatient Family Program

	Women		Men		Total	
	N	%	N	%	N	%
Completed treatment	22	65	4	57	26	65
Noncompleted	2	6	1	14	3	7
Transferred to chemically dependent program	8*	23	2**	29	10	24
Currently in treatment	2	6	0	0	2	5
Total	34	100%	7	100%	41	100%

*4 left against staff advice (they were unwilling to face their alcoholism) and 4 completed treatment.
**Both of whom left against staff advice (they were unwilling to face their alcoholism).

Table 3. Age Distribution

Years	Women N	Men N	Total N	%
15-21	4	1	5	15
21-30	10	0	10	32
31-40	5	2	7	22
41-50	3	1	4	12
51-60	4	1	5	17
Mean Age	33.3%	38.4%	35%	
SD	12.1	14.3	2.3%	
Range	38.5	39.5		

Table 4. Relationships to the Alcoholic

	N	%
Wife	15	48
Husband	2	6
Children		
Daughter of alcoholic mother	6	20
Daughter of alcoholic father	3	10
Daughter, both parents alcoholic	2	6
Son of alcoholic father	3	10
Total	31	100%

Table 5. Distribution of Family Program
Inpatients' Posttreatment

	Returned to Partner		Subsequently separated	
	N	%	N	%
Women				
Patients who completed treatment	17	74	5	83
Noncompleted treatment N.B. 2 patients currently in treatment	2	9	0	0
Men				
Patients who completed treatment	4	17	0	0
Noncompleted treatment	0	0	1	17
Total	23	100%	6	100%

The five men (two husbands, three sons) all converted anger and guilt feelings into physical symptoms. These included chest pain, partial paralysis of hands, severe headaches, severe back pain, and gastric symptoms. These symptoms resulted in lost time from work. They all also had consulted a physician or spent time in a hospital because of conversion anxiety or hypochondriacal conditions. All of them were over-achievers, perfectionistic, and successful in their fields of endeavor until, at times, they were immobilized by psychological stress.

All had been involved in Al-Anon -- three had extensive involvement but found themselves unable to utilize it sufficiently to deal with problems and feelings.

These men made significant changes in treatment, especially in dealing with repressed anger, of which they were extremely fearful. As they dealt with their feelings, their depression lifted and they were able to deal more realistically with their feelings of self-pity.

Subsequent MMPI followup, six months posttreatment of the four who completed, indicated that they have consolidated the progress that they made in treatment in dealing with fealings of guilt and anger. Their MMPI scores are within normal limits. They have had considerably less problems with conversion reactions and hypochondriacal complaints. This is supported by reports from their families and significant others. The family relationships of these men have improved in quality with the exception of one whose wife died in an alcohol-related accident. He, however, has continued to recover.

A group of women who have for many years been involved in relationships of a pathological nature, might be expected to need inpatient treatment. They would combine some of the features of both "the disturbed personality" and the "stress theory." (Edwards, Harvey, and Whitehead, 1973). Of the 26 women (22 completed, 2 left noncompleted, 2 currently in treatment), 15 of the group were wives of alcoholics, 3 of whom also had one alcoholic parent. Eleven were daughters and 2 of the daughters had parents, both of whom were alcoholic. Three of the daughters had also married alcoholics!

On admission, the <u>passively angry/victim type</u> was the most predominant feature of the MMPI profiles. Feelings of guilt and obsessive worry were also a very dominant feature. Fifteen of the women also had strong paranoid tendencies. The younger women, and daughters of alcoholics in their late teens and early 20's, were all lacking in self-esteem; even more so than the older women or wives (McLachlan, 1973). Several of these younger women

Table 6. Marital, Educational, and Vocational Background
of 31 Family Program Patients

	N	%
Married	16	50
Single	9	30
Divorced	3	10
Widowed	0	0
Separated	3	10
Total	31	100%
Left school before 16 years	7	20
Left school after 16 years	2	5
"O" or "A" levels	17	55
University/Degree	5	20
Total	31	100%
<u>Vocational Status</u>		
1. Professional/Tech. Admin.	10	30
2. Farmers/Farm Managers, Service/ Ind/Clerical	8	25
3. Skilled/Semi-skilled	1	5
4. Unskilled/Laborers	1	5
5. Housewife	7	20
6. Student	4	15
Total	31	100%

were also extremely introverted. Two had experienced psychotic episodes: one a daughter, as a child, and another just prior to admission to treatment. The latter temporarily needed chemotherapy. Fourteen of the group were also quite anxious and converted stressful emotional experiences into physical complaints.

At admission, 25 out of 26 women had profiles with several scales greater than two standard deviations above normal. Twenty-two women were evaluated during the followup posttreatment. Only 12 still manifested <u>one</u> scale greater than two standard deviations above normal. Six out of the 12 had one elevated scale which was related to repressed anger. Of the remaining 6 with an elevated scale, 3 manifested continuing impulsiveness, and 3 continued to show paranoid tendencies. All of these 22, at six-month followup, showed improvement in intra-extra personal relationships with family and friends. Notwithstanding this change, 5 of the women subsequently separated from their partners (3 husbands were continuing to drink; 2 were not, but a separation had been agreed upon).

All the women have continued to show improvement, and the quality of their lives has been enhanced. This is substantiated by family, significant others, and Al-Anon contacts.

Six-month followup of the daughters indicated significant improvement in the area of self-esteem. This occurred even if the parent was still drinking; although obviously in the area of family harmony and relationships there are still some difficulties. Where the parent is in recovery there is significant improvement after treatment, even where there had been a troubled relationship before the daughter's admission.

Obviously a group of 31 is a very small proportion of family members in relation to all families treated at Broadway Lodge -- there have been over 800 alcoholics and their families (conservatively 2,000 people) since the end of 1974.

This group represents people who have such profound problems that they need time and concentration on self to recognize their need for and ability to change. Intrapersonal improvement has been demonstrated by the MMPI. Extrapersonal growth has been demonstrated in the areas of dysfunction prior to treatment (repressed anger, guilt, over-achievement, perfectionism, fear, self-pity, obsessive worry, lack of self-worth, and introversion) by clinical observation and data confirmed at followup by family, significant others, personal physicians, and Al-Anon members.

CONCLUSION

These groups were not recovering with the help available. They had participated in Al-Anon, which they seemed unable to use or address the emotional needs of their condition. Some had tried professional or psychiatric counselling which had not addressed the significance of the family disease of alcoholism. Both helping agencies had only touched part of the problem. There seems to be a small group of people who need more than one-to-one counselling, or weekly conjoint family group therapy if they are to attain alleviation of the interpersonal disturbances and reduce the stress related to their involvement with the alcoholic. More research is needed in the area of followup of family members in treatment, whether it be in Al-Anon, conjoint family therapy, out-patient treatment, or treatment in a residential program. The family has an ongoing need for recovery, change, and growth in exactly the same manner as the dependent person.

The reader is reminded that the type of treatment most appropriate for the individual needs of each family member requires much more careful attention by all helping agencies.

REFERENCES

Black, C. "Children of alcoholics." Alcohol World. N.I.A.A.A., 4.1: 23-27, 1979.

Cork, R. M. The Forgotten Children: A Study of Children with Alcoholic Parents. Toronto: Alcoholism & Drug Addiction Research Foundation of Ontario, Canada, 1969.

Ditzler, J. M., "Alcoholism: A family illness." Nursing Times, 76(25): 1103-1105, 1980.

Edward, P., Harvey, C., and Whitehead, P. "Wives of alcoholics: A critical review and analysis." Quarterly Journal of Alcohol Studies, 34: 112-132, 1973.

Ellis, A. Guide to Rational Living. New York: Institute for Rational Living, 1961.

Fox, R. "The alcoholic spouse." In: V. W. Eisenstein (Ed.), Neurotic Interaction in Marriage. New York: Barie Books, 1956.

Gacic, B., Importance of the family and social network in the treatment of alcoholics. Proceedings of the 27th International Institute on the Prevention and Treatment of Alcoholism, Vienna, 1981.

Glasser, W. Reality Therapy: A New Approach to Psychiatry. New York: Harper & Row, Inc., 1965.

Jackson, J. K. "The adjustment of the family to the crisis of alcoholism." Quarterly Journal Stud. Alc., 15: 562-585, 1954.

Kammlier, M., and Laundergan, J. The Outcome of Treatment. Center City, MN: Hazelden Foundation, 1977.

Leite, E. When Daddy's a Drunk. Center City, MN: Hazelden Foundation, 1979.

McLachlan, J. F. C., Walderman, R. L., and Thomas, S. A. "A study of teenagers with alcoholic parents." Donwood Institute Research Monog., No. 3, Toronto, 1973.

Mason, M. "Relationships in recovery. Sexuality as an intimacy barrier." Alcoholism -- A Modern Perspective. Lancaster: M. T. P. Press, Ltd., Jan. 1982. (Proceedings of Alc.80, an International Conference on Alcoholism, University of Bath, England, 1980.)

Maxwell, R. The Booze Battle. New York: Praeger Publishers, Inc., 1976.

Orford, J. Alcoholism: New Knowledge & New Reponses. London: Croom Helm, 1977.

Pilat, J. M. Children of alcoholics. Needs and treatment intervention. Proceedings of the 27th International Institute on the Prevention & Treatment of Alcoholism, Viennea, 1981.

Rimmer, J. "Psychiatric illness in husbands of alcoholics." Quarterly Journal Studies on Alcoholism, 35: 281-283, 1974.

Ward, R. F., and Faillace, L. A. The Alcoholic and His Helpers. Study by Department of Psychiatry of the Baltimore City Hospitals, the Johns Hopkins University School of Medicine, and the Alcoholism Research Unit, Baltimore City Hospital, by grant from National Institute of Mental Health, 1981.

THE THERAPEUTIC COMMUNITY APPROACH TO REHABILITATION:

PERSPECTIVE AND A STUDY OF EFFECTIVENESS

George De Leon

Director of Research and Evaluation
Phoenix House Foundation
New York, New York 10023

INTRODUCTION

Since the 1960's the spectrum of drug abusers in America has widened. Opiate abuse and its catalytic effect on crime prevails particularly in the disadvantaged minority subcultures. Marijuana, polydrug, cocaine, and alcohol abuse interact malignantly with the development and health of youth in all classes. Adult misuse of licit and illicit substances is pervasive either for leisure or as aids to living.

Federal policy encouraged the development of a range of intervention approaches which could embrace individual differences among users in drug abuse patterns, life-styles, or motivation for a change. Thus, four major treatment modalities evolved: detoxification, methadone maintenance, outpatient settings, and drug-free residential therapeutic communities.

Not all of these approaches, however, constitute treatment or rehabilitation. This distinction illuminates some essential differences among the four modalities. As a treatment, the principal aim of detoxification is the elimination of a physiological dependence through a medically safe and relatively inexpensive procedure. A secondary objective is to refer the detoxified client to other modalities.

Methadone maintenance programs have been guided by the general view that opiate addiction is a recurring disease which may relate to physiological or metabolic anomolies, and that addiction to illicit substances assures involvement in a criminal life-style. The principal aim of methadone maintenance is to permit the addict to sustain a prosocial life-style while not being distracted by the illegal pursuit of narcotics. Abstinence from chemical dependency is not a primary goal, although it eventually occurs for some clients.

Outpatient drug-free centers include a diversity of ambulatory and day-care programs. Originally offered for relatively well-socialized opiate abusers, these outpatient settings evolved into a modality attracting nonopiate, alcohol, cocaine, and/or polydrug abusers.

The therapeutic community views abuse of any drug as self-destructive behavior, which is a reflection of personality problems and/or chronic

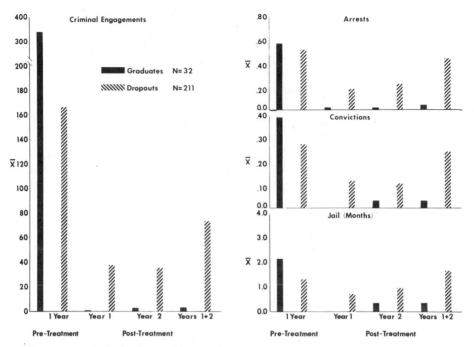

Figure 1. Criminal profiles of dropouts and graduates pre-treatment
and through two years followup.

deficits in social, educational, and marketable skills. Its antecedents lie
in socioeconomic disadvantage, poor family effectiveness, and in
psychological factors.

The principal aim of the therapeutic community is a global change in
life-style: abstinence from illicit substances, elimination of antisocial
activity, employability, prosocial attitudes and values. A critical
assumption, however, is that stable recovery depends upon a successful
integration of both social and psychological goals. Rehabilitation, therefore, requires multidimensional influences and training which, for most, can
only occur in a 24-hour residential setting.

Each modality has its view of drug abuse, each impacts the abuser in
different ways, and the effectiveness of each must be evaluated in terms of
its principal aims. These contrasts in approach underscore the rationale
for the ongoing research.

The effectiveness of the therapeutic community (TC) has been evaluated
primarily through followup studies. Recent reviews of this work are
contained in Bale (1979), Brook and Whitehead (1980), De Leon and Rosenthal
(1979), and Sells (1979). With few exceptions studies have focused upon the
"hard" variables of outcome, e.g., drug use-drug treatment, criminal
activity, employment, and education. TCs compare favorably with other
modalities on these measures (see, e.g., Burt et al., 1979; Sells, 1979).
Moreover, outcome status on these measures is positively related to time
spent in treatment (e.g., De Leon and Andrews, 1978; De Leon, Andrews,
Wexler, Jaffe, and Rosenthal, 1979; De Leon, Wexler, and Jainchill, 1982;
De Leon, 1984; Holland, 1978; Simpson and Sells, 1982).

These evaluations, however, have not addressed the social and
psychological goals of rehabilitation in drug-free residential settings.
The major aim of the present study, therefore, was to assess the

286

effectiveness of a single therapeutic community, Phoenix House, through a comprehensive followup investigation of psychological and social adjustment.*

METHOD

The followup sample consisted of male and female graduates (n = 53) and dropouts (n = 371) from the 1974 residential population of Phoenix House, a traditional therapeutic community in New York City. Dropouts were randomly selected by time in the program (<1 to > 17 months). Followup status two years after treatment was assessed through interview and agency records. Over 75% of the original sample completed a 4-hour interview that included psychological testing. There were no significant demographic, drug abuse, or time in program differences observed between the interviewed and uninterviewed clients. Only the interview findings are described in this paper.

Social adjustment profiles were obtained on all clients which comprised data on demography, social and family background, employment/schooling, drug use, and criminal patterns. Clients retrospectively traced their social adjustment from the year prior to treatment through all years of followup.

Status was assessed in two ways. First, criminality, drug, use, drug treatment, and employment were examined through 16 multiple outcome variables. Second, each individual received a global index based upon weighted combinations of criminality and drug use. Thus, a client was classified on a 4-point scale of success, from least favorable (1 -- any crime and any use of opiates or a non-opiate primary drug) to most favorable (4 -- no crime and no drug use). The most favorable criteria had to be maintained throughout followup to earn a best success index (No. 4).

Psychological profiles were obtained early in treatment and again at followup. Seven standardized psychological tests and scales were administered. All clinical and experimental scales of the MMPI; its reliability, validity, and wide applicability are extensively documented in the literature. Pathology Scales, validated in the psychiatric literature as indices of psychopathology, consisted of the Shortened Schizophrenia Scale (Sc), the Shortened Manifest Anxiety Scale (SMAS), the Beck Depression Inventory (BDI), and the Socialization Scale of the California Personality Inventory (Soc.). The Beta Intelligence test is a standard measures that is extensively used in every variety of population. It is constructed to minimize errors that arise from verbal and reading deficits. The Tennessee Self Concept (TSC) is a standardized test for assessing self-esteem and psychological disturbance. Norms and scores from various comparison groups have been published, and it has demonstrated sensitivity to detect psychiatric and self concept changes.

The analytical objectives in this study were (1) to quantify the relationship between social adjustment status at followup and time spent in program; (2) to correlate psychological status with success status at 2-years followup. Results for dropouts and graduates (completees) are described with respect to 10 outcome variables and the global success index.

OVERVIEW OF FINDINGS

Criminality: On all measures -- criminal engagements, arrest, convictions, and months in jail -- there was a significant decline in each year of followup compared to the year before treatment. Graduates showed little evidence of criminal activity, while for dropouts criminal involvement was at a significantly reduced level (Figure 1).

---

*Full details of this and other relevant research are contained elsewhere (De Leon, 1984, 1979, 1976).

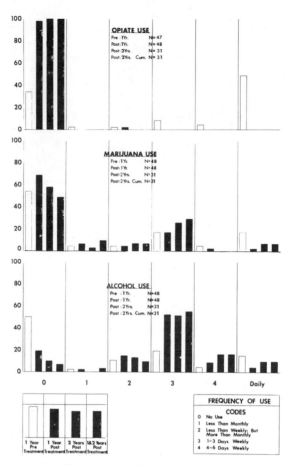

Figure 2. Frequency of drug use pretreatment and
through two years followup, for dropouts and graduates.

Drug Use: Opiate abuse was near zero for graduates and declined
significantly for dropouts. The proportions using marijuana and alcohol
showed modest increases for both groups but usually at a frequency of 1-3
days a week (Figure 2).

Drug Treatment: Across two years of followup no graduate entered drug
treatment. For dropouts the percentage of people in drug treatments
declined significantly compared to their pretreatment levels, but
percentage of time in drug treatment did not change.

Employment: On all measures of employment -- salary, job rank, months
working -- dropouts and graduates show significant improvement. The
graduate change was statistically greater (Figure 3).

Success Rates: Across two years of followup, 35% of the dropouts and
68% of the graduates maintained a best success index (#4), while 56% of the
dropouts and 94% of the graduates improved over their pretreatment status.

Time in Program: Among dropouts, length of stay in treatment was
highly associated with success and improvement rates posttreatment. The
rate of best success (#4) for those in the program less than 5 months, 5-12
months, and over 13 months were 18%, 32%, and 52%, respectively.
Improvement rates (positive change over pretreatment status) were 25%, 34%,
and 75%, respectively (Figure 4).

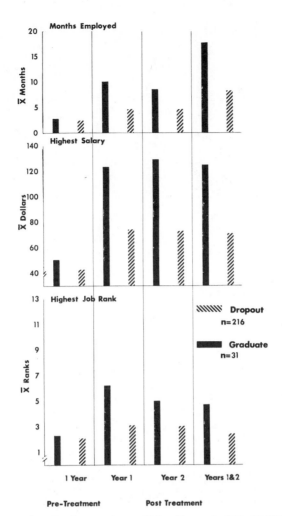

Figure 3. Employment profile through two years
followup, for dropouts and graduates.

Psychological Studies: The initial (Time 1) psychological profiles
reveal many signs of personality disorder, deviancy, poor self-esteem, and
low intellectual level. At two-year followup, overall psychological status
improved significantly across the entire sample but did not attain normative
or healthy levels. Maladjustment and dysfunction prevailed particularly
among shorter-staying dropouts. Improvements were greater and psychological
status was better for graduates, long-staying dropouts, and females.

Psychological Status and Success Status: Figure 5 shows that there is
a systematic and positive relationship between success status and
psychological adjustment at two-year followup. Among the two least
favorable groups (Index 1 and 2), psychological status was worse compared
with the two favorable success groups (Index 3 and 4). This relationship
was similar for dropouts and graduates, except that successful graduates
showed better psychological status than successful dropouts.

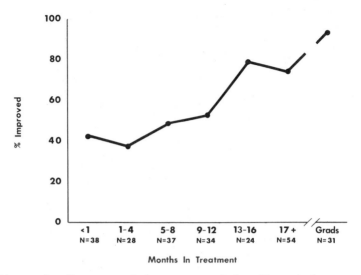

Figure 4.   Improvement in success status through two-years
            followup, by time in program.

SOME CONCLUSIONS AND IMPLICATIONS

The major findings of the present project provide convincing evidence for the effectiveness of the therapeutic community approach for drug abuse.   First, the systematic relationship between time in program and posttreatment success status firmly implies a relationship between "dosage" of treatment and outcome.   Second, success status related directly to psychological improvement at followup.   Those who made a better social adjustment revealed the best psychological change, a correlation which most directly reflects the global goals of rehabilitation in the theratpeutic community.

The relationship between treatment and client status at followup is firmly established in the present findings.   Still, however, treatment effectiveness is not proven in the rigorous sense of that word.   First, evaluation of other treatment and nontreatment factors remains to be assessed, particularly in terms of their influence on long-term followup. Second, further study must isolate the contribution of client factors, background and motivational, to length of stay and eventual outcome status. Third, despite the impressive psychological findings in the present studies, further research must focus more directly on treatment process components and their relationship to short- and long-term client change.   Fourth, differences between modalities and between programs with respect to treatment philosophy, program operation, staff experience, and clients, limit the generality of the treatment effectiveness conclusions obtained at Phoenix House.

Nevertheless, the present research represents the most comprehensive followup study to date of a single drug treatment program.   Moreover, Phoenix House is the largest therapeutic community nationally, and a most stable example of the traditional TC approach.   While the Phoenix therapeutic community is undoubtedly unique, the present research offers valid conclusions and hypotheses for further work in similar traditional TCs.

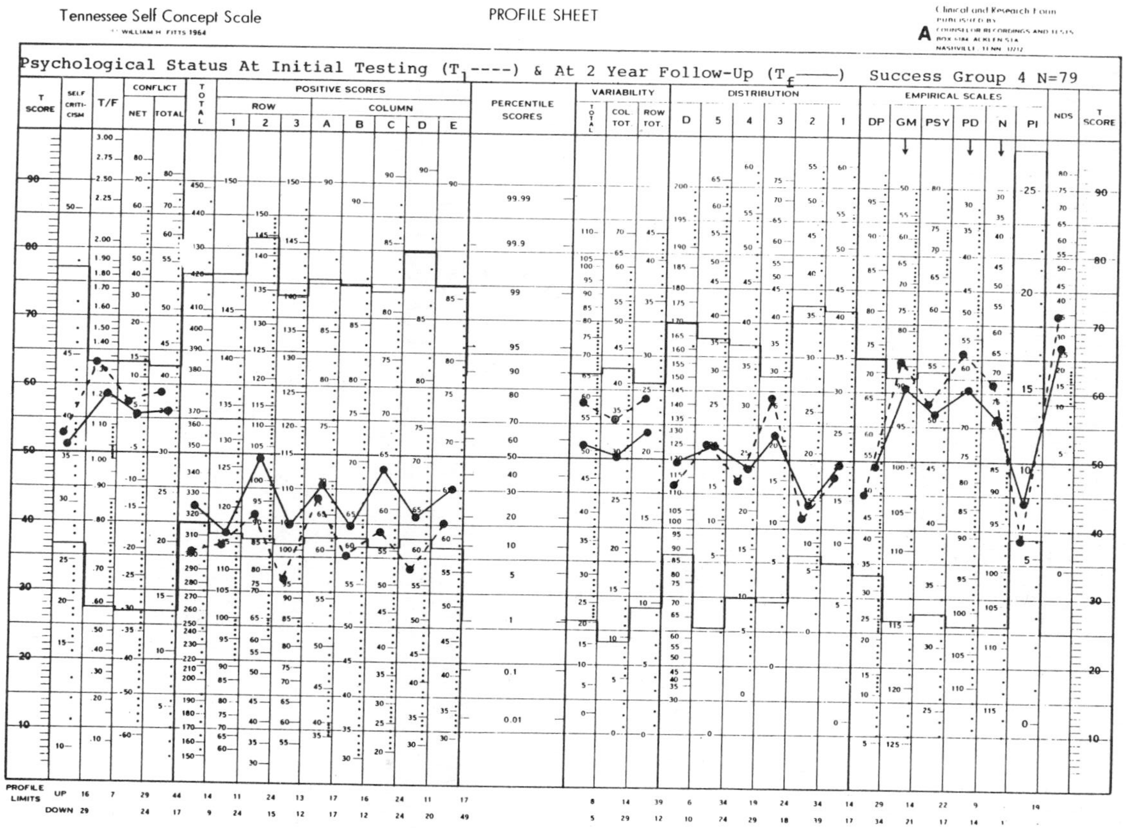

Figure 5. The TSC profiles in treatment (T_1) and at followup (T_f). Improvement is reflected in an elevation of the left segment (Positive Scales 1-E) and in a lowering of the right segment (Empirical Scales DP-NDS). Profiles are shown (A) for most favorable group (No. 4) and (B) for least favorable group (No. 1).

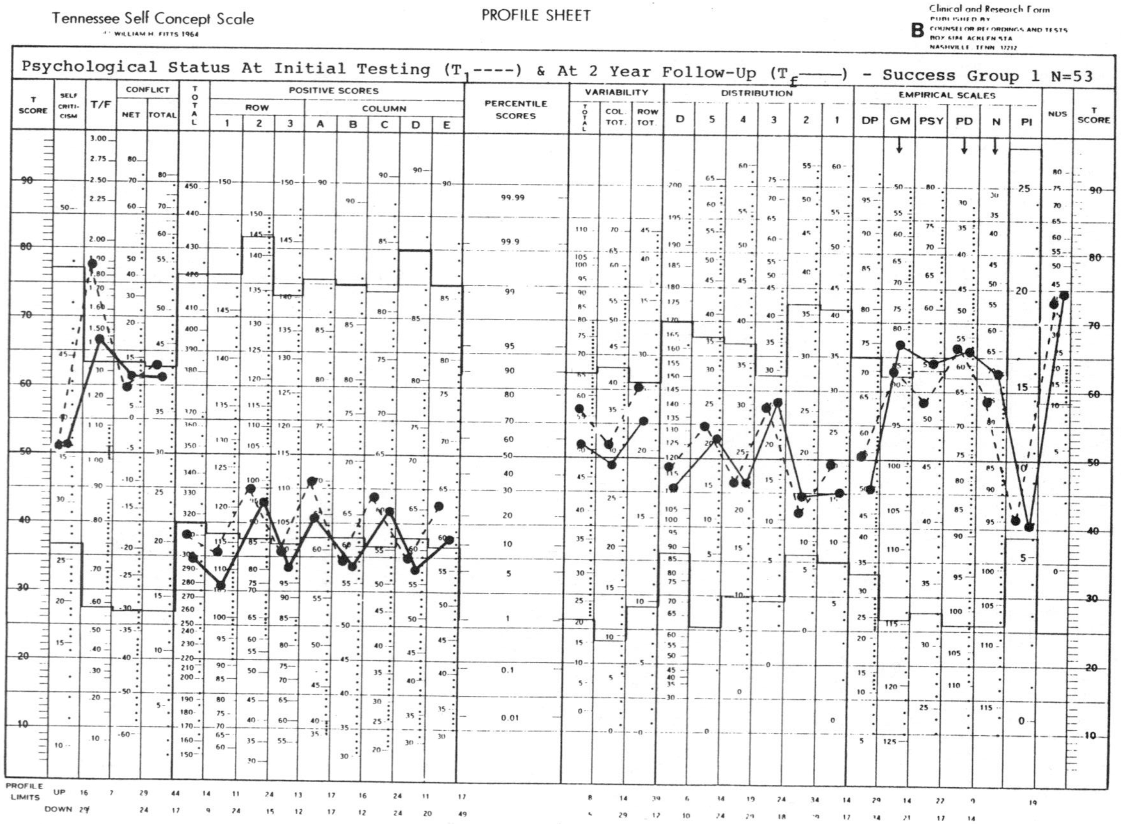

Figure 5. (Continued.)

REFERENCES

Bale, R. N. "Outcome research in therapeutic communities for drug abuse: A critical review 1963-1975." Int. J. Addict. 1979, 14(8): 1053-1074.

Burt, M. R., Pines, S., and Glynn, T. J. Drug Abuse - Its Natural History and Effectiveness of Current Treatments. Cambridge, Mass.: Schenkman, 1979.

Brook, R. C., and Whitehead, I. C. Drug Free Therapeutic Community. New York: Human Science Press, 1980.

De Leon, G. the Therapeutic Community: Study of Effectiveness. Rockville, MD: National Institute on Drug Abuse, Treatment Research Monograph Series, DHHS Pub. No. (ADM)84-1286, 1984.

De Leon, G. The Therapeutic Community: Study of Effectiveness. Final Report of Project Activities, NIDA Project No. DA-01228, November 1979.

De Leon, G. Psychological and Socio-Demographic, Profiles of Addicts in the Therapeutic Community. Final Report of Project Activities, NIDA Grant No. DA-00831, 1976.

De Leon, G. "Phoenix House: Psychopathological signs among male and female drug free residents." J. Addictive Disease. 1974, 1(2): 135-151.

De Leon, G., and Rosenthal, M. S. "Therapeutic communities." In Dupont, Goldstein, and O'Donnell (Eds.), Handbook of Drug Abuse. Washington, D.C.: National Institute on Drug Abuse, 1979.

De Leon, G., and Andrews, M. "Therapeutic community dropouts 5 years later: Preliminary findings on self reported status." In: B. Smith (Ed.), A Multi-Cultural View of Drug Abuse: Proceedings of the National Drug Abuse Conference, 1977. Cambridge, Mass.: 1978, pp. 369-378.

De Leon, G., Wexler, H. K., and Jainchill, N. "The therapeutic community: success and improvement rates 5 years after treatment." Int. J. Addict. 1982, 17: 703-747.

De Leon, G., Andrews, M., Wexler, H. K., Jaffe, J., and Rosenthal, M. S. "Therapeutic community dropouts: Criminal behavior 5 years after treatment." Am. J. Drug Alcohol Abuse, 1979, 6(3): 253-271.

Holland, S. "Gateway Houses: Effectiveness of treatment on criminal behavior." Int. J. Addict. 1978, 13(15): 369-381.

Sells, S. B. "Treatment effectiveness." In Dupont, Goldstein, and O'Donnell (Eds.), Handbook on Drug Abuse. Washington, D. C.: National Institute on Drug Abuse, 1979.

Simpson, D. D., and Sells, S. B. "Effectiveness of treatment for drug abuse: An overview of the DARP research program." Advances in Alcohol and Substance Abuse." 1982, 2(1): 7-29.

THE THERAPEUTIC COMMUNITY IN RETROSPECT AND PROSPECT

William B. O'Brien and Sy Halpern

Daytop Village, Inc.
New York, New York 10018

The 20th century therapeutic community movement is a phenomenon which owes its existence as much to the failure of the medical model as to any other reason. One could look back at some of the historical attempts at treating drug abuse with a sense of humor if the problem were not deadly. Thus, we see solutions such as hypnosis in 1890, heart tonics and vichy water in 1895, static electricity in 1903, insulin in 1931, and moving along, refrigeration therapy in 1941, and prefrontal lobotomies in the early 50's (Brill and Lieberman, 1972). Across the world traditional institutions have been confronted with growing numbers of drug addicts, alcoholics, and more recently a new breed, the polydrug abuser.

The proliferation of symptomatology among youth serves to underscore a key premise of the Therapeutic Community. Addiction itself, as well as drug misuse, are but symptoms. They are painful reminders of the world we live in; a world where morality, faith, and the institutions of home and family are under constant attack by disenchanted people. Morality and values may find little context in pharmacology or physiology. Nor can they provide roadmaps for progress when treatment as process, technique, or ideology is based largely on the concept that the addict/abuser's problem is primarily physical in etiology or nature.

These two factors -- morality and values -- can, however, to the extent that they are missing, account for the growing number of young and older people who are increasingly turning to simplistic chemical solutions, born of discouragement and despair. The eminent psychiatrist, Karl Menninger, has noted that too many of us have enjoyed great prosperity and scientific advances while closing our eyes to and forgetting the plight of our neighbors and environment as "we began to sink into a posture of apathetic inaction, the old time sins of acedia and anomie" (Menninger, 1973).

The therapeutic community success in restoring thousands of young substance abusers to drug-free productive lives is not a de novo phenomenon of the 20th century. Vexatious dilemmas have been the lot of mankind throughout recorded history. At the time of the American Revolution in 1777, David Hume, an English philosopher, noted that the first question regarding any cause is its necessity: "Whatever begins to exist, must have a cause for existence." The cause for the self-help movement might well be the historical inability of governments, social systems, and organized religion to deal with or ameliorate the day-to-day pain and suffering that has been the companion of man from time immemorial. Tacitus, a Roman historian, may have given the ultimate explanation for governmental or systematic myopia to the citizenry's plight when, some 2000 years ago, he commented: "Misserimam servitutem pacem apellant" (The most wretched slavery they call peace).

O. Hobart Mowrer, a noted psychologist, has done extensive research on the links between the therapeutic community of the 20th century and early Christian development (Mowrer, 1976). He points out that early Christianity was essentially a small group movement; albeit a movement with extraordinary appeal and power. There were no churches or cathedrals, but rather small groups of people meeting in homes which, of course, put a real limit on the possible size of such groups. Here we find the practice of Exomologesis, a Greek term which has today dropped out of common English usage. The term described a process which "involved complete openness about one's life, past and present, to be followed by important personal changes, with the support and encouragement of other members of the Congregation." The practice of mutual honesty, amendment of life, and growing involvement with the concern for others brought joy and freedom where there had been pain and suffering. The concept involved self-disclosure and confession of sin. This was followed by an appropriate announcement of penance, pleas for forgiveness, and plans for making restitution. A final period of friendly fellowship, Koinonia, closed the meeting. This general formula continued until the Council of Nicea (A.D. 325) when Constantine occasioned the ending of the requirement of open personal disclosure and replaced it with private confession to a priest. The concept of Exomologesis, interestingly enough, also surfaced in other places. Based upon material found in the Dead Sea Scrolls, it has been linked to the Qumran Community. The philosophy of mutual criticism is a salient feature of the 20th century therapeutic community: "They shall rebuke one another in truth, humility, and charity. Let no man address his companion in anger, or ill temper, or obduracy, or envy prompted by the spirit of wickedness. Let him not hate him (because of the wickedness of his uncircumsized heart), but let him rebuke him on the very same day, lest he incur guilt because of him. Furthermore, let no man accuse his companion before the Congregation without having first admonished him in the presence of witnesses."

The principles of Aedification Mutua (Mutual Edification) and Correptio Fraterna (Brotherly Correction) lie deep within the teachings of the New Testament and Gospels. There are many references to this throughout the Bible. In Thessalonians 5:11 we hear Paul exhort: "encourage one another and build one another up." The building process involves man's taking his fellows to task for their actions. In Colossians 3:16 Paul further enjoins: "teach and admonish one another." The regular confrontations that are a daily experience within the therapeutic community embody a precept in Hebrews 3:13: "but exhort one another everyday, as long as it is called 'today' that none of you may be hardened by the deceitfulness of sin."

Many centuries have passed since the Hebrew prophets preached the importance of a normal life in a world beset with an emphasis on material values. This focus is not, however, the exclusive domain of the Judeo-Christian world. There are parallel findings in Hinduism, Buddhism, Confucianism, and Islam. The ancient Hindu guide of souls had a vocabulary of the normal life comparable to that of the West. In the workship of Varuna there is a well-developed belief in a deity who is "all seeing, all judging, and merciful to a praying sinner."

This excursion into the past can very well provide a sense of continuity to many of the key concepts embodied within the 20th century therapeutic community and, moreover, provide some needed perspective in a century where humility has become an almost anachronistic trait.

Dr. Frederick B. Glaser has eloquently tied together the connection: "We may learn much from the history of the therapeutic communities. No wonder they seem different; they are the offspring of religion in a field dominated by science and medicine. No wonder they seem familiar; all of us have had more or less contact with religious principles. No wonder they place reliance upon individuals only recently emerged from the problems which they now confront; of such were Saint Mary Magdalene, Saint Paul, Saint Augustine."

If there is uniformity and great similarities to therapeutic communities around the world, it may well be that it derives from the unity of basic belief among the various religions they stem from. The rebirth, growth, and subsequent proliferation of therapeutic communities around the world in the past few decades represents one of the most fascinating social developments of our time.

The term "therapeutic community" as it is employed today refers to a phenomenon which emerged simultaneously in England and the United States in the middle of the 20th century (Jones, 1959). In England, Dr. Maxwell Jones was experimenting in humanizing institutional treatment settings both in Dingleton Hospital in Scotland and Henderson Hospital in London. The experimentation evolved into a method of organizing the social structure of a treatment setting to cultivate and take advantage of natural social components.

Filstead and Rossi have identified a number of organizational elements within the process:

First, there is a shift in the conceptualization of the patient role from a passive to an active agent in treatment. Second, the two class system of staff and patients, with its resultant authority pyramids, is flattened, and efforts to develop a sense of oneness through such procedures as daily community meetings, patient government, etc., are employed. Third, there is the shift in the staff's role. The traditional professional role is quite inappropriate to the therapeutic community. Therefore, in order to operate an effective therapeutic community, the professionals have to lose their professionalism. This means that the program (or the community) becomes the prime identity; the profession becomes secondary. Fourth, open communication between staff-patient, patient-patient, and staff-staff is essential. Finally, it is important to have the program resemble, as closely as possible, the real world "outside" the institutional setting.

The exciting early development of the "therapeutic community" in Europe, preceded by some 10 years its genesis in the United States. It is interesting to note that its English pioneers, while enthusiastic, were scientifically cautious and did not feel that they had come across a particular treatment method.

Thus, in 1959, Dr. Maxwell Jones was looking for further evidence of greater utility in the concept he was describing: "In conclusion I would like to stress that, in my opinion, the concept of a therapeutic community has value in the spheres of nurse and staff training generally, in opening up communications between all hospital personnel and patients, and in consequence of these factors improving patient management. It may and I believe does amount to a specific treatment method, but this awaits further development and proof."

In America in the late 1950's and early 1960's drug addicts had pitifully few choices if they were inclined to treatment. Although there were hospitals and prisons, not one of these choices was efficacious. Even the euphoria induced by drugs could not shroud the addict's acceptance of society's view that he was incurable. This dark view fed upon itself as hospitals and prisons released people of whom as many as 98% immediately fell into recidivistic patterns. Against this gloomy backdrop there appeared a ray of hope that was actually curing addicts without medical intervention. Charles Diedrich, a recovered alcoholic, had expanded his self-help group to include narcotic addicts and ultimately broke with AA to form Synanon and deal with addiction.

This was exciting news. In 1962, the National Institute of Mental Health awarded a grant to the probation department of New York City's Brooklyn Supreme Court to explore the possibility of starting such a program. Two professionals, a psychiatrist, Dr. Dan Casriel and a criminologist, Dr. Alexander Bassin, visited Synanon. They returned with a blueprint for a pilot project, Daytop Lodge, which came into being in 1963. Daytop Lodge was limited to 22 male probationers and was established on Staten Island. It was an experiment which failed in achieving its various treatment goals. Its participants, the addicts in treatment, continued with their street culture life-style -- their expertise at manipulation, conning, etc., and the professionals, who ran the program, continued with their focus and foreign models and techniques. A subculture quickly developed where the "patients" learned the degree of change that they would have to simulate in order to obtain rewards from their mentors.

Another source of problems was a lack of strong leadership. After the initial unsuccessful year an attempt was made to revitalize the program by bringing in a recovered exaddict from Synanon (David Deitch). His major impact was to offer a glimpse of an exaddict who had made it -- an active, available role model.

As in many other arenas of health and social intervention, it became apparent that the role-model concept was critical in any attempt at drug-free rehabilitation. Addicts who had believed that real help was impossible, that "once an addict, always an addict," were now stripped of that defense. The program began to work and treatment was provided almost exclusively by exaddicts. Most of the graduates were able to find employment either within Daytop or with other agencies who were anxious to share in the success that the program was experiencing.

Daytop's leadership has grown up in Synanon and brought with it the philosophy of the subculture. In essence, it is a view that saw the world as negative and even sick, and maintains that to return to that world would be tantamount to falling. This was no problem so long as graduates remained within programs as staff or were absorbed by a growing number of emerging programs. By the mid-sixties new therapeutic communities began to spring up across New York City, the Midwest, and the West. After a long period of competitiveness and uneasiness that had all the characteristics of "sibling rivalry," moves were made to bring the therapeutic community field together.

In 1975, seven therapeutic communities joined together to form Therapeutic Communities of America Incorporated (TCA), which would serve as the unified spokesman for the movement goals. In 1981, the organization is a vibrant, 120-member consortium. TCA represents over 12,601 treatment slots from New England to Seattle and Alaska, from Florida to California and Hawaii. An average of ten new therapeutic community members are added to the roster at each quarterly meeting.

Internationally, under the aegis of the International Council on Alcohol and Addictions (ICAA) based in Lausanne, Switzerland, therapeutic communities and the self-help, drug-free approach have conducted World Conferences beginning in 1976 in Sweden, followed by increasingly successful conferences in Montreal (1977), Rome (1978), New York (1979), and Amsterdam in 1980. Not the least of the benefits of the worldwide interaction of therapeutic communities is the training and exchange programs that have been operative.

On September 3, 1980, The World Federation of Therapeutic Communities (WFTC) was chartered by unanimous resolution of the Executive Council of the ICAA/TC Section in the Netherlands. That organization numbers among its members therapeutic communities in Canada, the Philippines, Malaysia, Great Britain, the Netherlands, Belgium, France, Sweden, Norway, Federal Republic of Germany, Italy, Thailand, Iceland, Ireland, Brazil, Panama, Venezuela, Algeria, Scotland, Puerto Rico, and South Africa, among others. Monsignor William B. O'Brien, first President of the WFTC, expressed that organization's philosphy in his initial President's Message:

> Therapeutic communities are all too familiar with the struggle to survive in an environment which is harshly adverse. We have seen the tragic results of an ever-expanding emphasis on the medical and the pharmacological. We are weary of simplistic, phantasmo-goric solutions dependent on chemistry to solve a dilemma that is not chemical. Somewhere along the road there has been far too much focus upon chemistry and far too little focus on the human dilemmas, i.e., the disintegration of the family and its attendant values. Said breakdown results in a monumental sea of turmoil for which the symptom of addiction represents only the tip of the iceberg.

The theory underlying the therapeutic community approach to substance abuse may be viewed from the perspective of Daytop Village, America's largest drug-abuse treatment program. Daytop's approach to substance abuse is that it is largely the result of the process and experiencing of alienation. Abuse behavior, resulting from alienation, is assumed to stem from or is exacerbated by faulty sociopsychological development. The individual, manifesting this kind of disorder, typically reacts to personal stress by withdrawing into a protective shell, viz., drug abuse. Such behavior is seen as immature and as reflecting felt-inadequacy or incompetence in dealing with stress. In response to these real or imagined shortcomings, the abuser, while using drugs as a defensive cover, tends to overtly compensate by developing an inflated self-image and a false sense of superiority. As a consequence, he does not relate openly and honestly about himself, attempts to manipulate others through a show of pretended dependency, and is unable to express real concern for others. Such behavioral strategies on the part of the individual become reinforced through his association and identification with others who are perceived as being similar.

As for these individuals whose drug-dependency seems to emanate mainly from social privation and alienation, the theory assumes a lesser degree of personality weakness, and more of a given debilitating social environment as the etiological basis. The abusers who would fall into this diagnostic category typically include individuals of minority ethnic groups who are subjected to racial, social, and economic discrimination and frustration. Such individuals appear to see their existential condition as hopeless and in response tend to seek a kind of oblivion from reality through the use of drugs.

The above concept is based on the clinical experience of the program, as well as some familiar psychodynamic concepts adhered to by the social work field. It is not intended as a formal theory nor does it claim a full understanding of the abuse dynamics. Its principal use has been in providing a conceptual framework out of which the therapeutic techniques are derived, tested, and used. Based on the aforementioned definition of drug abuse, the clinical theory of the program assumes the following: In most cases, individual abusers should be initially removed from their immediate natural environment which may be contributory to continued abuse behavior, and should subsequently reside within a controlled therapeutic setting which has the capability of partially keeping out negative environmental influences. During this residential period, personality and interpersonal dynamics underlying abuse can be improved through a regimen of varied therapeutic activities, thereby lessening and eventually eliminating drug-taking behaviors. Following this period, and as a means of testing the individual's growth derived from the residential experience, the individual gradually re-enters the outside community. And, during this latter period, the resident is given direct assistance in "basic survival" (i.e., counseling for future involvement in school, career development, and job placement).

Therapeutic communities reject the concept of the drug addict as a sick person who is automatically entitled to all the prerogatives of the ill in our society. Sick people are in hospitals where things are done for them: someone makes their bed, cooks their food, and gives them special concern, leniency, and forgiveness. Residents of a therapeutic community do for themselves, as work is considered an integral part of the process. We attempt to develop autonomy in young people who have not achieved that crucial developmental goal of adolescence. Residents learn to establish cooperative and workable relationships with peers, without being dominated by those peers, and preparing for a meaningful vocation. In the process they must develop a philosophy of life (e.g., a set of guiding moral beliefs and standards) which can lend some order and consistency to the many decisions and actions they will have to make and take in a diverse, changing, sometimes chaotic world.

There are two simple prescriptions which have been at the core of therapeutic community treatment: (1) no physical violence; (2) no narcotics or other chemicals and, by inference, no other shells under which to hide. Residents stay within the therapeutic community as a result of a strong group process, the genuine love and responsible concern that each member of the community gives to his brother and sister.

Therapeutic communities appear to answer basic human needs. Dr. Abraham H. Maslow succinctly summarized this in his reflections on a visit to Daytop:

It seems to me that there is a fair amount of evidence that the things that people need as basic human beings are few in number. It is not very complicated. They need a feeling of protection and safety, to be taken care of when they are young so that they feel safe. Second, they need a feeling of belongingness, some kind of a family, clan, or group, or something they feel that they are in and belong to by right. Third, they have to have the feeling that people have affection for them, that they are worth being loved. And, fourth, they must experience respect and esteem. And that's about it.... Could it be that Daytop is effective because it provides an environment where these feelings are possible?...

Isn't it a pity we're not all addicts, because if we were we could come to this wonderful place!

The therapeutic community places a heavy burden on the individual. It is an existential arena which calls upon people to be responsible for their own actions. Some visitors to therapeutic communities are taken aback with the apparent preoccupation with guilt -- a word that is looked upon as anathema to many involved in the helping professions. Words like moral values and ethics leave them with similar feelings of disquietude. The road to understanding the therapeutic community has to do with the aforementioned imponderables. I would suggest that they look to Erich Fromm, whose work Psychoanalysis and Religion may very well provide some answer to Karl Menninger's crucial question, "Whatever became of sin?"

The reaction to the awareness of guilt is not self-hate but an active stimulation to do better. Some Christian and Jewish mystics have even considered sin a prerequisite for the achievement of virtue. They teach that only if we sin and react to the sin not in fear but with concern for our salvation can we become fully human.

Despite these basic truths, successful treatment -- the outcome of a therapeutic community process -- is not a simple task. There is more to it than having an individual become drug free. People have problems with employment, socialization, housing, the legal system, training, and the day-to-day issues related to adapting and coping with life. In point of fact, there are a host of difficulties which await the recovered drug abuser. Any organization attempting rehabilitation in a significant sense is charged with realistically considering and hopefully fulfilling all these human needs.

Over the years therapeutic communities have chosen to become modern and sophisticated entities whose horizons have broadened as they cope with changing populations manifesting an ever increasing variety of addictions and social problems. Some of the therapeutic communities of the 1980's have widened their scope to include other segments of the acting-out disorder spectrum, e.g., adolescents, youthful alcoholics, and first-time offenders in the penal system, as well as prescription drug abusers. In an increasingly complex and intertwined society -- particularly in urbanized industrialized countries, the therapeutic community has of necessity developed links with health care institutions, mental health services, the courts, social welfare agencies, and other aspects of the intervention system.

REFERENCES

Brill, L., and Lieberman, L. Major Modalities in the Treatment of Drug Abuse. New York: Behavioral Publications, 1972.
Casriel, D., and Amen, G. Daytop. New York: Hill and Wang, 1971.
Devlin, C. The trial marriage in a modern therapeutic community: The relationship between the ex-addict and the paraprofessional and the traditional professional in drug-free therapeutic communities. National Drug Abuse Conference. San Francisco, California, May, 1977.
Fromm, E. Psychoanalysis and Religion. New York: Bantam Publications, 1950.
Geraigiry, J. Re-entry: A bridge between the therapeutic community and society. National Drug Abuse Conference. San Francisco, May, 1977.
Glaser, F. The Origins of the Drug-Free Therapeutic Community: A Retrospective History. Addiction Research Foundation, Toronto, Ontario.

Jones, M. "Toward a clarification of the therapeutic community concept." British Journal of Medical Psychology, 1959.

Lowinson, H. J., and Ruiz, P. Substance Abuse: Clinical Problems and Perspectives. Williams and Wilkins, 1981.

Maslow, A. H. The Farther Reaches of Human Nature. New York: Viking Press, 1971.

McNeil, J. T. A History of the Cure of Souls. New York: Harper & Bros., 1951.

Menninger, K. Whatever Became of Sin. New York: Hawthorn Brooks, Inc., 1973.

Mowrer, O. H. Morality and Mental Health. Chicago: Rand McNally, 1967.

Mowrer, O. H. "Therapeutic groups and communities." In P. Vamos and J. Devlin (Eds.), Proceedings of the First World Conference on Therapeutic Communities. Montreal, Canada: Portage Institute, 1976.

O'Brien, B. W. Drug Abuse and Alcoholism in Youth: A Perspective. Sixth International Institute on the Prevention and Treatment of Drug Dependence. Hamburg, Germany, June 28-July 2, 1976.

Rossi, J., and Filstead, W. The Therapeutic Community. New York: Behavioral Publications, 1973.

The World Federation of Therapeutic Communities. International Newsletter. Vol. 1, No. 1, Montreal, Canada: Portage Institute.

THE AVAILABILITY OF ALCOHOL AND ALCOHOL PROBLEMS:

SOME THEORETICAL EXPLANATIONS AND EMPIRICAL CONSEQUENCES[1]

Jerome Rabow and Ronald K. Watts

Department of Sociology
University of California, Los Angeles
Los Angeles, California

Stating that the availability of resources influences social life would indeed seem obvious. Cities and ports are built at the mouths of rivers, valleys attract nomads and urban villagers, and urban centers have served as gravitational pulls upon members of society. Yet, the availability of resources has not been systematically studied in social science research. The number of works utilizing this concept seems all too few. One such work involved the choice of friends, which was shown to be determined by position of residence in a housing complex rather than by personality or race (Deutsch and Collins, 1951). The destruction of resources brings to mind the death by dieselization of at least one community (Cottrel, 1951). Another study on prostitution reveals how such resources may influence commuting and residential patterns (Goldman, 1981). Recently the issue of availability has come to the fore in the area of handguns and violent crimes (Cook, 1981), while tourism has been linked with alcohol consumption (Watts and Rabow, 1981). After these observations are made, one may also, upon further reflection note that the building of a library or a bar in a neighborhood might not increase or modify the behavioral patterns of members located with proximal access to the new resource. The library and bar might lose in the competition with television or other resources for the time, money, and energy of community members. While there may not be immediate changes in the patterns of use of these newly available resources, the continuous encounter with these resources could effect changes in use over time. This suggests that the mere availability of resources does not lead to increased utilization. In the case of alcohol consumption, which has so many personal and social functions, its availability and utilization is perhaps even more problematic than the utilization of either natural or social resources.

Our objective in this paper is to outline and review some of the theoretical issues and empirical work that we have undertaken on alcohol

[1]This paper is based on research funded by the State of California, Department of Alcohol and Drug Programs. Further support was obtained from the U.C.L.A. Academic Senate. The two authors have worked closely for many years and authorship is alphabetical and indicative of our genuine collaboration. Completion of this paper was possible through a grant from The Wine Institute and The Wine Growers of California. We are grateful to Patricia E. Schneider of the institute for her support.

availability, alcohol consumption, and alcohol problems. For the past seven years, we have been trying to determine the impact of availability upon alcohol consumption and alcohol problems in the State of California. Our own efforts have to be placed in the context of the prior work that exists in this particular area, and it is to this task we first turn.

Currently, two major theoretical approaches are competing for dominance in the field of alcohol studies. They have been referred to as the sociocultural model (Pitman, 1965) and the distribution of consumption model (de Lint and Schmidt, 1970). Both have been recently critiqued (Harford, Parker, and Light, 1980). Briefly, the former has typically concentrated on explaining the observed differences in rates of alcoholism among various social and cultural groups, and in patterns of alcohol use by resorting to norms and differences in normative patterns among those groups (Bacon, 1943; Bales, 1944, 1946, 1962). Furthermore, research in the sociocultural model of explanation has usually followed the path of exploring ethnic and religious subgroup norms as primordial sources of learned drinking behavior (Glad, 1947; Snyder, 1958; Field, 1962; Lolli et al., 1958; Skolnick, 1958). The distribution of consumption model has provided an apparent alternative approach to the older sociocultural model. It focuses on the "statistical distribution of consumption" (Whitehead, 1975) and emphasizes the relationship between the mean level of per capita consumption and prevalence of cirrhosis of the liver mortality in a population (de Lint and Schmidt, 1970; de Lint, 1974).

Both of these models or approaches have generated contradictory recommendations for prevention of alcohol problems. The sociocultural model has given rise to the "integration of social norms approach," which seeks to make drinking a mundane aspect of everyday life, defined by prescriptive norms of moderation against excessive consumption (Plaut, 1976; Wilkinson, 1970). By introducing children at an early age into responsible drinking, it is argued that they will, as they mature, come to accept drinking as an ordinary part of everyday life and eventually the mystique surrounding drinking will be lost. These policies would tend to encourage consumption by a wider number of people, thus increasing the overall level of consumption, but resulting in fewer excessive drinkers and excessive drinking episodes.

The distribution of consumption model, on the other hand, suggests that by lowering the overall level of consumption in a population, the rate of cirrhosis of the liver mortality will, of necessity, decline. Policies aimed at reducing the overall level of consumption, however, appear to contradict the objectives of the sociocultural approach (Whitehead, 1975; de Lint, Schmidt, and Popham, 1975). While Whitehead (1975) has called for an integrated approach to prevention, there is some evidence that the two approaches may not be contradictory after all but rather are two different ways of interpreting the same phenomena. We believe that it is possible to give a "sociological interpretation" of the Ledermann (Ledrmann, 1956) log normal curve, upon which is based the distribution of consumption model. Beauchamp has also suggested that the curve may be a graphical representation of "conforming behavior" to norms regarding drinking. Beauchamp states:

> Of course there is no single set of norms regarding alcohol use; there is wide variation between the sexes, ages, re-
> ligions, ethnic groups, and strong urban-rural differences.
> Despite this variation, however, the overall structure of
> drinking norms is still supportive of the social order --
> the dominant norms restrict alcohol use; heavy or alcoholic
> drinking is not the norm for any group of consequence.
> (Beauchamp, 1980).

It is an interesting intellectual problem as to why these two theories have been regarded as separate and distinct. In describing the existence of a log normal curve, are we not establishing a rate of consumption as a social fact? When we discover normative prescriptions and proscriptions, are we not also looking at social facts? If, for the moment, we assume that one kind of sociology is interested in the relationship of social facts to each other, it would seem to be a very natural progression to establish rates and curves for different groups. However, the careful establishment of rates appears less important than the fact that groups differed and that these differences could be approached through analysis of normative influences. When we approached this research four years ago, we believed these were indeed competing hypotheses. Our work over the years has now led us to the point where we see these hypotheses as complementary, or alternative ways of looking at the same phenomena. Given the data base that we have worked with, the theories seem to be equal in significance. For now we will assert that the availability and the cultural models are supportive of each other. Since we have reviewed the corpus of theoretical and empirical work on availability and consumption elsewhere (Rabow and Watts, 1983), we will focus exclusively on our own research on alcohol availability and alcohol problems, noting what we feel to be some significant contributions.

Our first effort was to study cross-sectional data for counties and cities in the State of California. We studied availability by analyzing two types of off-sale premise sites where only alcohol beverages were purchased and five types of on-sale premise outlet rates in relation to social area characteristics. We utilized four dependent variables: public drunkenness arrest rates, misdemeanor and felony drunk driving arrest rates, and cirrhosis of the liver mortality in 51 of 58 counties (Rabow and Watts, 1982). We further extended the level of analysis to 213 cities in California (Watts and Rabow, 1983). Since California is a state with practically no local control over licensing, much of the local variation in hours and the like was eliminated.

Employing a social area approach first utilized by Donnelly (1978), we found that certain types of alcohol beverage outlets were correlated both with specific social area characteristics and particular types of alcohol problems for 51 counties in California. We tested several hypotheses derived from previous research regarding social status, urbanization, minority status, household composition, and family structure in relation both to the physical availability of alcoholic beverages and to the four alcohol-related problems. The hypothesis that physical availability was related to indicators of a alcohol problems was also tested.

The results of the county analysis provided support for most of the hypotheses tested. Indicators of status, which included income, education, and occupation, were generally related inversely to alcohol-related problems. In particular, the percent of persons in poverty was found to be directly related to public drunk arrests, and both misdemeanor and felony drunk driving arrests after the controls for urbanization and median income were made. However, social status was found to be virtually unrelated to cirrhosis mortality rates.

While we found little support for the hypothesis that urbanization is positively related to alcohol problems, we did find moderates evidence for the hypothesis that minority status is positively associated with alcohol problems. Public drunk arrests were significantly correlated with the percentage of Black household population, the percentage of Spanish-American household population, and the percentage of other non-White population but unrelated to percentage of foreign-born or native

of mixed population parentage. Felony drunk driving arrests were unrelated to any of the minority status variables. After controls for
urbanization and income were made, percentage Spanish-American was the
only population significantly related to misdemeanor drunk driving arrests.
When we controlled for urbanization and income, cirrhosis mortality was
significantly related to three of the four race/ethnicity indicators, with
all zero-order correlations increasing after removing the effects of
urbanization and income.

A hypothesis relating household composition and family structure to
alcohol problems found little support in the county data. It was expected that when individuals were integrated into relatively more traditional family-oriented settings, alcohol-related problems would be reduced.
For the most part, the household composition and family structure variables,
including percentage of households made up of husband-wife families,
percentage of divorced or separated males and females, sex ratio, female-
headed households, and youth and aged dependency ratios showed few consistent relationships with arrest rates. This was also true of indicators of familism, such as percentage of single-person households and
percentage of large households with six or more persons.

However, cirrhosis mortality rates were found to have an inverse relationship with traditional family structure. Only two of the nine
correlations for cirrhosis mortality and the household composition and
family structure variables were insignificant (sex ratio and aged
dependency ratio), while most of the significant coefficients were significant at p < .01 level. The familism indicators provided strong
evidence that cirrhosis mortality is related to traditional/nontraditional
families. The percentage of large households (six or more persons) was
significantly and inversely related to cirrhosis after controls at the
p < .05 level and was positively related to single-person households at
the p < .001 level after controls, confirming recent national findings for
389 cities (Gove and Hughes, 1980).

Relatively few studies have concentrated on specific types of
alcohol beverage outlets in relation to social area characteristics in
any comprehensive way. Previous research on the correlates of alcohol
control policies by the state have generally been linked to comparisons
between "monopoly" and "license" states (Jellinek, 1974; New York State
Moreland Commission on the Alcohol Beverage Control Law, 1963). The
Moreland Commission did give attention to the number of liquor sales
outlets per unit of population. Other studies have concentrated on
consumption and selected socioeconomic and demographic characteristics
and their relationship to cirrhosis mortality at the state (Terris, 1967)
and county (Tokuhata, Digon, and Ramaswamy, 1971; Donnelly, 1978) levels.
Though Donnelly (1978) pursued the social area approach with a number of
sociodemographic indicators, he did not consider varying outlet rates
across county jurisdictions. The works by Smart (1977a,b), Parker, Wolz,
and Harford (1978), and Harford et al. (1978), are useful are useful
starting points for the consideration of physical availability, although
only a few sociodemographic variables were used in their analyses.

Rabow and Watts (1982) considered a wide range of sociodemographic
variables in relation in seven types of on- and off-sale premise outlets,
and found that specific outlet types were related to specific socioeconomic and demographic characteristics. On-sale premise outlets located
in bona fide eating places were expected to be positively related to
social class and urbanization, but inversely related to minority status,
household composition, family structure and familism, and unemployment.
On the other hand, outlets such as off-sale premises, general (all types
of beverages sold), on-sale premise beer bars (beer only), and on-sale

premises, general (all types of beverages sold), were expected to be inversely related to social class and positively related to minority status, nontraditional household composition, low familism, and other variables indicating social disorganisation. The results of the analysis for California counties confirmed most of the hypotheses about the relationships between specific outlet types and sociodemographic characteristics. Thus, restaurants with on-sale premises licenses were positively related to social rank and either inversely or nonsignificantly related to race and ethnicity. The restaurant outlets were also inversely related to structural features such as unemployment. On-sale premises licenses associated with eating were positively and strongly related to women's labor force participation at the $p < .001$ level after controls for urbanization and income were made. Furthermore, the household composition and family structure variables, as well as the familism variables, showed support for the idea that social areas with traditional family structures have relatively fewer on-sale premises restaurant outlets.

Although on-sale premises general outlets were unrelated to social class after controlling for urbanization and income, off-sale premises, general outlets were found to be positively related to social class variables. After controls for urbanization and income, beer bars remained substantially correlated with percentage of families in poverty at the $p < .001$ level. Less support was found for the hypothesized positive relationships between on-sale premises, general outlets, on-sale premise beer bars, off-sale premises, general outlets, and the race/ethnicity variables. Only after controls were made were significant relationships between off-sale premises, general, on-sale premises, general, and both percentage of population Black and percentage of population other non-White, revealed. Beer bars were significantly related to the percentage of Spanish American population after controls were introduced. This lends support to recent research that suggests that beer is the beverage of choice among Spanish-Americans (Technical Systems Institute, 1977). Solid support was found for the hypothesized inverse relationship between traditional family structure and off-sale premises, general outlets, but both beer bars and on-sale premises, general outlets provided little support for the hypothesis.

Familism was related to off-sale premises, general and on-sale premises, general outlets in the hypothesized way. However, beer bars were related to familism in an inverse manner. Of all outlet types considered, beer bars showed the strongest correlation with the percentage of persons living in crowded dwelling units (1.01) or more persons per room).

We also investigated the relationship between the seven outlet types and alcohol problems. Of the seven outlet types, only beer bars were significantly related to both public drunk arrests and misdemeanor drunk driving arrests after controls for urbanization and income were made. Felony drunk driving appeared to be related to several outlet types, while cirrhosis mortality was related specifically to off-sale premises, general and on-sale premises, general outlets.

A similar analysis was carried out on data for 213 cities in California (Watts and Rabow, 1983). Corroboration was obtained for most of the county results with some exceptions. Both of our studies at the county and city levels were hampered by reliance on taxable retail sales for packaged liquor stores, a less than adequate measure of consumption.

In sum, we analyzed cross-sectional data from a variety of sources on physical availability of different types of outlets and sociodemographic characteristics. We used a social area ecological framework to study distributions of structural groupings such as age, sex, social class or

rank, racial-ethnic composition, household composition and family structure, and their net effects on selected alcohol problems. We found at
both county and city levels of analysis relationships between particular
types of physical availability, particular types of social area characteristics, and particular types of alcohol problems. We employed correlation, partial correlation, and regression techniques to analyze our data.
In almost every case specific types of outlet availability were implicated
in specific types of alcohol problems when controlling for selected sociodemographic characteristics. A major assumption is that differentially
distributed social area characteristics can serve as "proxies" for varying
subgroup norms and normative patterns, the direct measures of which were
unavailable.

Similarly, we found that specific social area characteristics representing social class, minority status, traditional versus nontraditional
family structures, and other structural features such as unemployment,
women's labor force participation, and physical density were variously
implicated in selected alcohol problems net of the effects of differing
levels of outlet availability.

We concluded, as had Donnelly (1978), that social area analysis offers
a most promising tool for developing integrated and comprehensive models for
testing a number of hypotheses in the sociocultural model.

Other promising areas of research on types of availability are being
pursued. Rabow, Schwartz, Stevens, and Watts (1982) investigated
"dimensions" of availability through a pilot survey of 580 Southern
California residents. Questions concerning price considerations, physical availability, "propinquity," sociability or obligation to serve,
purchasing patterns, and sociodemographic data have revealed that sociability appears to have the greatest net effect on individuals' frequency
and quantity of consumption. Work by Neuman and Rabow (1986) using
aggregate data on the physical distribution of outlets in the respondents'
immediate environment, revealed that the efficient and effortless purchase
of alcohol contributes to only a small amount of the variance in consumption. However, food stores account for more than half the sample's
purchases. In another study Rabow and Neuman (1984a) found that the
opening of a single new outlet in a community provided mixed results. A
survey revealed no shifts in alcohol consumption while the examination of
garbage indicated an increase in alcohol consumption. Finally, "time" may
be another form of availability. College students' drinking patterns were
shown to begin on Thursday night with increased acceleration until a peak
was achieved Saturday night (Rabow and Neuman, 1984b).

A second direction that our work took was to focus on the licensing
of alcohol outlets. ABC's (Alcohol Beverage Control), up until a couple
of years ago have been one of the more neglected arena in the alcoholism
area and it was to this area that we turned. We began a study of the
licensing and enforcement process in the State of California Alcoholic
Beverage Control (CABC). Rabow, Johnson-Alatorre, and Watts (1980)
employed a mail survey sent to all 161 ABC investigators in the 25 State
of California ABC offices designed to assess investigators' decisions to recommend issuance or denial of alcoholic beverage licenses. The return rate
was 81%. The instrument gathered background data on investigators and attitudinal vignettes developed by the authors. The three vignettes involved:
(1) a license application that violated a proximity to residence rule;
(2) a license application that involved an undue concentration rule; and
(3) a license application that involved an applicant's moral character.
Vignettes were varied as to the number of community protests. Findings
indicated that investigators, while sensitive to community protests, will

often <u>recommend</u> a license when there is a rule violation. Investigators also tended to perceive the administrative law judge as more likely than themselves to recommend issuance. Investigators would personally prefer to be more severe in cases of moral character than they publicly recommend. The effectiveness of formal rules designed to facilitate the community's interest was minimal. The rules did not present barriers to obtaining an alcohol license, even though investigators are personally concerned about community protests. However, investigators are responsive to protests, moral character, and to illegal applications (Rabow, 1986).

CONCLUSIONS AND SUGGESTIONS FOR FURTHER RESEARCH

This review of the literature on alcohol availability and consumption and the problems which result from the abuse of alcohol has attempted to show that the phenomena relating to alcohol and drinking behavior are complex and multifaceted. Although the issue of availability has received increasing attention in recent years by many researchers interested in the improvement of the public's health, the tendency has been to promote the distribution of consumption model over the older sociocultural model as the way to study policies aimed at preventing the growth of and hopefully reducing alcohol problems. Though the focus of this review has been on various issues of availability and consumption and alcohol problems, we have tried to show that the single-minded attachment to one or the other of the approaches is not productive. A truly realistic model of complex alcohol phenomena will necessarily employ a host of variables in a multivariate framework. Some attempts have been made in this direction. Furthermore, cross-sectional studies, while important in determining statistical relationships at one point in time, need to be supplemented with sophisticated new time-series modeling techniques. There is even the possibility of employing "cross-sectional time series" models which follow a number of variables for any number of jurisdictions over time. The rapid advancement in statistical programming has made this entirely feasible.

The mathematical and statistical techniques are available. However, the data currently are not available, at least not in any comprehensive way. Efforts should be made to develop a national epidemiological system of alcohol-related information that could be integrated with a comprehensive set of sociodemographic information such as that first employed by Donnelly (1878).

Science proceeds cumulatively. Research is needed to isolate those factors which show the greatest promise for conscious manipulation by collectives. Alcohol problems result from a combination of cultural, sociological, psychological and genetic factors. Progress at any of these levels will improve health and the quality of life in advanced industrial societies.

REFERENCES

Bacon, S.D. "Sociology and the Problems of alcohol: Foundations for a sociological study of drinking behavior," <u>Q. J. Stud. Alcohol</u>, 1943, 4: 402-445.

Bales, R.F. "The fixation factor," In R.F. Bales, <u>Alcohol Addiction: A Hypothesis Derived from a Comparative Study of Irish and Jewish Social Norms</u>. Unpublished doctoral dissertation, Cambridge: Harvard University, 1944.

Bales, R.F. "Cultural differences in rates of alcoholism." <u>Q. J. Stud. Alcohol</u>, 1946, 6: 480-499.

Bales, R.F. "Attitudes toward drinking in the Irish culture." In

D.J. Pittman and C. Snyder (Eds.), <u>Society, Culture and Drinking Patterns</u>. New York: Wiley and Sons, 1962.

Beauchamp, C.E. <u>Beyond Alcoholism: Alcohol and Public Health Policy</u>. Philadelphia: Temple University Press, 1980.

Cook, P.J. "The effect of gun availability on violent crime patterns." <u>The Annals of the American Academy of Political and Social Sciences</u>, 1981, 455: 63-80.

Cottrell, W. "Death by dieselization: A case study in the reaction to technological changes." <u>Am. Soc. Rev.</u>, 1951, 16: 358-365.

Deutsch, M., and Collins, M.E. <u>Interracial Housing</u>. Minneapolis: The University of Minnesota Press, 1951.

de Lint, J. "The prevention of a alcoholism." <u>Prev. Med.</u>, 1974, 3: 24-35.

de Lint, J. and Schmidt, W. "Estimating the prevalence of alcohol consumption and mortality data." <u>Q. J. Stud. Alcohol</u>, 1970, 131: 957-964.

de Lint, J., Schmidt, W., and Popham, R.E. "The prevention of alcoholism: Epidemiological studies of the effects of government control measures." <u>Brit. J. of Addict.</u>, 1975, 70: 125-144.

Donnelly, P.G. "Alcohol problems and sales in the counties of Pennsylvania: A social area investigation." <u>J. Stud. Alcohol</u>, 1978, 39: 848-858.

Field, P.B. "A new cross-cultural study of drunkenness." In D.J. Pittman and D.R. Snyder (Eds.), <u>Society, Culture and Drinking Patterns</u>. New York: Wiley and Sons, 1962.

Glad, D.D. "Attitudes and experiences of American-Jewish and American-Irish male youth as related to difference in adult rates of inebriety." <u>Q. J. Stud. Alcohol</u>, 1947, 8: 406-472.

Goldman, M.S. "<u>Gold Diggers and Silver Miners: Prostitution and Social Life on the Comstock Lode</u>. Ann Arbor: University of Michigan Press, 1981.

Gove, W., and Hughes, M. "Reexamining the ecological fallacy: A study in which aggregate data are critical in investigating the pathological effects of living alone." <u>Social Forces</u>, 1980, 58: 1157-1177.

Harford, T.C., Parker, D.A., and Light, L. "Normative approaches to the prevention of alcohol abuse and alcoholism." <u>Proceedings of a Symposium, 1977</u>. Washington, D.C. Department of Health Education and Welfare, 1980.

Harford, T.C., Parker, D.A., Paulter, C., and Wolz, M. "Relationship between the number of on-premise outlets and alcoholism." <u>J. Stud. Alcohol</u>, 1978, 40: 1053-1057.

Jellinek, E.M. <u>Recent Trends in Alcoholism and Alcohol Consumption</u>. New Haven: Hillhouse, 1947.

Ledermann, S. "Alcool, Alcoolisme, Alcoolisation." Vol. 1. "Donnees Scientiffiques de Caractere Physiologique, Economiglue et Social. Paris Institut National e'Etudes Demographiques, Travaux et Documents 1956.

Lolli, G., Serianni, E., Golder, G., Balboni, C., and Mariani, A. <u>Alcohol in Italian and Italian-Americans</u>. New Brunswick, N.J.: Rutgers Center of Alcohol Studies, 1958.

Neuman, C.A., and Rabow, J. "Buyer use of physical availability: Consumer and drinking behavior." 1986, <u>Int. J. Addictions</u>.

New York State Moreland Commission on Alcohol Beverage Control Law: The Relationship of Alcoholic Beverage Control Law and the Problems of Alcohol. Study Paper No. 1, 1963.

Parker, D.A., Wolz, M.W., and Harford, T.C. "The prevention of alcoholism: An empirical report on the effects of outlet availability." <u>Alcoholism: Clinical and Experimental Research</u>. 1978, 2: 339-343.

Pittman, D.J. "Alcoholic beverages in culture and society: An overview." In <u>Alcohol and Accidental Injury: Conference Proceedings</u>. Washington, D.C.: U.S. Department of Health, Education and Welfare, 1965.

Plaut, T.F.A. <u>Alcohol Problems: A Report to the Nation</u>. New York: Oxford University Press, 1967.

Rabow, J. "Alcohol beverage licensing practices in California: Autonomy and control in a regulatory agency" 1986.

Rabow, J., Johnson-Alatorre, R., and Watts, R.K. The regulation of public welfare: The failure of balancing interests in alcohol licensing. Dept. of Alcohol and Drug Abuse, State of California, 1980.

Rabow, J. and Neuman, C.A. "Garbaeology as a method of cross-validating interview data on sensitive topics." Soc. and Soc. Research, 1984a, 68: 480-497.

Rabow, J. and Neuman, C.A. "Saturday night live: Chronicity of alcohol consumption among college students" Substance and Alcohol. Actions/Misuse. 1984b, 5: 1-7.

Rabow, J., Schwartz, C., Stevens, S., and Watts, R.K. "Social psychological dimensions of alcohol availability: The relationship of perceived social obligations, price considerations and energy expended to the frequency, amount, and type of alcoholic beverage consumed." Int. J. Addiction. 1982, 17: 1259-1271.

Rabow, J. and Watts, R.K. "Alcohol availability, alcohol sales, and alcohol-related problems." J. Stud Alcohol. 1982, 43: 767-801.

Rabow, J., and Watts, R.K. "The role of availability in alcohol consumption and alcohol problems." In M. Galanter (Ed.), Recent Address in Alcoholism, 1983, 1: 285-302.

Skolnick, J.H. "Religious affiliation and drinking behavior." Q. J. Stud. Alcohol, 1958, 19: 452-470.

Smart, R.G. "The effect of two liquor store strikes on drunkenness, impaired driving and traffic accidents." J. Stud. Alcohol, 38: 1785-1789.

Smart, R.G. "The relationship of availability of alcoholic beverages to per capita consumption and alcoholism rates." J. Stud. Alcohol, 1977b, 38: 891-896.

Snyder, C. Alcohol and the Jews: A Cultural Study of Drinking and Sobriety. New Brunswick, N.J.: Rutgers Center of Alcohol Studies, 1958.

Technical Systems Institute: Drinking Practices and Alcohol Related Problems of Spanish Speaking Persons in Three California Locales. Sacramento, California, The California Office on Alcoholism.

Tokuhata, G., Digon, E., and Ramaswamy, K. "Alcohol sales and socio-economic factors related to cirrhosis of the liver mortality in Pennsylvania." Health Serv. Ment. Health Reps., 1971, 86: 253-264.

Watts, R.K., and Rabow, J. "The role of tourism in measures of alcohol consumption, alcohol availability and alcoholism." J. Stud. Alcohol, 1982, 43: 797-801.

Watts, R.K., and Rabow, J. "Alcohol availability and alcohol related problems in California cities." Alcoholism: Clinic and Experimental Research, 1983, 7: 47-58.

Whitehead, P.C. "The prevention of alcoholism: Divergences and convergences of two approaches." Addictive Diseases, 1975, 4: 431-443.

Wilkinson, R. The Prevention of Drinking Problems: Alcohol Control and Cultural Influences. New York: Oxford University Press, 1970.

EVALUATION AS A TOOL FOR THERAPEUTIC PLANNING

Pritam S. Bhatia

Graduate School of Social Work
University of Arkansas
Little Rock, Arkansas   72204

INTRODUCTION

In the past 150 years, there has been an increasing emphasis on the understanding, diagnosis, and treatment of alcoholism. In reviewing the history of this progress, it becomes self-evident that the treatment of choice for the alcoholic individual is invariably related to the theoretical framework in which the problem is defined. In the 17th century, for example, intemperate drinking was simplistically viewed as some form of moral weakness. With this unicausal attitude, most "treatment" approaches employed physical punishment, incarceration, and/or social condemnation.

During the last of the 18th and early 19th centuries, the writings of Drs. Benjamin Rush and Thomas Trotter introduced the concept that "inebriety" was an illness rather than some kind of immoral or criminal behavior (Jellinek, 1960). This point of view reached fruition in the mid-1900's with a "new approach" called epidemiological that described alcoholism as a disease entity. Although this increasingly focused attention on the physiological facets of alcoholism, the epidemiological approach still represented a sophisticated unicausal theory. Thus, the concomitant treatment plan was directed toward the physical control of the patient in hospitals or in institutions (rather than jail) as well as some form of physiological control (e.g., Antabuse).

In recent years, the purely epidemiological unicausal model has been modified by the increasing awareness of the social-cultural factors involved in alcoholism. For example, the 1951 World Health Organization definition of alcoholism is more nearly a social one, in keeping with the greater acceptance of socially dysfunctional roles within the compass of ""ill roles" by society (Keller, 960). In the same vein, Blane, Overton, and Chafetz have demonstrated that the diagnosis of alcoholism is a function of social perceptions, dependent upon both cultural and personal biases (Blane et al., 1963). These authors have pointed out that some cultural and subcultural expectations are far more accepting of alcohol intake (and subsequent alcoholism) than are other cultures. This theoretical focus has been the basis for the treatment concept of placing the patient in a "subcultural" milieu whose expectations discourage alcohol intake. This approach has been instrumental in the development of the "halfway house" idea and also in an implicit aspect of the Alcoholics Anonymous movement.

Another relatively recent development has been the psychological and psychiatric interest in the understanding and treatment of alcoholism. The major theoretical contribution from these professions was a delineation of the psychodynamic factors and behavioral learning principles as these issues apply to uncontrolled drinking. The resultant treatment models for the alcoholic utilize individual and group psychotherapy to enhance the patient's awareness of the intrapsychic conflicts that motivate his intemperance. A concerted effort has also been made to control the alcoholic's drinking patterns via the principles of behavior modification (e.g., aversion therapy). Again, these treatment techniques represent a relatively narrow focus because the social-cultural factors often were minimized in the isolation of the psychology laboratory and therapist's office. Consequently, the insights and improvements obtained in therapy were not often to be generalized in the patient's day-to-day contact with his environment.

It can be seen that diagnostic formulations vary according to their purposes. Treatment, prognosis, etiology, or some other confusion of purpose has occurred in the nomenclature of alcoholism with the result that the accuracy and usefulness of the current diagnostic scheme is minimal. For the most part, a unicausal approach to diagnosis has been employed. That is, a number of potentially relevant, individual symptoms, behaviors, etc., are identified, but they are subsumed under one criterion and are considered to be indicative of one trait or phenomenon, i.e., alcoholism. Such an approach is not conducive to the understanding of the problem of alcoholism. Rather, emphasis should be placed upon the interactional process between diagnosis, treatment, and the availability of treatment.

To date, research has shown that the unicausal model for diagnosis does not adequately explain the phenomena of alcoholism. Alcoholism must be considered in a progression; that is, if the drinking continues, an individual will not only continue to experience more symptoms, but will experience an increased severity of such symptoms. Mendelson and Mello conclude that their data "strongly support the notion that the perpetuation of drinking behavior is related to complex social environmental factors rather than alcohol dosage per se" (Mendelson and Mello, 1966). Pattison and others point out that "alcoholism may be best described as a psychosocial behavior syndrome which is dependent on a variety of social and cultural variables. Changes in patterns of drinking behavior may result no so much from changes in personality variables as from changes in social and cultural variables" (Pattison, Heady, Gleser, and Gottschalk, 1968).

In the same vein, other studies suggest that abstinence as a factor of treatment outcome appears to be relatively independent, or there is adjustment in other areas such as work, psychological and emotional stability, or assumptions of adequate roles. Cohen indicates that a disease model, which is based on a progressive illness concept, is an oversimplification of a complex problem, the acceptance of which will lead us to believe that the cure is both simplistic and unidimensional (Cohen, Lielison, and Faillace, 1971).

Talcott Parsons points out that alcoholism can be more fully understood within the context of a multilevel concept (Parsons, 1958). This includes a strictly physiological aspect that pertains to neither personal discomfort nor to any social role (such as a slowly developing cirrhosis). Thus, alcoholism may be defined as a disease with multiple conceptual levels; some symptoms are primarily physiological, others primarily social. Recognition of this can avert futile efforts to conceptualize all alcohol-related problems within a unicausal framework. Research suggests that it seems more logical to approach the problem of alcoholism with a classification of symptoms based upon a multilevel scheme. The description of all potentially relevant factors or dimensions associated with alcoholism is a prerequisite for the effective treatment of the syndrome.

Jellinek has observed that the etiquette of American alcoholism literature requires that we "acknowledge that physio-pathological, cultural, and social elements have a role in the genesis of alcoholism" (Jellinek, 1960). Some authors have preferred to concentrate on all interpretation of alcoholism as a form of social deviance. In accepting this, there is an overt admission that the abuse of alcohol arises out of a matrix of behavioral and symbolic transactions that constitute social reality. Becker has reviewed several specific meanings of "social deviance" and commented on each. One can take a quite radical sociological view of deviance, considering deviant those, and only those, actions which are typically called deviant by others in society (Becker, 1963). In this view, no behavior is deviant in and of itself; it is deviant only when someone in a position to make a definition stick and have weight, defines it as deviant. The question of what is deviant becomes, then, a political question: Who has the power in a given situation to label something as deviant and have others accept this definition?

Although agreement among researchers appears to be the increase, there is still a myriad of definitions of alcoholism and problem drinking found in the literature. While most of these are not conflicting, strictly speaking, neither are they exactly comparable. That is, a prevailing estimate based on one definition usually cannot be compared meaningfully to another estimate using a different frame of reference and another population. However, if there is not a single definition acceptable to all researchers, certain elements of definitions regularly turn up in many otherwise different formulations. Central among these shared elements are: (1) excessive intake; (2) concern about own drinking; (3) disturbance of social and economic functioning; and (4) loss of control. While these common elements have become part of the alcoholism literature, they have not been organized in such a way as to lend themselves to a model of measurement. The lack of a uniform approach creates a need for consistency and predictability for diagnosis and treatment.

The Multiphasic Matrix for the Diagnosis of Alcoholism (MMDA) was designed to aid counselors in their initial interviews of clients, to help obtain uniform diagnostic data in regard to these shared elements, and to develop a treatment plan. The MMDA should also provide an overall evaluation of the patient's drinking problem that is quantifiable, yet is unique to the individual.

The MMDA is meant to serve as a tool for the trained counselor. It is not intended to supplant or stifle the counselor's clinical impressions. Rather, the matrix may help clarify impressions that often arise as the preconscious level about factors that should be involved in a diagnosis of alcoholism.

THE MMDA AS A DESCRIPTIVE BEHAVIORAL MODEL

The MMDA was planned as a model of human behavior and, therefore, may be used as a research tool. In the behavioral sciences, the development of models has become widely accepted as a means of studying complex behavior. The use of a model provides a way of ordering large amounts of data so that all facets of an area of inquiry receive attention.

Representing the real work via a model can be a complex process, because reality -- particularly human behavior -- is diverse and cannot be completely visualized or understood. In addition, even when a model may

Table 1

	"Abstinence" Score 0	"Mild" Score 1	"Moderation" Score 2	"Serious" Score 3	"Severe" Score 4
Category 1 Amount of alcohol ingested per day (averaged)	None	1-5 ounces 1-5 cans of beer 1-5 glasses of wine 1-5 ounces of whiskey (1-3 "shots")	6-10 ounces 6-10 cans of beer 1 pt.-1 fifth of wine 6-10 ozs. of whiskey (4-7 "shots")	11-20 ounces 11-20 cans of beer 1-2 fifths of wine 1 pt.-fifth of whiskey	20+ ounces 20+ cans of beer 2 fifths of wine 1 fifth of whiskey
Category 2 Drinking pattern	None	Drinking an occasional weekend party/social gathering near or to the point of inebriation; before dinner cocktail, after work beer, etc.	Drinking at regular weekend parties plus occasional weekday drinking (e.g., TGIF get-together) near or to point of inebriation	Evidence of daily drinking near or to point of inebriation	Drinking occupies a major time investment; binge drinking during previous month
Category 3 Physical symptoms related to drinking	0 symptoms (see scoring guide)	1-10 symptoms (see scoring guide)	11-25 symptoms (see scoring guide)	26-40 symptoms (see scoring guide)	40+ symptoms (see scoring guide)
Category 4 Social Interactions (previous month)	13+ composite score (see scoring guide)	10-12 composite score (see scoring guide)	7-9 composite score (see scoring guide)	4-6 composite score (see scoring guide)	0-3 composite score (see scoring guide)

	0 incidents (see scoring guide)	1–3 incidents (see scoring guide)	4–6 incidents (see scoring guide)	7–9 incidents (see scoring guide)	10+ incidents (see scoring guide)
Category 5 Loss of personal control related to drinking (previous month)					
Category 6 Occupational adjustment	No problems on job; steadily employed; accepts occupational responsibility (home, school, daily living)	Minor work problems; alcohol effects job performance (home, school, daily living) to a minor degree; occasional work day missed because of hangover	Noticeable work problems; poor efficiency reports or warnings; has been absent 2+ days/month from job (home, school, daily living)	Significant job problems; poor job attendance/frequent job changes because of drinking; minimal efforts to fulfill occupational responsibility (home, school, daily living	Chronically unemployed, little or no effort to find employment; takes little or no responsibility for him/herself at home, school, daily living (e.g., drops out of school
Category 7 Family relationships	Solid family relationships (or non-alcohol related family stress)	Basic family relationships maintained; some transient alcohol-related problems (e.g., argument about specific incident of drinking)	Distinct marital/family discord; obvious strain in family relationships related to drinking; threats of divorce, police neighbors, etc., aware of family discord	Serious marital/family problems; Divorce/separation related to alcohol problem; family ties are vague, distant, or broken	No meaningful family ties; patient is basically alone with his drinking problems

succeed in encompassing most of the aspects of reality, it then becomes
difficult to cope with the available statistical data.

Consequently, researchers attempting to construct a model of behavior
usually begin by simplifying reality. While it is realized that the
viewpoint expressed by this process may only "approach" reality as
experienced by every individual, a model, nevertheless, has economic
advantages, in terms of human effort, and of available funds. As Forester
(1961) pointed out, "A model, compared to the real system it represents, can
yield information at a lower cost. Knowledge can be obtained more quickly
and for conditions not observable in real life.

The purpose of any model is to describe, predict, and explain the
behavior under investigation. A descriptive model yields a frame of
reference in which observations made at a certain point in time can best be
understood and explained. On the other hand, a predictive model provides
order from current observations in such a way as to make interpretations of
other situations and the anticipation of future behavior possible.

The MMDA is a descriptive model. It attempts to simplify the complex
realities of the alcoholism syndrome into a matrix of seven broad categories
that describe the major features of any drinking problem. Within these
categories, then, aspects of the problem are scored according to their
severity.

THE MMDA AS A PREDICTIVE BEHAVIORAL MODEL

As a secondary goal, the MMDA was developed to include a potentially
predictive capacity. For example, if a series of descriptive measurements
were made of an individual over a period of time, the response pattern might
establish a predictive value for the model.

A series of in-depth discussions with experienced alcoholism counselors
was undertaken to delineate the specific kinds of behavior that should be
included in the seven diagnostic categories of the MMDA.

Each of those seven diagnostic categories was then further divided into
levels of progressive pathological functioning. Five levels of increasing
symptom severity were developed, including: (1) abstinence; (2) mild; (3)
moderate; (4) serious; (5) severe.

Parameters of the MMDA were carefully designed so that progression
within each category would be feasible and changes over a period of time
could be measured. Because of the nature of the alcoholism syndrome, the
resultant matrix contains some interrelationships and overlap to accommodate
the most realistic appraisal of the experience. For example, "hospital
ization" was included in different categories based on the fact
that there are often two different purposes for hospitalizing an alcoholic
patient. One is to provide medical attention for the patient's physical
problems, i.e., gastritis, drinking-related injuries, or others; and the
second is the use of the detoxification facility as an external control of
the individual to stop him from drinking. In the latter case, the medical
considerations are usually secondary to the control of the socially
unacceptable behavior.

Another example of matrix interrelationship is seen in categories four
and seven where, although both parameters deal with interpersonal
relationships, in one case it is with the family and, in the other, outside
of the family. Details of the MMDA appear in Table 1.

Categories Described

The seven categories -- and the rationale underlying their development -- are described in the following pages:

1. <u>Quantity of alcohol ingested daily</u>: The assumption is made that the severity of a drinking problem is directly related to the quantity of alcohol taken every day. The amount included in each cell is operationally defined in ounces and is then translated into comparable amounts of beer, wine, and whiskey. Because this scale is administered to clients considered to have drinking problems, a score of zero is arbitrarily assigned to total abstinence. This score is not meant to represent "normal" drinking or to conceptualize any proposed standard for drinking as practiced in American society.

The number of ounces shown in each matrix cell should represent the average amount of alcohol the client consumes daily. A counselor utilizing the MMDA should carefully evaluated the client's (or significant others') statements regarding the amount of alcohol consumed during a typical recent week. The average daily consumption may then be estimated.

This information would then be scored in the following manner (see Table 1):

For example, if a patient reports having a couple of beers every night after work, plus drinking about a pint or so of whiskey at a weekend party, then the counselor would calculate that beer consumption equalled 10 ounces per week, and whiskey consumption equalled 20 ounces per week, all for a total of 30 ounces of alcohol per week, or 4 ounces per day. In this case, the client would receive a score of 1 for category one.

But if a patient states that he spends most evenings drinking beer with friends and becomes intoxicated on wine and vodka each weekend, the counselor may assume six to eight beers are being consumed each evening, or 35 ounces per week; this is in addition to consumption of about two quarts of wine and a fifth of vodka on weekends, for an additional 60 ounces. Thus, the total weekly consumption of alcohol would be 95 ounces, or 13 plus ounces per day. This client should then be given a score of 3 for this category.

A counselor may have occasional difficulty in determining the average amount of alcohol ingested in recent weeks by the client because there appears to be a downward trend in the quantity being consumed. For example, a patient who has been drinking heavily may appear to be "tapering off," so that his report would show only a few beers within the two to three days before counseling. Should this trend not appear temporary or situational (i.e., lack of money), the counselor should take into account the consumption trend in estimating the level of severity for this category.

Drinking Time Measured

As drinking becomes an increasing problem for an individual, s/he tends to spend an increasing amount of time at this activity. Category Two measures -- on a time continuum -- a person's investment of time in drinking from a minor to a major degree.

2. <u>Drinking patterns</u>. Definitions of severity are derived from the judgments made by the counselor about the amounts of time the client is investing in drinking in order to reach some discernible effect, such as drinking near to or to the point of inebriation. In scoring this category,

the counselor should make a judgment of the patient's general drinking patterns over the past month.

The scoring in this category for <u>binge drinkers</u> may show marked fluctuations from month to month (i.e., from a score of 0 to a score of 4). Operationally, a "binge" is considered as a period of uninterrupted drinking, or maintenance of a high level of intoxication, for three days or more. If the counselor finds one or more binges during the previous month, category two should be scored as 4.

Another potential problem area may result when a patient claims drinking only to relax, with never being really inebriated. In this case, the counselor must make a judgment as to the likely occurrence of the degree of intoxication. Additional information from friends, spouses, or others may prove helpful in making this determination.

Scoring the four levels in category two involves the following:

A score of 1 is meant to include patterns where the major drinking activity is limited primarily to occasional weekend parties or social gatherings. Also included at this scoring level would be the individual who spends an hour every evening after work in a bar, but consumes only two to three beers, with little or no behavior indicative of ingestion. In this situation, one can assume that time was invested more in socializing than in gaining some effect from the alcohol. The same rationale would apply to a couple that routinely spends an hour talking and drinking cocktails before dinner.

A score of 2 indicates that the individual's drinking pattern includes frequent, if not regular, weekend parties where alcohol is an important facet of the socializing process. In addition, the patient may occasionally imbibe near to, or to the point of, intoxication on a weekday (i.e., a get-together after work).

A score of 3 is important, because the discernible feature of this level is almost daily drinking. For example, the patient may spend almost every evening watching TV while consuming six bottles of beer. Included at this level would be the after-work drinker who routinely spends an hour or more in a bar before going home and whose subsequent behavior reflects the intake of alcohol.

A score of 4 should be given when the patient appears to focus most of his or her effort into drinking. In some cases, an individual may hold a regular job but manage to continue drinking throughout the day. Binge drinkers would be included at this level also.

Physical Problems Scored

Past research has demonstrated that adverse physical effects become pronounced as an individual ingests an inordinate amount of alcohol over a prolonged period of time. Category three is designed to measure the progressive severity of physical effects as the drinking problem increases.

3. Physical sequelae. There are broad individual differences in the development of physical problems related to drinking. This makes it difficult to delineate a specific symptom complex at any particular level of severity. To surmount this scoring problem, the severity levels of category three are determined by the number of physical symptoms a patient has manifested during the previous month.

For example, if a patient reports having to settle his nervousness with a morning drink two or three times a week, this would be scored as 10 symptoms (see Table 2). It should be noted that the list does not include permanent physical problems such as peripheral neuropathy, parethesia, cerebellary dysfunction, or speech impairment.

A counselor should be cognizant of the more several alcoholism symptoms and record them on an intake data sheet; but these symptoms should not be scored in category three. The reasoning is that the MMDA is a clinical and research instrument focused upon evaluating the drinking problem and measuring any therapeutic changes that occur. Because the more severe physical problems noted above cannot be changed, they are considered outside the purview of the MMDA matrix.

Table 2. MMDA Scoring Guide -- (Category Three).

---

1. Hangover (headaches, nausea, and general malaise).
2. Morning after shakes.
3. Morning after "nervous stomach" (with or without vomiting).
4. Missed meals while drinking.
5. Need for a morning drink to "settle nerves."
6. Physician has warned patient to stop or limit his drinking.
7. Physical injuries related to drinking (cuts, bruises, sprains, broken limbs, etc.).
8. Blackout periods while drinking.
9. Hallucinatory experiences or delusional ideas.
10. Hospitalization or emergency room contact related to drinking (gastritis, pancreatitis, cirrhosis, malnutrition, DT's, etc.)

Self-Gratification Categorized

For the fourth category, the assumption is made that, as the drinking problem increases, an individual's emotional energies become more inner-directed and involvement in social interaction diminishes. Instead, the alcoholic's energies become focused upon self-gratification. He or she becomes less able to invest time or involvement in things, events, or other people outside of the immediate need for gratification. This parameter is intended to define a continuum for meaningful emotional commitments; for example, close friendships to superficial commitments, such as with alcohol-related associates.

4. Social interaction. Defining the levels of severity for social
interactions present some rather baffling problems because it is difficult
to distinguish between the relative effect of a client's drinking problem and
the effect of that client's basic personality upon the developing pattern of
social interaction.

While it is generally assumed that alcohol abuse adversely affects
social interactions, it is also known that the subtly intertwined
motivations of alcoholism and basic personality factors cannot be completely
sorted out. Consequently, category four was determined upon the basis of
the patient's general behavior rather than what could be assumed to be
purely alcohol-related behavior.

It is recognized, nevertheless, that an estimation of an individual's
social relationships will require some difficult judgments by the counselor.
The general frame of reference should be the evaluation of the client's
"inner-directed" versus his or her "outer-directed" emotional energy.
Within this context, social interactions should be judged and scored
according to the following levels:

Friends

Loss or withdrawal from friends should be scored 0; maintenance of a
circle of friends should be scored 2; and gaining new friends would receive
a score of 4.

Involvement in Nondrinking Social Activities

A determination should be made of the number of events the client
attended over the past month and scored accordingly:

Events	Score
0-1	0
2-4	2
3-7	3
8+	4

Meaningfulness of Social Contact

In each of the four areas, a score of 0 would be given if few or none
of the traits are observed; a score of 1 would represent "average"; a score
of 2 would be given if a considerable degree of the trait is recognized (see
Table 3 for details).

It is assumed that a counselor will recognize the extremes of
meaningful versus nonmeaningful interactions for the patient, and the
concept of "average" would be somewhere in the middle of the continuum from
meaningful to none. The possible maximum scoring for this section would be
8.

The total range of scoring possible for category four would be 0 to 16,
and the level of severity recorded would be based upon the following guide:

Category four score	Equals matrix score of
0-3	4
4-6	3
7-9	2
10-12	1
13+	0

Table 3. MMDA Scoring Guide -- (Category Four)

1.	Friends		
	Loss or withdrawal from		Score 0
	Maintains same group		Score 2
	Gains friend(s)		Score 4
2.	Social activities		
	0-1		Score 0
	2-4		Score 2
	5-7		Score 3
	8+		Score 4
3.	Meaningfulness of social contact		
	(a) Self-awareness/insight	Min.	Score 0
		Avg.	Score 1
		Max.	Score 2
	(b) Expression of feelings	Min.	Score 0
		Avg.	Score 1
		Max.	Score 2
	(c) Shares personal experiences	Min.	Score 0
		Avg.	Score 1
		Max.	Score 2
	(d) Lack of defensiveness	Min.	Score 0
		Avg.	Score 1
		Max.	Score 2
	Range of total score	0-16	
		0-3	Score 4
		4-6	Score 3
		7-9	Score 2
		10-12	Score 1
		13+	Score 0

Table 4. MMDA Scoring Guide -- (Category Five)

1. Argumentative while drinking
2. Mauldin behavior (e.g., crying jags)
3. Overly gregarious behavior
4. Incidents of physical fighting
5. Sexual promiscuity
6. Noninterest in one's personal appearance
7. Driving while intoxicated arrest
8. Personal or property damage while drinking (e.g., burning hole in clothing or mattress, damaging car)
9. Arrest for disorderly conduct, public nuisance
10. Fugue-like state
11. Larcenous behavior (theft, burglary, armed robbery)
12. Suicide attempts
13. Impulsive spending or gambling while drinking
14. Placement in detoxification facility

Personal Control Scale

It is generally recognized that alcoholism progressively interferes with an individual's ability to adjust to the accepted social standards or to exert appropriate control over personal behavior. In category five, the degree to which drinking results in maladaptive social behavior is defined in terms of external controls such as arrests or hospitalizations, among others.

5. <u>Loss of personal control</u>. Loss of personal control due to drinking manifests itself in any number of ways, depending upon the individual's basic personality. Because of personal factors, it would be impossible to establish a system of relative severity for the various types of maladaptive social behavior. In determining levels of severity for category five, therefore, both the different types of noncontrol and the number of occurrences over the past month should be noted. Examples of the loss of personal control may be found in Table 4.

In scoring this category, the counselor should exercise care in differentiating between a patient's general behavior and loss of personal control as a result of drinking. Many of the characteristics listed may represent behavior that is typical or usual for a particular patient, whether drinking or not. For example, a client may have been careless about his or her appearance, regardless of drinking.

An added consequence of an individual's loss of personal control is the necessity of society's placing some kind of constraint on the person. Items 7, 9, and 14 in Table 4 are examples of this kind of external control. When evaluating these items, a counselor should note whether or not an arrest resulted in a relatively short incarceration, for example 1-5 days, which would be considered as one incident of losing control.

However, should a situation result in a week or more of incarceration, the episode would be scored 10. In evaluating item 14, a counselor should note whether the admission to a detoxification facility was voluntary or involuntary. If the client voluntarily sought admission, this should be scored as hospitalization under category three; if the admission was involuntary, it should be scored 10 in category five.

In evaluating item 10 (a "fugue-like" state), the counselor should distinguish this from "black-outs" (category three). When the amnesic state also involves the patient's moving about the community and behaving in an uncharacteristic manner, this will be considered "fugue-like" and scored in category five.

The level of severity for category five is based on the total number of loss-of-control incidents that occurred for the client over the previous month:

Incidents	Score
0	0
1-3	1
4-6	2
7-9	3
10+	4

Job Problems Scored

The ability to provide for one's self and family diminishes as alcohol becomes the dominant factor in one's life. Category six reflects the degree to which increased drinking has interferred with the ability to earn a living. This category also includes a continuum of efficiency as mother and homemaker. In a broad sense, this category measures one's capacity to accept and fulfill adult responsibilities.

6. Occupational adjustment. For a majority of patients, the level of severity can be determined by a review of work history. The more that drinking impairs one's ability to continue working, the greater the severity score in this category. It must be remembered that what is being measured is a patient's willingness to accept the responsibility of bringing his paycheck home (this is included in category five).

In the case of patients who are housewives, their capacity to fulfill the role of homemaking should be evaluated (i.e., preparing meals, cleaning, doing the laundry, caring for the children). School-age patients are evaluated on their area of primary responsibility -- attending school and completing homework.

When the patient is a retired senior citizen, welfare recipient, or a disabled individual, the usual definition of occupational responsibility must be broadened. In this case, category six will be scored on the basis of an individual's capacity to be a useful member of his living community and to handle adequately the responsibilities of daily living. For example, a patient's ability to shop, keep house, help more infirm neighbors, keep appointments at the clinic, could all be utilized.

In making a judgment of this sort of responsibility, what must be evaluated in the degree to which a drinking problem interferes with the patient's functioning. The counselor should exercise some care to differentiate between nonfunctioning due to alcoholism and unwillingness to accept occupational responsibility because of a pervasive dependent attitude.

Family Dysfunctioning

The assumptions underlying this subscale are similar to the rationale outlined in category four. A person's emotional commitments are usually stronger and more enduring for family relationships than for other social interactions. However, as alcohol becomes a more insistent force in a person's life, even the familial bonds begin to deteriorate. Category seven measures the extent of marital and/or family dysfunctioning as a result of excessive drinking.

7. Family relationships. Although the scoring of this category is relatively straightforward, two areas of potential scoring difficulty must be considered.

Marital and family relationships are scored together. In some situations, there may be distinct differences between the client's relationship with a spouse and a relationship with parents. In scoring, the counselor should make a judgment as to the relative importance of relationships for the client. For example, the client may have frequent contact with parents who condone the client's behavior, whereas the spouse with whom the client lives may be upset by drinking. In this example, it would seem that the marriage relationship has more emotional impact on the client, and category seven should be scored on this basis.

Infrequently, it will be found that the client has no family ties (neither marital nor blood relations). If this circumstance appears to bear no relationship to the drinking behavior, a score of 8 (nonapplicable) should be arbitrarily assigned.

Counselor Judgment Essential

In completing the MMDA, one must not lose sight of the primary purpose of the matrix: To gain an accurate description of the patient's drinking problem. Each category should be viewed as a continuum. The symptoms listed in the different levels of any category should serve only as guidelines. A patient need not manifest all of the characteristics listed in order to be classified at a particular level.

It is hoped that the counselor's clinical judgment will not be hampered or "locked in" by these specific criteria. For example, one often finds that a patient's report included behavior appearing in two or three levels of severity. The counselor must determined which most accurately depicts the patient's current functioning in that particular category of behavior.

Also, it must be noted that, in completing the MMDA, diagnostic judgments should be made only on problems that are alcohol-related. For example, a patient may have a physical problem stemming from a birth injury. This would not be an alcohol-related physical problem and should not be considered in scoring category three.

A second consideration in filling out the MMDA is that, occasionally, personal considerations will motive a patient to exaggerate or minimize some facets of his or her drinking problem. The MMDA need not necessarily be completed solely on the basis of the patient's report. In an effort to gain accurate data, the matrix should reflect a composite of whatever information is available to the counselor (for example, reports from the spouse, family physician, hospital social worker, probation officer, etc.).

Obviously, administration of the MMDA will require a great deal of judgment. Counselor experience will play an important role in the final evaluation of a client, even when the MMDA is used. This instrument will provide a useful tool for counselors and, with the counselors' input, it will provide a more graphic representation of the stages and influences of alcoholism upon an individual than has been possible previously. With such an evaluation it is possible to plan treatment more scientifically, realistically, and meaningfully while being able to maintain therapy in an individualized manner. For example: If a client comes in for treatment, the counsel will use the MMDA to assess baseline information outlined in the MMDA's seven categories. This input results in a category score as well as an overall cumulative score, which tells the counsel where the client is starting from. The counselor then can project a treatment plan for each individual client in terms of that client's needs. As treatment progresses over time (approximately at one-month intervals) the client is reassessed and the treatment plan is either continued as started or modified according to the client's current needs. Each reassessment is used as a comparison to the prior assessment. The category scores and the cumulative scores are compared to determine if there has been progress with the client on a monthly basis. As the MMDA is geared for monthly assessment, the counselor can quickly determine short-term as well as long-term treatment impact.

It seems evident that, aside from the MMDA's worth as a diagnostic instruments, its structure can be beneficial in designing an intake form that will be maximally efficient.

Following this line of reasoning, the MMDA also could be employed as a technique for evaluating the efficacy of an agency's current intake form. For example, if the MMDA could not be completely scored on the basis of available information from the intake data, it would point out the need for revising their current intake form. Such a structuring device would benefit the alcoholism counselor in upgrading his case record maintenance and in formulating his written summaries.

Because of the MMDA's ability to differentially diagnose and monitor the individual alcoholic client, this matrix lends itself to an eventual cost-benefit framework. With the ongoing use of the MMDA, an agency will be in the position to evaluate the effectiveness of current treatment programs so that maximum cost benefits can be achieved. Similarly, innovative treatment modalities can be explored and evaluated via the MMDA in relatively short periods of time. In this way, nonproductive programs can be weeded out without inordinate expenditures, and monies can be invested in treatment approaches demonstrating more promise. Perhaps a somewhat intangible cost-benefit also can be achieved in terms of gaining more efficient intake information, better case record maintenance, and the use of the MMDA to help focus the direction of the therapeutic thrust.

A final future consideration is the long-range applicability of the MMDA as a predictive instrument in addition to its function as a short-range descriptive model. For example, by obtaining a series of MMDA scores on a monthly basis, the mathematical model of "Markovian chains" can be employed. In this way, the probabilities of a client's future behavior can be predicted from an analysis of his current functioning as reflected in his MMDA scores.

Although the MMDA was designed as a diagnostic instrument, one must note some potential limitations to this scale. One basic consideration must be the variations among cultural groups and social classes in drinking patterns and associated social values. The present study focused upon one relatively homogeneous subculture, a predominantly Black, inner-city, lower socioeconomic group. Other groups may manifest quite dissimilar patterns, which may show up particularly in the correlations among the categories of the model.

Another issue is the potential for bias in client reports regarding their own behaviors. For example, the various facets of drinking behavior have been carefully quantified and yet the counselor often must rely on the client's report as factual, which may result in some distortion of the data. The MMDA was intended to provide a broader description of the various facts of alcoholism. Even with the inclusion of seven behavioral categories, this still may not yield a totally comprehensive view of this complex syndrome. Another possible limitation is that the MMDA requires a training period and considerable practice before it can be optimally utilized by the counselors. Further, the evaluation and scoring of the seven categories often takes additional time from the counselor's already busy schedule. When one takes these limitations into account, it is still believed that the MMDA reflects a significant step forward in the diagnostic conceptualizing and describing of alcoholism problems. Its use permits treatment planning in a reliably scientific, yet humanly individualized way. This diagnostic tool also permits the treatment agents to become more sensitive to and remain aware of training which they may need to undergo in order to more effectively meet the treatment needs of their clients during particular points of time.

REFERENCES

Becker, H. S. Outsiders. New York: Free Press, 1963.
Blane, H. T., Overton, W. F., and Chafetz, M. E. "Social factors in the diagnosis of alcoholism." Quarterly Journal of Studies on Alcohol, 1963, 24: 640-663.
Cohen, M., Lielison, I., and Faillace, L. (Eds.) "The modification of drinking of chronic alcoholics." Recent Advances in the Studies of

  Alcoholism. Washington, D. C.: U. S. Government Printing Press Office,
  1971.

Forester, W. J. *Industrial Dynamics*. Cambridge, Mass.: MIT Press, 1961,
  p. 49.

Jellinek, E. M. *The Disease Concept of Alcoholism*. New Haven: College and
  University Press, 1960.

Keller, M. "Definitions of alcoholism." *Quarterly Journal of Studies on
  Alcohol*. 1960, 21: 125-134.

Mendelson, J. H., and Mello, N. K. "Experimental analysis of drinking
  behavior of chronic alcoholics." *Annals*. New York Academy of
  Sciences, 1966, 133: 828-845.

Parsons, T. "Definitions of health and illness in the light of American
  values and social structures." In E. G. Jaco (Ed.), *Patients,
  Physicians and Illness*. New York: Free Press, 1958.

Pattison, E. M., Heady, E. G., Gleser, G. C., and Gottschalk, L. A.
  "Abstinence and normal drinking: An assessment of changes in drinking
  patterns in alcoholics after treatment." *Quarterly Journal of Studies
  on Alcohol*, 1968, 29: 610-633.

ASSESSMENT OF PREVENTION EFFORTS:

A FOCUS ON ALCOHOL-RELATED COMMUNITY DAMAGE

Richard M. Earle

University of Evansville
Evansville, Indiana   47702

Since 1970 and the passage of a national alcoholism law (PL 91-616) and the creation of the National Institute on Alcohol Abuse and Alcoholism (NIAAA) there has been a flurry of research activity and program development concerned with the prevention of alcohol-related problems (USDHEW, 1971, 1974, 1978, 1981). However, the question of how to measure the incidence of alcoholism and alcohol abuse in a given community has been a serious omission among published reports in the field (Staulcup, Kenward, and Frigo, 1979). Selden Bacon, an internationally recognized expert in studies on alcohol, writes that for "the past 78 years there have been many claims for prevention programs but few people can think of even one such attempt which can provide the slightest evidence that such a prevention in fact incurred" (Bacon, 1978, p. 1143).

The purpose of this paper is to identify 16 measures sensitive to the incidence of alcoholism and alcohol abuse specific to a modern, metropolitan community, which under examination might facilitate assessing the effectiveness of community-wide primary prevention efforts.

Briefly, primary prevention is a term borrowed from the public health literature and refers to lowering the incidence of a given problem "by altering susceptibility or reducing exposure for susceptible individuals" (Mausner and Bahn, 1979, p.9). Within the purview of this paper, primary prevention refers to planned intervention that prevents the occurrence of an undesirable or unwanted event, behavior, or phenomenon. The terms alcohol abuse and alcoholism refer to drinking behavior "that on a continuous basis, interferes with adequate functioning in any significant area of a person's life" (Krimmel, 1971, pp. 15-16). The concept of community is defined as a large, complex network of formal organizations collected in one geopolitical area. Using this conceptualization of community, for example, Albert Reiss, Jr., and Harold Wilensky have defined modern society as a system of "complex

---

The research on which this article is based was funded by the Scientific Advisory Council of the Distilled Spirits Council of the United States and the National Institute on Alcohol Abuse and Alcoholism grant #5T32 AA06139-02.

territorial organizations whose populations are more-or-less integrated by
economic, legal, military, and political institutions and by the media of
mass communication and entertainment" (Reiss and Wilensky, 1972, p. i).

INTERORGANIZATIONAL COORDINATION

The strategy used to collect information for this paper uses concepts
employed in organization analysis that might aid in understanding processes
one might use as an effective alcoholism-prevention agent, perhaps working
as a community planner and developer (Earle, 1981). Specific attention
is concentrated on the unique perspectives of several major community
institutions and how they are affected and must react to alcohol abuse and
alcoholism within their organizational environments. This orientation stem
from the theory that the social setting and interorganizational coordinatio
can undergird a viable community-wide strategy to effect primary prevention
of alcohol problems (Zinberg and Fraser, 1979).

An insightful discussion of interorganizational coordination as it
relates to primary prevention of alcoholism and alcohol abuse appears in th
Proceedings of the Tripartite Conference on Prevention--a 1976 meeting of
United States, Canada, and British professionals in the fields of
alcoholism, drug abuse, and mental health. On several occasions throughout
the published proceedings, various participants called for an interorganizational coordination strategy. Draper, for example averred that:

> Public policies and the way in which government agencies carry
> them out are part of the social environment just as the norms that
> govern individual behavior. So is the behavior of private
> organizations and professions in the position to identify and
> treat alcohol problems whether they are concerned with medicine,
> social work, education or law. The task of the agency that wishes
> to provoke a broad attack on alcohol problems is to enlist all of
> these bodies or groups as working allies. (Draper, 1977, p. 26
> emphasis added)

Earle (1981) has written an extensive literature review and discussion on
interorganizational coordination as an effective strategy to effect primary
prevention of alcoholism.

In this paper, a list of 16 social indicators of alcohol-related damage
is described as being useful for testing by community organizations.
Boguslaw and Vickers (1977) define the sociological term "social indicators"
as:

> Statistics, statistical series, and other forms of evidence that
> enable us to assess where we stand and where we are going with
> respect to our values and goals and to evaluate specific programs
> and determine their impacts. (p. 59)

The 16 social indicators described in this paper are designed to guide
interorganizational behavior in a particular community by presenting
important information to brief executive officers (CEOs -- meaning people ir
administrative and management positions) about the stated goals of lowering
the incidence of alcohol abuse and alcoholism. These indicators were
developed in consultation with a variety of experts and administrators in
the greater metropolitan area under study to insure their pertinence to that
community (see Table 1). All opinions of CEOs were incorporated into the
final list. If reaffirmation of the indicator's importance was provided by

330

Table 1. 16 Social Indicators of Alcoholism and Alcohol Abuse

Social indicator	Current level in the community
I. Number of people treated for alcoholism in medical Alcoholism Treatment Centers (ATCs)	7651 people in one year
II. Average age of patient treated for alcoholism in ATCs	41 years old
III. Number of Alcoholics Anonymous, Al-Anon, and Alateen groups	226 groups
IV. Average age of members of Alcoholics Anonymous	33 years
V. Number of family problems or family breakups that are alcohol-related	4676 families in one year
VI. Number of industrial personnel problems related to alcohol abuse	540 employees in one year
VII. Number of physicians that list the treatment of alcohol-related health problems as one of their specializations	15 physicians
VIII. Number of people receiving care from Salvation Army for chronic alcohol problems	320 people in one year
IX. Incidence of liver cirrhosis mortality	269 deaths in one year
X. Rate of diagnosis of liver cirrhosis in major hospitals (International Classification of diseases, Adapted 571.0)	807 Cases in one year
XI. Number of police arrests for Driving While Intoxicated (DWI)	2121 arrests in one year
XII. Per capita consumption of alcoholic beverages in the community	23.2 gallons of beer, .8234 cases of distilled spirits and .4984 gallons of wine
XIII. Incidence of alcohol-related criminal activity (crimes committed while drinking)	1975 crimes in one year
XIV. Alcohol-related financial loss to the total output of the metropolitan economy (due to poor work performance, absenteeism, tardiness and injuries on the job).	$880 million in one year
XV. Number of alcohol-related highway fatalities and personal injuries	1788 fatalities and personal injuries in one year
XVI. Number of "problem drinkers" in community bars (as defined by owner or manager of bars)	375,000 problem drinkers in one year

an additional CEO, the reaffirmation assumed the form of agreeing that the indicator merited attention in a policy-setting/organizational decision-making process.

Each social indicator (see Table 1) defines a specific objective, e.g., find the number of family problems occurring in the community that are due to alcohol abuse. Numerous library, telephone, and in-person inquiries were then conducted to identify and refine a specific technique of annually monitoring the social indicators and securing figures for a recent 12-month period.

The criteria for a useful social indicator was that it be: (1) specific to the community under examination; (2) available for public scrutiny; (3) collectable on a annual basis; (4) likely to remain available for at least the next 10 years; and, (5) financially feasible given the research budget.

In addition, all of the social indicators had to be of some obvious and immediate value to at least one major organization in the community. There was no expectation that any single indicator would be considered important to every organization nor was there any pretense of discovering an indicator or set of indicators that could be proven scientifically valid or reliable for all organizations in the community. The autonomous judgments of each major organization that comprised the community were the sole judges of the usefulness of each indicator. The intent of creating the social indicators was to facilitate interorganizational coordination concerning alcohol-related community damage and evaluate the primary prevention effort.

While creating the social indicators and correlative statistics, the temptation was to generate figures by popular formula or borrow figures from interested government and private organizations. It was felt, however, that those organizations that traditionally put the most effort into generating statistics and formulas for statistical reports were the same organizations that had a vested interest in the result. Therefore, in all cases except for social indicator XIV, regarding alcohol-related financial loss,[1] primary local sources were used. This procedure also stimulated more wide-spread community participation and focused attention on the social problem of interest thereby increasing the likelihood of useful innovation and local applicability of results. The effort resulted in some figures that were far less that ideal, but nevertheless demonstrated the feasibility of gathering valuable data from primary sources. Thus, this study generated more credible data than previously found in the literature.

SOCIAL INDICATORS

The remainder of the paper consists of a detailed discussion of 16 social indicators designed to meausre alcohol-related damage in one particular midwestern city. In each case the discussion explains: (1) the possible importance of each social indicator to one or more types of urban organization, e.g., private enterprise, religious, social welfare, medical, governmental; (2) how information was obtained for each of the social indicators; and (3) how to interpret fluctuations over time in the figures generated for each indicator (in light of our interest in community-wide primary prevention of alcohol abuse and alcoholism). In contrast to general theories that purport to apply equally to all communities, the 16 social indicators attempt to measure progress toward primary prevention among one complex network (one community) of organizations. Other communities can modify this set of indicators in accordance with their own perceptions of what constitutes an important indicator. Care was taken to incorporate types of community organizations that are common to most modern societies (e.g., social welfare, governmental, business, religious, and medical).

I. Number of People Treated for Alcoholism in Medical Alcoholism Treatment Centers (ATCs)

Most organizations that defined alcohol-related problems as health problems (Carmody, Mesard, and Page, 1977), would have some interest in this indicator.

---

[1]The financial loss indicator was difficult to compile locally due to lack of public access to local financial and medical records. This indicator was included in the study, however, because of its perceived importance within the business community. Over time, perhaps, local sources of information could be located or developed.

Table 2. Alcoholism Treatment Centers (ATCs)

Number of admissions to ATC over past year.

Number of hospital discharges with specified diseases over 12-month period using Hospital Adaptation of ICDA: H-ICDA International Classification of Diseases and Operations (2nd ed.).

Number discharges	Code number	Description	Remarks
	302.4	Acute alcoholism ("simple alcoholism")	Please report primary and secondary diagnoses combined for all classifications
	313.0	Episodic excessive drinking	
	313.1	Habitual excessive drinking	
	313.2	Alcoholic addiction (chronic alcoholism)	
	313.9	Other and unspecified alcoholism (acute) NOS	
	571.0	Cirrhosis	
	Total		

Nine[1] medical centers within one metropolitan area were asked to complete the form shown above (see Table 2). Optimum expected ratings (assuming an effective prevention program is operating) for this social indicator over a 10-year period would show a marked and steady increase in the number of people treated over the first five to seven years, followed by a gradual decline. This trend is expected due to the increase in sensitivity and general awareness of alcohol-related damage in the community. A saturation point is anticipated, however, and then a gradual decline is predicted, indicating acutal reduction in the incidence of alcohol abuse.

II. Average Age of Patient Treated for Alcohol Problems in Alcoholism Treatment Centers

This indicator holds some significance for those organizations in the community who can make the assumption that as problems drinkers are identified and treated, presumable with some success (cf. Francek, 1980; McGuirk, 1980; Trice, 1980), earlier in life, they spend fewer years of their total lifetimes abusing alcohol. Therefore, over the long term (two decades or more), one might be able to effect less alcohol-related community damage through a program of early identification and treatment.

Two community institutions were contacted for an estimate of average age. The first was a major medical center that had grouped data covering a six-month period (January through June, 1979). The 230 patients were age categorized 17-20, 21-25, 26-30, etc., to 61-65, 66-80 years. Precise ages of youngest and oldest patients were obtained by an "enlightened guess" on

---

[1]Nine medical centers represent 100% of the area's major medical centers that have alcoholic treatment units.

333

the part of the director of the Alcoholism Unit at the medical center. Blalock's formula for computation of the mean from grouped data was used to attain the figure 40.858 years (Blalock, 1972). Secondly, a large religious organization in the community was contacted regarding the average age of clergy that are diagnosed as having sufficiently serious alcohol problems to require treatment in a medical facility. Their figure was almost identical to that of the medical center.

Dr. Richard Bates, a well-known primary-care physician in Michigan with a long-time interest in alcohol-related health problems, averred the theory that as successful ATCs treat progressively younger patients over the years, which has been the case for Bates as well as many other professionals in the field, they begin to impact "closer to the center of the disease." Thus a gradual lowering of the average age of patients treated for alcohol problems in ATCs is proposed to be a mark of successful primary prevention within the total community under examination.[1]

III. Number of Alcoholics Anonymous, Al-Anon, and Alateen Groups

This number was obtained from telephone conversations with volunteers working for Al-Anon Family Groups in the community. The indicator is useful because, even by one of its chief critics, Alcoholics Anonymous is hailed as "primary representative of alcoholics and recovered alcoholics in our society" (Tournier, 1979, p. 230). As with social indicator I above, the number of AA, Al-Anon, and Alateen groups monitored is optimally expected to rise over the first decade of a successful implementation of a prevention strategy. Subsequent decades of effective prevention programming and planning should witness a gradual leveling off and ultimate decrease in the number of groups.

IV. Average Age of New Members of Alcoholics Anonymous

The central Alcoholics Anonymous (AA) office furnished an estimate of the average age of Alcoholics Anonymous members. The Information and Referral operator stated that "AA members are aged 13 and up," and estimated the average age of a novice Alcoholics Anonymous member in 1979 to be as much as ten years younger than a decade earlier. Unfortunately, AA does not have a useable (for research purposes) data base for making comparisons -- over time -- or even "educated guesses." Perhaps AA could be encouraged to do so if they were convinced there was a sound rationale supporting this procedure.

As with social indicator II, above, the rationale supporting this is that if alcohol-related problems are recognized and dealt with early, they are more likely to be treated successfully. Also, this trend toward earlier identification and treatment results in fewer years of abusing alcohol during the alcoholic's lifetime. Thus, again, the optimum statistical trend, in light of our (previously defined) interest in primary prevention, would be manifest in a gradually decreasing average age of new AA members across the community.

V. Number of Family Problems or Family Break-ups that are Alcohol-Related

Much recent research in the alcohol field considers family problems, including family violence, marital separation, or divorce to be reasonably common alcohol-related life problems (USDHEW, 1979). The occurrence of such problems should be automatically investigated for possible alcohol

[1] R. C. Bates, a primary care physician, personal communication, April, 1979.

involvement by the therapist or physician involved (Schuckit, 1979). A recent government report supports this position by clearly highlighting the influence of alcohol abuse in family problems across the nation (USDHEW, 1981). Likewise, it has been found recently that major organizations in American communities put great emphasis on the importance of these alcohol-related family problems (Earle, 1981).

For purposes of the present research project, a Community Service Directory, published by the United Way, was consulted for a list of all agencies in the community classified as providing "marriage counseling" services. All 18 agencies were included in a telephone survey. The director of the agency or a head counselor was asked to give an approximation (best guesstimate -- without researching the question[1]) of the number or percentage of all the families contacted (i.e., agency contacts) per year who have alcohol-related problems. This included instances where alcohol is a contributing factor and not only those families where alcohol is obviously the primary cause of problems. None of the agencies kept precise records of this information. Most responses were couched in terms resembling: "about ___% of all cases involve some alcohol-related problems per year." Agency directors were then asked how many cases they work on in one year. Thus, the total number of families in the metropolitan area identified as having alcohol-related problems by family counseling agencies was estimated at 4,675.

Of the 18 agencies contacted, one was too new to have generated any statistics, two had been combined to form one agency, and two agencies simply refused to provide any information (one of them cited "breach of confidentiality" as its reason and the other gave no reason). Of the 14 agencies that provided figures, the lowest percentage given was 4% and the highest was 73%. Five of the 14 cited the figure 20% and 10 of the 14 gave percentages that fell in the range from 13% to 35%. Only three agencies supplied figures of 50% or more alcohol-related cases and only one agency gave an estimate of less than 13%. Therefore, about 25% seemed a reasonable figure to use.

The results from this cursory telephone survey can be compared with an independent finding reported by the St. Louis Regional National Council on Alcoholism, Inc. The Council publishes a flyer "Facts on Alcoholism" which states that "25% of family problems, disruptions and divorces are due to alcoholism."

The reader should keep in mind that alcohol-related problems are usually concealed. Under such conditions whether or not social service agencies correctly identify such families and this issue as a problem is questionable. Recently published literature reveals growing professional interest and sensitivity to this issue, (Ehline and Tighe, 1977; Regan, 1978). Interestingly, one agency director, who conducted a thorough study of 150 case records over a 12-month period (notwithstanding my request for his "best-guess approximation"), reported the highest rate of alcohol-related cases (73%). Moreover, he mentioned that state-wide records revealed a rate of 78% alcohol-related problems among all cases

---

[1]Several earlier attempts had been made to collect more precise figures from these family counseling agencies, but the agencies consistently refused to cooperate with the research effort. Their reasoning cited the lack of accurate case records kept by their staff detailing the role played by alcohol abuse in family problems. Therefore, requesting an "approximation" clearly asked for something more estimable, i.e., sacrificing some degree of verifiable accuracy for greatly enhanced attainability of data.

handled by the agency since the formation of his organization. Most directors offered information for this social indicator based on their personal knowledge of typical agency clientele (instead of reporting data from a formal inquiry) -- which was acceptable because they were often counselors themselves. In light of this, the 25% figure seems to err on the conservative side, which buttresses the argument advocating for the potential importance of the problem to a variety of community organizations.

Given the creation of an effective primary prevention program, this indicator (representing the amount of family alcohol-related problems throughout the community) would be expected to increase over the short run (5-7 years) and then gradually decline as the "preventative" impact becomes apparent.

VI. Number of Industrial Personnel Problems Related to Alcohol Abuse

The importance of EAPs (Employee Assistance Programs) within the business community has been articulately expressed in a two-part article written by Bobbi Linkemer (1976), Assistant Editor of St. Louis Regional Commerce & Growth Association (RCGA). The following brief excerpt from one of Linkemer's articles summarizes the business-person's perspective:

> There are an estimated 136,000 alcoholics in the St. Louis metropolitan area, and since four other lives are directly affected by every alcoholic, well over 500,000 people (or about 20%) of this area's population are touched by the disease....Through a vehicle called Employee Assistance Programs (EAPs), many companies are becoming the catalyst for saving the lives and livelihoods of their own employees....Companies like McDonnel Douglas, Missouri Pacific, Chrysler, Mallinckrodt, General Motors, eighteen of the Bell systems, Eastman Kodak, duPont, and hundreds of others obviously agree. Each of these firms has an Employee Assistance Program (EAP) -- either management or labor initiated or established and operated by a private counsulting firm, such as Personnel Performance Consultant, which is currently setting up an EAP at Monsanto. (pp. 73-74)

In the course of this research project, six major Employee Assistance Programs (EAPs) in the community were contacted and asked how many employees with alcohol-related problems they were able to identify during the last 12-month period.

As with the indicator citing the number of people treated in medical Alcoholism Treatment Centers (ATCs), this indicator would be expected to increase dramatically during the first 5-7 months of an effective prevention program, then slowly but continuously to decline.

VII. Number of Physicians that List the Treatment of Alcohol-Related Health Problems as One of Their Specializations

The significance of this social indicator stems from the premise that alcoholism is a disease and should be treated in medical setting by a physician. Assuming the operation of an effective prevention program, the optimal expected change of the number of physicians treating alcohol-related health problems, within a given community, should increase over a 10-15 year period to meet the demand for increased treatment and ultimately decrease over a period of 20-25 years as the primary prevention program has its effect. Other types of caregivers -- psychiatrists, psychologists, counselors -- are excluded from this social indicator, but may be acknowledged in other social indicators, i.e., Employee Assistance Programs.

Documenting community-wide figures for this social behavior was accomplished by personal interviews with three community physicians, the Executive Director of a local National Council on Alcoholism, and a social work faculty member at a community college that had a special interest in alcoholism and alcohol abuse treatment.

VIII. Number of People Receiving Care from Salvation Army for Chronic Alcohol Problems

This social indicator is an attempt to secure an extimate of the size of the "alcoholic bum" population in the community. Although there are other facilities that treat "skid row bum" types, these people usually "make the rounds" among available facilities in the metropolitan area. By counting "first time" people treated in one major facility such as the Salvation Army over a one-year period, one can obtain a representative figure. This approach to calculating a figure was originally suggested by an alcoholism-treatment social worker at a state hospital, alcoholism services unit. There is, of course, no way to obtain a perfectly accurate census of this population.

Data for this social indicator was obtained from personal interviews with Salvation Army personnel at their three treatment/residential centers in the community. Annual totals of "first time" people treated at Salvation Army facilities were available covering a four-year period (1975-1978). The figure used (320 people treated) is a simple arithmetic mean of the four annual totals available.

It is expected that even the most effective community-wide primary prevention program would have very little effect on the size of this population within the first decade of operation, but succeeding decades could be expected to witness a measurable decline.

IX. Incidence of Liver Cirrhosis Mortality

Since there is considerable controversy within the alcoholism field as to the usefulness of this indicator, a review of the literature is included in the discussion below.

The argument against it was articulated during a personal interview with Selden Bacon,[1] who stated that possibly only 40% of late-state alcoholics develop cirrhosis -- and earlier stages rarely have it. Also, it usually requires many years of heavy drinking to die from alcoholic liver cirrhosis and obviously cirrhosis can be caused by non-alcohol-related factors. Moreover, heavy continuous drinking does not guarantee the development of cirrhosis. Using this rationale, it was suggested that perhaps cirrhosis not be used as an indicator.

On the other side of the controversy, it is argued that alcoholics suffer from cirrhosis more often than other people. The importance of the indicator was emphasized as recently as 1978, when a British group of medical researchers studying the causes of death in persons under the age of 50 cited alcoholic liver cirrhosis as a significant contributor ("Deaths Under 50," 1978).

A lengthy rationale in favor of using this indicator was articulated in a nutritional study conducted at the Research Laboratories of the Finnish State Alcohol Monopoly.

[1]S. Bacon, personal communication, September, 1978.

One fairly concrete measure of the prevalence of alcoholism is the incidence of diseases that result directly or indirectly from the constant abuse of alcohol. More commonly than other people, alcoholics suffer from liver cirrhosis, pancreatitis, cancer of the mouth, pharynx, and esophagus, and disorders of the central nervous system like the Wernicke-Korsakoff syndrome, polyneuropathy, and pellagra. Brunn and co-workers have concluded on the basis of abundant epidemiological and clinical evidence that, after cardiovascular diseases, liver cirrhosis is the most common cause of death in alcoholics and that constant abundant use of alcohol is the prime factor in the genesis of alcoholic cirrhosis. Consequently, the mortality rate from cirrhosis is a reliable indicator of the prevalence of alcoholism even though other causes of cirrhosis exist. (Pekkanen and Forsander, 1977, p. 83)

In addition, one of the highlights stemming from the Third Special Report to the U.S. Congress on Alcohol and Health from the Secretary of Health, Education, and Welfare, June, 1978, is that "studies of international alcohol statistics demonstrate a high correlation between the per capita level of consumption and the rate of cirrhosis deaths" (p. xi).

Also it is worth noting that this indicator has a long history of use in the United States -- making comparative data relatively easy to obtain. In the early 1940s, E. M. Jellinek began to monitor the size of the alcoholic population by noting the statistics on liver cirrhosis deaths. More recently, in 1979 both the 100th Edition of the Statistical Abstracts of the United States (1979) and the Book of American Rankings (1979) use measurements very similar to the original Jellinek formula to estimate the number of alcoholics in the United States.

The figure used for the present study, 269 deaths, was provided by the State Center for Health Statistics and includes a five-county metropolitan area.

It is expected that a community prevention program would have little or no effect on this indicator for the first 10-15 years. Subsequent decades, however, would be expected to witness a decline in the rate.

X. Rate of Diagnosis of Liver Cirrhosis in Major Hospitals

Korsten and Lieber (1979) have clearly expressed the critical role played by cirrhosis in American urban communities. They write that:

Alcoholic liver injury and its complications are of enormous significance in terms of morbidity and mortality. This is apparent in New York City where alcoholic cirrhosis now ranks as the third or fourth leading cause of death in males (35-45 years) and in Canada where it is the most rapidly increasing cause of death in those over age 25.

Although early human and animal studies suggested the primacy of malnutrition in the production of alcoholic liver injury, recent experimental work has clearly demonstrated the direct, hepatotoxic effects of alcohol. It is now clear that the alcoholic cannot fully prevent the development of liver injury by adequate or even supplemented nutrition. (p. 24)

For purposes of the present study, seven major hospitals (some of the same medical centers contacted for social indicator number I) were contacted to determine the rate of diagnosis of liver cirrhosis. Figures were provided by Medical Records Departments.

Fluctuation in the rate of this indicator is expected to rise and fall dramatically with the degree of sensitivity or awareness of the problem on the part of physicians or other health care professionals and the degree of acceptability (or social stigma) within society of personally admitting to having a drinking problem. Over a 10-year period, however, a successful prevention program would be expected to raise the rate of this indicator considerably over the first 5-7 years and later witness a slow decline.

XI. Number of Police Arrests for Driving While Intoxicated (DWI)

Perhaps the single-most, obviously important indicator of the level of alcohol abuse in the mind of the public is the number of problems caused by the drunken driver. Studies of people arrested for DWI indicates that the majority are indeed chronic abusers of beverage alcohol -- not social drinkers who have no history of alcohol abuse (Barni, 1980). Unfortunately, the number of arrests can fluctuate dramatically, however, with a change in public policy regarding arrests (Korsten and Lieber, 1979), rendering the indicator less useful than its broad public appeal might indicate.

The figure used in this study, 2,121 DWIs, comes from local police department records for the greater metropolitan area.

It is expected that using this indicator over the short term (less than 5 years) will not be useful in evaluating a prevention program. However, over a long period of time (perhaps 10 years or more) this indicator will warrant attention as an evaluation instrument.

XII. Per capita Consumption of Alcoholic Beverages in the Community

This social indicator of alcohol abuse and alcoholism has been a highly controversial one for many years within the alcohol field. When the National Institute on Alcohol Abuse and Alcoholism (NIAAA) made the decision to favor using consumption levels as an indication of alcohol abuse they were strongly criticized by the Alcohol and Drug Problem Association of North America (ADPA) as being "neo-prohibitionistic." Joining in the criticism of NIAAA's decision was the National Coalition of Adequate Alcoholism Programs, which is another major national constituency (St. Louis NCA, 1977/1978). The points made to buttress this position included the statement that drinking, per se (with the exception of minors or pregnant women), has not been proven harmful to one's health and that for many people drinking in moderation seems to have some positive effects (The Alcoholism Report, 1977).

The argument in favor of using the per capita consumption rate as a social indicator is perhaps best summarized by Jan DeLint (1977). After marshalling an impressive amount of literature to support her position, she defends the following statement:

> Since the relaxation of alcohol control has undoubtedly facilitated the proliferation of alcohol use in society and therefore higher rates of excessive use and related problems, it would seem reasonable in the context of public health to try to stabilize these trends and to control alcohol availability. (p. 437)

Also, in an independent research report, Pekkanen and Forsander (1977) aver that:

From the point of view of public health, it is the level of total consumption that is important for when the total consumption level of alcohol rises, the number of alcoholics increases and, consequently, the incidence of disorders caused by alcohol. (p. 91)

Clearly, further discussion of this social indicator is appropriate; therefore, it was included among the list of 16 social indicators.

Figures for this indicator stem from the following sources: (1) gallons of beer as reported by the United States Brewers Association, Inc.; (2) cases of distilled spirits from Liquor Industry Marketing 1979 (p. 151); and (3) gallons of wine as per Automated Professional Services Company (Automated Professional Services, 1979).[1]

The decision was made to monitor all three beverage types independently for conceptual clarity and ease of replication. Twenty-five years from now Americans might switch their consumption patterns from one type of beverage to another -- maintaining a constant gallons-per-person rate. But drinking a gallon of beer is not the same behavior as drinking a gallon of whiskey and concomitant (increase physiological, if not psychological and social/cultural) damage.

The beer and distilled spirits information was converted to a per capita rate by industry sources but the wine consumption figures had to be converted by the investigator. Calculation of the rate for this indicator was made by dividing total gallons of wine sold in the metropolitan area (761,306 gallons) by total number of people 21 years of age (the state's legal minimum drinking age) or over 1,527,400 people).[2]

Using the NIAAA policy guideline, the optimal rate change for this indicator would be to hold the per capita alcohol consumption rate at its present level.

XIII. Incidence of Alcohol-Related Criminal Activity

Gerson and Preston (1979) have tried to strengthen a relationship between alcohol abuse and violent crime similar to the one that DeLint and Schmidt (1971) tried to establish between per capita consumption and general alcohol problems. Gerson and Preston explain that:

The ability to control for other variables in a regression analysis makes it possible to control, for example, for a bias of more frequency police patrols and, hence, greater reporting in certain neighborhoods. Regardless of which combinations of variables describing neighborhoods and police activity are entered into the equation ahead of alcohol, the rate of alcohol consumption in licensed establishments affects the rate of violent crime in the district. (p. 311)

John (1978), however, emphasizes several caveats regarding studies that have been conducted in this area. His opinion, based on a thorough

---

[1]This report was concerned with case sales and was divided by wholesaler, supplier, brand, and container size and was copyright dated 1979. Automated Professional Services explained that: "This report is compiled from the records of the Liquor Control Department, State of Missouri. It covers all shipments to wholesalers during the full calendar month. It does not include direct shipments to military installations."
[2]Information about the total adult population of the metropolitan area was obtained from the local chamber of commerce.

literature review, is that one cannot identify a causal relationship between alcohol abuse and violent crime. Notwithstanding this view, John does support the use of this as a legitimate social indicator. He writes that:

> Regardless of the above-mentioned limitations about research data and clearly defined terms, many researchers have independently reached the general conclusion that people who use alcohol excessively seem more likely than others to be involved in acts of violent crime and, more specifically, in criminal homicide. (p. 9)

Information for the present study was obtained from the State Department of Probation and Parole. The Department estimates that about 60% of all criminal cases they handle involve alcohol abuse as an obvious, major contributing factor.

The optimal rate for this indicator is a gradual average decline when viewed over a period of a decade or more. Significant fluctuations are not expected to be correlated with the overall goals of a primary prevention project.

XIV. Alcohol-Related Financial Loss to the Total Output of the Metropolitan Economy

Berry (1976) has written a brief report providing a rationale for this indicator. He explains that:

> Alcohol abuse generates costs to society within the contexts of lost production (including premature mortality), health care, motor vehicle accidents, fire, crime, and certain social responses to alcohol abuse. The economic cost is manifested in either of two forms. Alcohol abuse can adversely affect productivity, causing society actually to lose some potential production. Or it may be necessary to produce goods and services (for example, health care) to cope with the consequences of alcohol abuse, thus incurring the opportunity cost implied when scarce resources are used in the production of one good or service rather than another. (p. 621)

Any estimate of the economic costs of alcohol abuse will be difficult to determine for many reasons. One is the strong social stigma attached to the problem. A business consultant interviewed in connection with this research project, for example, thought that asking about the drinking habits of community businessmen was tantamount to "asking about their private sex lives. It's too personal, and too sensitive an issue to broach," he declared. Several business executives simply referred the researcher to the Director of Personnel for their organization, or refused to admit they had had to deal with "that problem." If the local business/financial community does monitor the extent of problem drinking, those records are not made public. The damage, however, seems to be privately acknowledged and therefore is well worth trying to measure -- due to its presumed magnitude.

In addition, there are powerful and complex economic factors that must be addressed in any thorough examination of this social indicator. Clearly, there are large corporate profits directly linked to the promotion and sale of alcoholic beverages. Manufacturer sales, leisure industries, advertising, and other business-related activities stand ready to gain great economic advantage stemming from both widespread "social" drinking and "heavy" drinking. One spokesperson for the alcoholic beverage industry was quoted as saying, "When the economy is bad, not as much liquor is consumed, it's as simple as that" (St. Louis Post-Dispatch, 1977).

The importance of this social indicator, however, in light of the present focus on primary prevention, was referred to in a packaged training course titled "Alcohol and Federal Employees: A Training Course for Supervisors." The course was constructed by a Federal Occupational Health Task Force and an integral part of the course targeted costs to private industry. The script of the seminar read as follows:

> Reliable information indicates there are over 2.5 million untreated cases of alcoholism in industry, each costing the employer approximately 25% of the employee's annual salary. This then is a total cost to industry in the U. S. of approximately 4.25 billion dollars. (Occupational Health Task Force, 1973, p. 12)

Using this formula, the Task Force was able to estimate the loss to private industry within the community under study at $68 million -- assuming an alcoholic employee population of 40,000 people.

The figures used for this indicator in the present study was derived as follows:

1. According to the Regional Commerce and Growth Association the total output of the economy for the metropolitan area in 1978 was $46.3 billion. This figure is not identical to the Gross Regional Product, but "it is the closest approximation available" according to a RCGA spokesperson.[1]

2. According to the U. S. Department of Commerce, U. S. Industry and Trade Association (ITA), the Gross National Product (GNP) for 1978 was $2,107.6 billion.[2]

3. According to the National Institute on Alcohol Abuse and Alcoholism (Alcoholism and Alcohol Education, 1979), alcohol abuse and alcoholism cost the United States over $40 billion per year.[3]

4. One could then say that the metropolitan community's share of the national economy is about 2.2% ($46.3 billion divided by $2,107.6 billion). If the metropolitcan community had 2.2% of the economic loss due to alcohol abuse and alcoholism, they had 2.2% of $40 billion lost ($880 million) due to alcohol-related damage in 1978.

The above formula is far from ideal. However, its inclusion serves to highlight the need, as perceived by many community organizations, for some measurement of financial loss.

---

[1] Telephone conversation with RCGA Staff, June 1979.
[2] Ibid.
[3] This figure varies depending on one's method of calculation. For example, a much higher estimate was suggested by Dr. Samuel E. Guze, Head of the Psychiatry Department and Vice-Chancellor for Medical Affairs at Washington University in St. Louis. He placed the cost of alcohol-related damage to the U. S. between $45 and $70 billion per year during a presentation of an "Alcoholism Forum" sponsored by the George Warren Brown Continuing Education Program and the St. Louis Area National Council on Alcoholism, 9-10 October 1979. An article about the Forum was published in Links Newsletter (November 1979), page 5, available from the George Warren Brown School of Social Work, GWB Alumni Association, Washington University, St. Louis, Missouri 63130.

It is expected that as a primary prevention project succeeds, the amount of financial loss throughout the community will decline.

XV. Number of Alcohol-Related Highway Fatalities and Personal Injuries

The intoxicated automobile driver is generally thought to be a danger to him or herself as well as others on the nation's highways. The Missouri Division of Highway Safety sponsored a study in which blood alcohol content was analyzed of drivers killed in highway crashes. They report that:

Of the 378 drivers analyzed, 60.1% had been drinking. 46.8% had a blood alcohol level of over .10, which is the legal presumption level of intoxication in the State of Missouri. (Alcohol Study, 1977/1978)

Figures for this social indicator were provided by the Department of Public Safety, Missouri Division of Highway Safety and calculations were made as follows: (1) The most recent information available for this research project was from July 1977 through June 1978. During this period, there were a total of 16,363 crashes involving either personal injuries or fatalities. (2) According to figures available from the Missouri Division of Highway Safety, 927 (July 1977 through December 1977) plus 861 (January 1978 through June 1978) or a total of 1,788 crashes occurred involving fatalities or personal injuries "where one or more drivers had been drinking."[1] This amounts to about 11% of all crashes.

The spokesperson for the Department of Public Safety emphasized that these numbers only include "proven incidences" -- that is, only where the police officers at the scene had filed a report and an analysis of blood alcohol content had been performed immediately. But he added that officers do not test everyone for blood alcohol content. Many drivers are left untested even though they may be staggering or otherwise acting intoxicated because the aftereffects of having the accident will often cause accident victims to manifest inebriate-like behavior.[2] Therefore, this is thought to be a very conservative way of estimating the actual number.

Additional evidence that official statistics are misrepresenting the actual incidence of the problem stems from an award-winning series of articles published in the St. Louis Post-Dispatch on drunken driving in which it was revealed that eight out of every ten drunken-driving defendants whose cases came to court in 1979 "walked out of court with nothing on their driving record to indicate and alcohol-related traffic offense" (Rose and Malone, 1980, p. 4C).

Assuming that an effective prevention program is operating in the community, this indicator is expected to decline measurably over the first 5-7 years and continue to decline slowly over subsequent five-year blocks of time. A caveat regarding this indicator, however, is that the rate depends in part on recommended police procedures and officer training regarding suspected alcohol problems. Without specific orders from the police officers at the scene of the accident, none of the necessary tests are conducted to determine intoxication levels of accident victims.

XVI. Number of "Problem Drinkers" in Community Bars

Numerous investigators have addressed the question of a possible

---

[1]Personal correspondence with Department of Public Safety, Missouri Division of Highway Safety, March 1979.
[2]Ibid.

important relationship between the rate of alcohol abuse and alcoholism and the type, size, character, and number of drinking establishments in any given community (Brunn et al., 1975; Popham et al., 1975; Smart, 1977; Parker and Harman, 1978; Parker, Wolz, and Harford, 1978). Recently, Harford et al. (1979) concluded a study concerning the relationship between alcoholism and the number of on-premise outlets by arguing that "more attention be given to the development of a comprehensive model of alcohol availability" (p. 1057).

The procedure used to collected data for this social indicator was constructed as follows: (1) There are about 3,000 bars in the metropolitan area under study (information from a local bartenders union). (2) Each bar serves an average of 500 (different) people in one week. (3) Out of an estimated population of 1,500,000 bar customers, 375,000 (25%) are problem drinkers.

These figures were drawn from conversations with 14 bartenders (two bartenders from each of seven types of bars, listed below) who had been working in bars for several years. Essentially, two important questions were asked during these interviews: (1) Approximately how many different people does your bar serve in one week? and (2) In your opinion, how many people that you serve can you diagnose as "problem drinkers?" (Give number in average week or percentage.) A definition of "problem drinker" was provided as follows:

1. The guy who comes in for lunch, orders two martinis, then forgets about lunch, orders two more martinis and leaves.

2. The guy who is loud and disturbs other customers.

3. The guy who gets drunk, then goes out and beats up his "old lady" or runs someone over with his car.

4. The guy whose drinking becomes a problem while at work.

5. The guy who has physical (health) problems related to alcohol.

6. Any combination of the above.

Several problems were encountered during this series of interviews. Responses varied widely and many bartenders did not know their clientele well enough to determine who qualified as a problem drinker. Some bartenders had difficulty estimating the average number of different people served in a one-week period. Several bartenders were suspicious of the motivations behind the research project or were cautious about how they responded if bar customers were within hearing range of the conversation with the researcher.

All bars in the metropolitan area were classified by "type" -- Country Club, Hotel, Convention, Ball Park, Fancy Restaurant, Skid Row, and Community or Neighborhood bars -- and two bars from each "type" were surveyed. The percentage of all bar customers that were labeled problem drinkers varied from 2% to 100%. In no instance did a bartender claim to have had no experience with problem drinkers.

The means of computing a figure for this indicator, notwithstanding its heuristic and innovative qualities, should not be expected to accurately measure the impact of a prevention program unless it can be done annually over a minimum of a 10-year period. Also, it should be remembered that drinking patterns within the metropolitan community, whether problematic or not, are totally ignored by this social indicator if they occur off the

premises of a bar. And, finally, this indicator ignores female problem
drinkers, yet the literature points to the possibility that "today women may
be more at risk for developing alcohol problems than in the past" (Bourne
and Light, 1979, p. 94).

SUMMARY AND FUTURE APPLICATION

The value of this list of social indicators is a function of its
ability to mobilize existing community organizations toward a common goal of
primary prevention. It is based on a strategy of using a broadly
constructed interorganizational communication network (incorporating both
alcohol-specific and non-alcohol-specific organizations) to establish and
maintain a primary prevention community program. Although the focus of this
study was not on testing for the degree of empirical validity and
reliability of the 16 social indicators, care was taken to identify
indicators that would be of interest to organizations commonly found in most
American communities, e.g., educational, business, social welfare, mass
media, religious, and medical. There is recent support within the alcohol
literature for this research orientation (Plant, 1979).

Also, it should be noted that data for the social indicators were never
collected over a full 10-year period during the research project. Rather,
the intent of this paper has been to:

1.   Construct a framework for collecting data.

2.   Secure one set of figures (covering a 12-month period) to
      demonstrate the feasibility of data collection.

3.   Ultimately to produce a mechanism for measuring the increase or
      decrease of alcohol misuse (including alcoholism, alcohol abuse,
      and problem drinking) that would be accessible in the greater
      metropolitan community under study.

Additionally, it should be emphasized that none of the indicators are
sufficiently robust to stand alone or be useful indicators over less than a
10-year period. [Some authors have suggested a minimum of a 25-year time
frame (Bacon, 1978) for the purposes of evaluating a primary prevention
project.] While collecting data over a 10-year period (a minimum time frame
suggested by Bacon) or more, it would be expected that some indicators would
be sensitive to short-term advertising campaigns, etc., that would increase
their raw scores over a period of a few years. But this is predicted to
level off after a period of five to seven years and decline in subsequent
years if a successful prevention technique is operating in the community.
The above caveats deserve to be highlighted because they provide important
guidelines for interpretation and analysis of the indicators.

There are, potentially, an infinite number of social indicators of
alcohol-related damage in urban communities -- depending on the particular
community in question, the important interests of any particular
organization or network of organizations within the community and the point
in time at which the social indicators are operating. For example, none of
the 16 social indicators measures alcohol-related illnesses such as
gastrointestinal problems, ulcers, nerve damage, or unplanned suicide
attempts. Also, it is likely that families seeking help in which "alcohol"
is a factor would consult with religious leaders (priests, rabbis,
ministers) and not "orthodox" marriage counseling agencies in the community.
In certain communities (other than the one studied in the present project)
the incorporation of these and other perspectives may easily be included in
the evaluation instrument, and perhaps should be. Care should be taken,

however, to use a source of information that will remain clearly
identifiable and constant through the requisite 10-year minimum time frame.

Using this research orientation toward the assessment of primary
prevention policy and planning relative to alcohol abuse, this paper has
underscored the feasibility of:

1. establishing numerous primary local sources to monitor prevention
    activities;

2. maintaining these sources as a cohesive entity (incorporating all
    16 independent measures to formulate one message to a large group
    of organizational decision-makers about the prevention activities)
    covering a period of at least one decade.

3. suggesting the potential usefulness of this approach to other
    modern urban communities.

REFERENCES

Alcohol Study, Drivers Killed, 1977 and 1978, Missouri Traffic Crashes.
    Unpublished report available from Department of Public Safety, Missour
    Division of Highway Safety, Jefferson City, Missouri.
Alcoholism and Alcohol Education, 82 (May 28, 1979).
The Alcoholism Report, 7 (September 14, 1979). Published semimonthly by JSL
    Reports, 1264 National Press Building, Washington, D. C.
Automated Professional Service Company, Cumulative Year-to-Date Report of
    Wine and Spirits Shipments and Missouri Wholesalers,1979. Kansas City,
    MO: P. O. Box 7245, 64113.
Bacon, Selden D., "On the prevention of alcohol problems and alcoholism."
    Journal of Studies on Alcohol, 1978, 39, 1143.
Barni, E. "ARTOP Advances." The Key, 1980, March, 2-4.
Berry, R. E. "Estimating the economic costs of alcohol abuse." The New
    England Journal of Medicine, 1978, 295-621.
Blalock, H. M., Jr. Social Statistics, 2nd ed. New York: McGraw-Hill,
    1972.
Boguslaw, R., and Vickers, G. R. Prologue to Sociology. Santa Monica, CA:
    Goodyear, 1977.
Bourne, P. G., and Light, E. "Alcohol problems in Blacks and women." In J.
    H. Mendelson and N. Mello (Eds.), The Diagnosis and Treatment of
    Alcoholism. New York: McGraw-Hill, 1979.
Bruun, K., Edwards, G. M., Lumio, M., Makela, K., Pan, L., Popham, R. E.,
    Room, R., Schmidt, W., Skog, O. J., Sulkunen, P., and Osterberg, E.
    Alcohol Control Policies in Public Health Perspective. Helsinki:
    Finnish Foundation for Alcohol Studies, Vol. 25, 1975.
Carmody, A. P., Mesard, L., and Page, W. F. Alcoholism and Problem
    Drinking: 1970-1975. A Statistical Analysis of VA Hospital Patients.
    Controller Monograph No. 5. Washington, D. C.: Reports and Statistics
    Service, Office of Controller, Veterans Administration, 1977.
"Deaths under 50." Medical Services Study Group of the Royal College of
    Physicians of London. British Medical Journal, 1978, 2, 161-62.
DeLint, J. "Alcohol Control Policy as a Strategy of Prevention: A Critical
    Examination of the Evidence." In J. S. Madden, R. Walker, and W. H.
    Kenyon (Eds.), Alcoholism and Drug Dependence: A Multidisciplinary
    Approach. New York: Plenum Press, 1977.
DeLint, J., and Schmidt, W. "Consumption Averages and Alcoholism
    Prevalence: A Brief Review of Epidemiological Investigations."
    British Journal of Addiction, 1971, 66, 97-107.
Draper, R. A. "Social and Environmental Factors in Prevention." Summary
    Proceedings: Tripartite Conference on Prevention. U.S. Department of

Health, Education, and Welfare, DHEW Publication No. (ADM) 77-484.
Washington, D.C.: U.S. Government Printing Office, 1977, 23-28.

Earle, R. M. Primary Prevention of Alcoholism: Interorganizational
Coordination Approach. Doctoral dissertation listed in Dissertation
Abstracts International, 1981, 41(8). Order number 8103676.

Ehline, D., and Tighe, P. O. "Alcoholism: Early Identification and
Intervention in the Social Service Agency." Child Welfare, 1977, 56,
584-592.

"Fewer Hard Liquor Drinkers Indication. `Unhealthy Economy'." St. Louis
Post Dispatch (September 1, 1977).

Francek, J. L. "Occupational Alcoholism Programs: Challenge and
Opportunity." Alcohol Health and Research World, 1980, 4, 2-3.

Gerson, L. W., and Preston, D. A. "Alcohol Consumption and the Incidence of
Violent Crime." Journal of Studies on Alcohol, 1979, 40, 307-312.

Harford, T. C., Parker, D. A., Pautler, C., and Wolz, M., "Relationship
Between the Number of On-Premise Outlets and Alcoholism." Journal of
Studies on Alcohol, 1979, 40, 1057.

John, H. W. "Alcoholism and Criminal Homicide: An Overview." Alcohol
Health and Research World, 1979, 2, 8-13.

Judge, C. S. The Book of American Rankings. New York: Facts on File,
1979.

Korsten, M. A., and Lieber, C. S. "Hepatic and Gastrointestinal
Complications of Alcoholism." In J. H. Mendelson and N. K. Mello
(Eds.), The Diagnosis and Treatment of Alcoholism. New York:
McGraw-Hill, 1979.

Krimmel, H. Alcoholism: Challenge for Social Work Education. New York:
Council on Social Work Education, 1971, pp. 15-16.

Linkemer, B. "Business' Billion Dollar Hangover." St. Louis Commerce
Magazine, November, 1976, pp. 72-76.

Linkemer, B. "In the Forefront of the Fight against Alcoholism." St. Louis
Commerce Magazine, December, 1976.

Links Newsletter, November, 1979. George Warren Brown School of Social Work
Alumni Association, Washington University, St. Louis, MO 63130.

Mausner, J. S., and Bahn, A. K. Epidemiology: An Introductory Text.
Philadelphia: W. B. Saunders Co., 1974.

McGuirk, T. R. "Evaluation and Development of Employee Assistance
Programs." Alcohol Health and Research World, 1980, 4, 17-21.

Occupational Health Task Force of the Human Resources Committee of the
Federal Executive Board of Greater St. Louis, Alcohol and Federal
Employees: A Training Course for Supervisors. November, 1973.

Parker, D. A., and Harman, M. S. "The Distribution of Consumption Model of
Prevention of Alcohol Problems: A Critical Assessment." Journal of
Studies on Alcohol, 1978, 39, 377-399.

Parker, D. A., Wolz, M. W., and Harford, T. C. "The Prevention of Alcoholism: An Empirical Report on the Effects of Outlet Availability."
Alcoholism" Clinical and Experimental Research, 1978, 2, 339-343.

Pekkanen, L., and Forsander, O. "Nutritional Implications of Alcoholism."
Nutritional Bulletin, 1977, 20, 93.

Plant, M. A. "Estimating Drinking Patterns and Prevalence of Alcohol-
Related Problems." British Journal on Alcohol and Alcoholism, 1979,
14.

Popham, R. E., Schmidt, W., and DeLint, J. "The Prevention of Alcoholism:
Epidemiological Studies of the Effects of Government Control Measures."
British Journal of Addiction, 1975, 70, 125-144.

Regan, J. M. "Services to the Families of Alcoholics: An Assessment of a
Social Support System." Doctoral dissertation, Brandeis University,
1978. Social Work Research & Abstracts, 1978, 14, 67-68. University
Microfilms International No. 7821713.

Reiss, A. J., Jr., and Wilensky, H. L. "Forward." In C. Perrow, Complex
Organizations: A Critical Essay. Glenview, IL: Scott, Foresman and
Co., 1972.

Rose, L. J., and Malone, R. "Post-Dispatch Series Is Honored By Group."
    St. Louis Post Dispatch, August 13, 1980.
Schmidt, W., and Popham, R. E. "The Single Distribution Theory of Alcohol
    Consumption: A Rejoinder to the Critique of Parker and Harman."
    Journal of Studies on Alcohol, 1978, 39, 400-419.
Schuckit, M. A. "Treatment of Alcoholism in Office and Outpatient
    Settings." In J. H. Mendelson and N. K. Mello (Eds.), The Diagnosis and
    Treatment of Alcoholism. New York: McGraw-Hill, 1979.
Smart, R. G. "A Note on the Effects of Changes in Alcohol Control Policies
    in the Canadian North." Journal of Studies on Alcohol, 1979, 40,
    908-13.
Smart, R. G. "The Relationship of Availability of Alcoholic Beverages to
    Per Capita Consumption and Alcoholism Rates." Journalof Studies on
    Alcohol, 1977, 38, 891-896.
St. Louis Area National Council on Alcoholism. "NIAA Criticized for
    `Neo-Prohibitionist' Strategy." The Key, December 1977/January 1978.
Staulcup, H., Kenward, K., and Frigo, D. "A Review of Federal Primary
    Alcoholism Prevention Projects." Journal of Studies on Alcohol, 1979,
    40, 943-968.
Tournier, R. E. "Alcoholics Anonymous as Treatment and as Ideology,"
    Journal of Studies on Alcohol, 19679, 40, 230.
Trice, H. M. "Applied Research Studies: Job Based Alcoholism and Employees
    Assistance Programs." Alcohol Health and Research World, 1980, 4,
    14-16.
U. S. Department of Commerce, Bureau of Census, Statistical Abstract of the
    United States (100 edn.). Washington, D. C.: U. S. Government
    Printing Office, 1979.
U. S. Department of Health, Education, and Welfare, Alcoholism Prevention:
    Guide to Resources and References, DHEW Publication No. (ADM) 79-886.
    Washington, D. C.: U. S. Government Printing Office, 1979.
U. S. Department of Health, Education, and Welfare, First Special Report to
    the U. S. Congress on Alcohol and Health from the Secretary of Health,
    Education, and Welfare, December 1971. Washington, D. C.: U. S.
    Government Printing Office, 1971.
U. S. Department of Health, Education, and Welfare, Second Special Report to
    the U. S. Congress on Alcohol and Health from the Secretary of Health,
    Education, and Welfare, June 1974.
U. S. Department of Health, Education, and Welfare, Third Special Report to
    the U. S. Congress on Alcohol and Health from the Secretary of Health,
    Education, and Welfare, June 1978. Washington, D. D.: U. S. Govern
    ment Printing Office, 1978.
U. S. Department of Health and Human Services, Fourth Special Report to the
    U. S. Congress on Alcohol and Health from the Secretary of Health and
    Human Services, January 1981. Washington, D. C.: U. S. Government
    Printing Office, 1981.
Zinberg, N. E., and Fraser, K. M. "The Role of the Social Setting in the
    Prevention and Treatment of Alcoholism." In J. H. Mendelson and N. K.
    Mello (Eds.), The Diagnosis and Treatment of Alcoholism. New York:
    McGraw-Hill, 1979.

METABOLISM THROUGH ALCOHOL DEHYDROGENASE IS NOT

A PREREQUISITE FOR FATTY LIVER INDUCTION BY ALCOHOLS

R. Nordmann, F. Beaugé, M. Clément, and J. Nordmann

Service de Biochimie de la Faculté de Médecine de Paris-Ouest
et Institut National de la Santé et de la Recherche Médicale
(INSERM) U 72, 45 Rue des Saints-Pères, 75270
Paris Cedex 06, France

INTRODUCTION

It is well known that the most common manifestation of alcohol hepatotoxicty is the occurrence of a fatty liver, that is to say, an increase in the hepatic triacylglycerol level. Such an increase can be reproduced experimentally even by a single ethanol administration. It is still subject to debate whether the fatty liver induced by such administration is due to ethanol metabolism, resulting in an enhanced production of reducing equivalents in the hepatocyte and thereby inhibiting fatty acid oxidation. However other experimental data, such as those of Kalant, Khanna, Seymour, and Loth (1975), have brought arguments in favor of the role of ethanol per se, acting through an unspecific stress action, independent from ethanol metabolism.

We have previously shown (R. Nordmann, 1979; Beaugé, Clément, Nordmann, and Nordmann, 1980) that n-propanol and isopropanol, when administered acutely to the rat, share two common properties with ethanol, i.e., (a) their metabolism through the alcohol dehydrogenase (ADH) pathway; (b) their ability to induce a fatty liver.

The aims of the present research were to determine whether the oxidation through ADH is a prerequisite for the fatty liver inducing property of an alcohol. We investigated, therefore, the effects on liver lipid disposal of t-butanol, an alcohol which is not a substrate for ADH (von Wartburg, 1971).

The main abnormalities which can lead to a fatty liver are:

1. An increase in the blood free fatty acid level.

2. An increase in the hepatic uptake of these fatty acids.

3.   A decrease in the mitochondrial fatty acid oxidation.

4.   An increase in the esterification of acyl-CoA to triacylglycerols

5.   A decrease in the output of these triacylglycerols related to an
     impaired, very low density lipoprotein (VLDL) biosynthesis and/or
     secretion.

We studied the influence of t-butanol administration on these five
parameters and compared the results with those concerning ethanol,
n-propanol and isopropanol that we have previously reported (Beaugé et al,
1979, R. Nordmann, 1980).

METHODS

     t-Butanol was dissolved in water (25% by volume) and administered by
gastric tube at the dosage of 25 mmol/kg body weight to female Wistar rats
fasted during the overnight period immediately preceding the experiments.
An equal volume of water was administered to the control rats.

     Blood-t-butanol levels were determined according to Thurman and Pathman
(1975).

     The techniques used for the determination of blood glucose and FFA,
blood and liver triacylglycerols (TAGs), hepatic lactate, and pyruvate, as
well as for the measurement of hepatic palmitate uptake, $^{14}CO_2$ production
from $(1-^{14}C)$ palmitate and $(1-^{14}C)$ palmitate incorporation into blood and
liver TAGs are the same as those previously described (Beaugé et al., 1979)

RESULTS

     Table 1 indicates that the rate of t-butanol oxidation in the intact
rat is very limited. This is shown by the slow rate of its disappearance
from the blood following its administration.

     This administration is followed by an increase in the liver
triacylglycerol level, which represents about 350% of the control value at
the 20th hour (Table 1).

     The blood-free fatty acid (FFA) level is increased at the 5th hour. It
is interesting to note that it is accompanied by an increased blood glucose
level (Table 1).

     The hepatic fatty acid uptake measured by the uptake of $U-^{14}C$ palmitate
(Abrams and Cooper, 1976) is increased. The percentage of injected dose
recovered in the liver 5 min. after the palmitate injection is 11.5 + 2.3
(n = 9) and 7.3 + 0.5 (n = 9) in t-butanol treated and control rats,
respectively (p < 0.01).

     The mitochondrial fatty acid oxidation (as measured by $^{14}CO_2$ production
from ^{14}C-palmitate in liver slices) is unaffected by t-butanol. The
cytosolic $NAD^+/NADH$ ratio is also unaffected as shown by the absence of
modification of the lactate/pyruvate ratio following its administration
(results not shown).

     t-Butanol administration results in an increase of ^{14}C-palmitate
incorporation into liver TAGs, contrasting with a strong decrease of
palmitate incorporation into blood TAGs. Both abnormalities are highly
significant at the 5th, as well as at the 20th, hour following t-butanol
administration (Table 2).

Table 1. Influence of t-Butanol Administration (25 mmol/kg body wt., p.o.) on Various Blood and Liver Substrates in the Rat

Time	Animals	Blood				Liver	
		t-Butanol (mM)	Glucose (mM)	Free fatty acids (mM)	Triacylglycerols (mM)	Wet weight (g)	Triacylglycerols (µmol/g wet wt.)
2h	Control	--	3.94 ± 0.27(5)	0.87 ± 0.07(5)	0.68 ± 0.07(5)	2.40 ± 0.14(5)	8.7 ± 2.0(5)
	Treated	13.24 ± 0.50(4)	4.05 ± 0.44(5)	0.92 ± 0.06(5)	0.68 ± 0.01(5)	2.62 ± 0.14(5)	14.9 ± 1.5(5)
			p > 0.05	p > 0.05	p > 0.05	p > 0.05	p < 0.01
5h	Control	--	4.33 ± 0.27(6)	0.75 ± 0.03(6)	0.74 ± 0.10(6)	2.78 ± 0.28(6)	7.1 ± 1.7(6)
	Treated	12.57 ± 0.34(4)	5.22 ± 0.50(6)	0.98 ± 0.11(6)	0.58 ± 0.12(6)	3.08 ± 0.16(6)	13.3 ± 3.2(6)
			p < 0.01	0.01 < p < 0.02	p > 0.05	p > 0.05	p < 0.01
20h	Control	--	3.83 ± 0.22(6)	0.73 ± 0.11(6)	0.76 ± 0.09(8)	2.69 ± 0.06(6)	8.5 ± 1.9(8)
	Treated	11.35 ± 0.28(6)	5.72 ± 1.55(6)	0.78 ± 0.08(6)	0.56 ± 0.09(8)	2.98 ± 0.18(6)	28.5 ± 4.6(8)
			p < 0.01	p > 0.05	p < 0.01	p > 0.05	p < 0.01

Results are given as means + S.E.M. for the number of determinations in parentheses.

351

Table 2. Radioactivity Incorporated into Liver and Blood Triacylglycerols 1 hr after 1-^{14}C-Palmitic Acid Injection.
Influence of t-Butanol Administration (25 mmol/kg, p.o.)

Time (h)	Liver (10^{-2} cpm/g wet wt.)		Blood (10^{-2} cpm/ml)	
	Control	Treated	Control	Treated
5	517 ± 25	634 ± 43 (p < 0.01)	10 ± 2	5 ± 1 (p < 0.01)
20	456 ± 35	1253 ± 276 (p < 0.01)	42 ± 9	21 ± 6 (p < 0.01)

Animals were injected intraperitoneally with albumin-bound $(1^{-14}C)$ palmitic
acid (2.5 µCi/100 g body wt.) 1 hour before sacrifice. The number of hours
shown in the table indicates the time elapsed between t-butanol administration and sacrifice of the animals. Results are given as means + S.E.M. for
6 determinations.

DISCUSSION

t-Butanol shares with ethanol, n-propanol, and isopropanol the ability
to induce a fatty liver when acutely administered to rats (Beaugé et al.,
1979). The increase in the liver TAG level is of special magnitude in the
case of isopropanol and t-butanol, alcohols which disappear slowly from the
blood, whereas ethanol and n-propanol are rapidly cleared from the blood
(Beaugé et al., 1979; R. Nordmann, 1980). The high alcohol blood level
following t-butanol administration may play a role in the ability of this
alcohol to induce, as well as ethanol, an alcohol dependence syndrome
(Mc Comb and Goldstein, 1979).

The comparison of the effects of t-butanol administration on liver
lipid metabolism reported here with those following the administration of
either ethanol, n-propanol, or isopropanol previously reported (Beaugé et
al., 1979) allows for the following comments.

The blood FFA level studied 2 hours after the alcohol administration is
increased only in the case of isopropanol, whereas such an increase is found
at a later stage of intoxication following t-butanol administration.

The hepatic fatty acid uptake is increased after the administration of
any of the four alcohols studied, this increase being particularly marked
with ethanol.

The fact that mitochondrial fatty acid oxidation, which is inhibited in
the case of ethanol, n-propanol, and isopropanol (Beaugé et al., 1979), is
unaffected by t-butanol seems related to the fact that this alcohol is not
metabolized through ADH and therefore does not alter the NAD^+/NADH ratio.

When considering the incorporation of ^{14}C-palmitate into TAGs, it
appears that administration of either ethanol or n-propanol results in an
increased incorporation of the label in both the liver and blood TAGs. In
the case of t-butanol and isopropanol, on the contrary, palmitate
incorporation is increased in the liver TAGs but strongly decreased in the
blood ones.

It is interesting to consider that this difference in the action on
fatty acid incorporation into the blood TAGs of ethanol and propanol on one
side, isopropanol and t-butanol on the other, is accompanied by parallel
findings in the blood TAG level. As a matter of fact, ethanol or n-propanol
administration results in an increased blood TAG level, whereas a decrease

is found after isopropanol and after t-butanol. In the latter case, this decreased blood TAG level is specially apparent at a late stage of intoxication.

Summarizing our results concerning the comparison of the acute effects of ethanol, n-propanol, isopropanol, and t-butanol on liver disposal, it appears that any of the alcohols studied enhances the fatty acid supply to the liver by increasing the hepatic fatty acid uptake.

Isopropanol and t-butanol induce furthermore an increase in the blood free fatty acid level accompanied by an enhanced blood glucose level. Both result probably from an unspecific stress action leading to an increase in catecholamine release.

An inhibition of fatty acid oxidation contributes to the fatty liver induced by ethanol, but not by t-butanol.

An enhanced fatty acid incorporation into liver triacylglycerols is apparent after administration of any of the alcohols studied. This enhancement is likely to result from an increased activity of phosphatidate phosphohydrolase, an enzyme which catalyzes the rate limiting step of this metabolic pathway and the activity of which is enhanced by glucocorticoids (Brindley et al., 1979). An unspecific stress action of the alcohol administered seems thus to play a prominent role in the enhanced fatty acid esterification.

A decreased VLDL biosynthesis and/or secretion contributes to the intensity and duration of the fatty liver induced by either isopropanol or t-butanol. It seems related to the high and long-lasting permeation of the organism by these alcohols.

In conclusion, an alcohol such as t-butanol, although not metabolized through ADH, is able to induce a fatty liver. It cannot be excluded that formaldehyde, which can be produced by microsomal oxidative demethylation of t-butanol (Cederbaum and Cohen, 1980), could play a role in this induction. However the fact that the Km for such an oxidative demethylation is as high as 30mM suggests that such a role cannot be essential. It appears therefore that the fatty liver induced by t-butanol administration results mainly from an unspecific stress action which impairs both the fatty acid supply to the liver and the biosynthesis and/or secretion of VLDL. Such an unspecific stress action, due to the alcohol itself, also contributes to the fatty liver inducing ability of ethanol which, by its metabolism, affects at the same time liver fatty acid oxidation.

These findings contribute to our understanding that the hepatoxicity of ethanol results (a) from the alcohol itself and (b) from its hepatic metabolism. This hepatotoxicity cannot therefore be prevented either by an inhibition of alcohol dehydrogenase or opposite, by an activation of alcohol metabolism in the liver. The only way to diminish the incidence of alcoholic liver disease appears thus to reduce the permeation of the organism by alcohol. This should be achieved either by the reduction of the amount of alcoholic beverages consumed or by the inhibition of the intestinal ethanol absorption.

SUMMARY

The alcohols which were previously reported to induce a fatty liver (ethanol, n-propanol, isopropanol) are all oxidized by ADH. The aims of the present research were to determine whether oxidation by this metabolic pathway is a prerequisite for the fatty liver inducing property of an alcohol. The effects on liver lipid disposal of the acute administration of t-butanol, an alcohol which is not a substrate for ADH, were therefore studied in the rat.

The per os administration of 25 mmol t-butanol per kg body weight is followed by a 3.5-fold increase in the liver TAG level contrasting with a decrease in the blood TAG level. The results of the study of the five most important parameters which could contribute to this fatty liver show that it is related both to an enhanced fatty acid supply to the liver, an enhanced fatty acid incorporation into liver TAGs, and a very significant decrease in VLDL biosynthesis and/or secretion. These findings are similar to those previously found after administration of isopropanol, which, like t-butanol, results in a long-lasting permeation of the organism by the alcohol injected. Contrary to the action of isopropanol, t-butanol does not inhibit liver fatty acid oxidation. This finding seems related to its lack of effect on the NAD^+/NADH ratio.

It can be concluded that t-butanol, an alcohol not metabolized through ADH, is nevertheless able to induce a fatty liver which seems to result mainly from an unspecific stress action. This strengthens the concept that the fatty liver, resulting from ethanol administration, is related both to the metabolism of ethanol through the ADH pathway (resulting in an excessive production of reducing equivalents which inhibits liver fatty acid oxidation) and to a stress action due to a direct effect of the alcohol per se. It seems therefore hopeless to prevent ethanol toxicity on liver metabolism either by inhibiting ADH (which would increase the direct ethanol effects) or by activating ethanol metabolism through ADH (which would favour the disturbances related to ethanol metabolism). The only way to diminish the incidence of liver damage would be to reduce the amount of alcohol beverages consumed or to inhibit the absorption of ethanol in the gastrointestinal tract.

REFERENCES

Abrams, M. A., and Cooper, C. "Quantitative analysis of metabolism of hepatic triglyceride in ethanol-treated rats." Biochem. J., 1976, 156: 33-46.

Beaugé, F., Clément, M., Nordmann, J., and Nordmann, R. "Comparative effects of ethanol, n-propanol and isopropanol on lipid disposal by rat liver." Chem. Biol. Interactions, 1979, 26: 155-166.

Brindley, D. N., Cooling, J., Burditt, S. L., Pritchard, P. H., Pawson, S., and Sturton, R. G. "The involvement of glucocorticoids in regulating the activity of phosphatidate phosphohydrolase and the synthesis of triacylglycerols in the liver. Effects of feeding rats with glucose, sorbitol, fructose, glycerol and ethanol." Biochem J., 1979, 180: 195-199.

Cederbaum, A. I., and Cohen, G. "Oxidative demethylation of t-butyl alcohol by rat liver microsomes." Biochem. Biophys. Res. Comm., 1980, 97: 730-736.

Kalant, H., Khanna, J. M., Seymour, F., and Loth, J. "Acute alcoholic fatty liver. Metabolism or stress." Biochem. Pharmacol., 1975, 24: 431-434.

Lieber, C. S., and De Carli, L. M. "Metabolic effects of alcohol on the liver." In C. S. Lieber (Ed.), Metabolic Aspects of Alcoholism. Lancaster: MTP Press, 1977.

Mc Comb, J. A., and Goldstein, D. B. "Quantitative comparison of physical dependence on tertiary butanol and ethanol in mice: correlation with lipid solubility." J. Pharmacol. Exp. Ther., 1979, 208: 113-117.

Nordmann, R. "Metabolism of some higher alcohols." In C. Stock and H. Sarles (Eds.) Alcohol and the Gastrointestinal Tract. Paris: INSERM ·Editions, 1980, Vol. 95, pp. 187-206.

Thurman, R. G., and Pathman, D. E. "Withdrawal symptoms from ethanol: evidence against the involvement of acetaldehyde." in K. O. Lindros and C. J. P. Eriksson (eds.), The Role of Acetaldehyde in the Action of Ethanol. Helsinki: Finnish Found. Alcohol Stud., 1975, pp. 217-231.

von Wartburg, J. P. "The metabolism of alcohol in normals and alcoholics: Enzymes." In: B. Kissin and H. Begleiter (Eds.) The Biology of Alcoholism, Vol 1: Biochemistry. New York, Plenum Press, 1971, pp. 63-102.

CEREBRAL ATROPHY IN CHRONIC ALCOHOLIC PATIENTS

Meir Teichman

Institute of Criminology, Tel-Aviv University
69978 Tel-Aviv, Israel

Steve Richman and Eric W. Fine

Department of Psychiatry
Albert Einstein Medical Center, Daroff Division
Philadelphia, Pennsylvania

The purpose of the study was to assess the occurrence of gross cerebral atrophy in alcoholic patients and to investigate the relationships between cerebral atrophy and two variable which are considered to be major determinants in the production of brain damage. The variables are (1) the quantity-frequency of alcohol intake; and (2) the duration of abuse.

It is widely known, and well documented, that chronic abuse of alcohol leads to various somatic and psychological disorders, among which the organic brain syndrome is considered to be the one with the most serious and crippling effects. There are a number of well recognized clinical neuropsychiatric consequences that are directly related to alcohol abuse and/or withdrawal (e.g., delirium tremens, stupor, hallucination, etc.) (Albert, Butters, and Levin, 1979; Courville, 1955; Fox, Ramsey, Huckman, and Proske, 1976; Freund, 1973; Lee, Moller, Hardt, Haubek, and Jensen, 1979; Peterson, 1976; Miller and Orr, 1980. Brewer and Perrett (1971) suggested that gross cerebral atrophy is more common among alcoholics than was previously thought, and that the level of pathology might be directly related to the intensity and the duration of abuse. Summarizing his review of scientific literature on brain damage in chronic alcoholics, Ron (1977), like Brewer and Perrett (1971), stated that "brain damage may be commoner than is thought and may be undetected because no adequate search is made."

Researchers applied several methods in their search for evidence of the existense of organic brain syndrome among alcoholics. Ron (1977) pointed out three different domains of inquiry; psychological, neuroradiological, and neuropathological. The first two areas of investigation were applied in the present study.

Draper (1978) in a recently published review about the alcohol brain damage syndrome concluded that such a "syndrome can be regularly identified by psychometric means." However, while it is possible only to infer about brain damage from psychological data, the introduction of computerized axial tomography (EMI scanner) offers a safe, noninvasive and reliable neuroradiological diagnostic method. Greitz (1975) recently demonstrated the accuracy of this technique. He reported that the tomograms were found to be 100% accurate in the diagnosis of cerebral atrophy.

We assume the "dose/duration-effect" relationship between cerebral atrophy and the alcohol consumption will be manifested by the increased severity of brain damage as measured by computerized tomography and by the Memory-For-Design Test.

METHODS

Subjects: 105 alcoholic patients, who were admitted to the alcoholism Unit, the Department of Psychiatry, Albert Einstein Medical Center, Daroff Division, Philadelphia, during a designated period (Dec. 1979 to Feb. 1980) served as subjects. 83.52% were males and 16.48% were females. Their ages ranged from 22 to 68 years (X = 49.11, S.D. = 10.91). All the patients were informed that we are investigating the effect of alcohol on the brain and their agreement was requested. None of the admitted patients refused. Five patients were omitted due to known previous psychiatric illness.

PROCEDURE

Two days after admission, the patient was interviewed and the quantity-frequency of alcohol consumption and the duration of drinking as well as of problem drinking were assessed. This information was validated in a second interview which was conducted 6 days after admission.

Following the detoxification phase, the patient was referred for a brain scan as part of his/her standard medical evaluation. The tomograms for the alcoholism unit were regularly performed by the Department of Radiology, Pennsylvania Hospital, Philadelphia. On the same day, the patient's psychomotor performance was tested

MEASUREMENTS

1. The measure of alcohol consumption was the Quantity-Frequency Index which represents the average ounces of alcohol intake per day over the past month (Eaglestone and Mothershead, 1974). The index is the sum of the frequencies with which beer, wine, and liquor are consumed during a typical day during the past month, multiplied by each beverage's absolute alcohol content.

2. The duration of alcohol consumption was determined by (a) the number of years the patient had been drinking regularly, (b) the number of years of recognized problematic drinking. These two figures were the patient's subjective assessments. Nevertheless, when possible, they were validated through other family members.

3. The computerized tomograms were evaluated by two staff members of the Department of Radiology, Pennsylvania Hospital for gross cerebral atrophy in relationship to the patient's ages. The findings were thereafter rated on a 6-point scale according to the severity of the cerebral atrophy. 0 represents no atrophy; 1 stands for a minimal damage, 3 represents a mild level of atrophy; while 5 stands for severe damage.

4. The psychomotor performance was measured by the Memory-For-Design Test (Graham and Kendall, 1960). The test (MFD) consists of 15 designs, each printed in black on a five inch cardboard square. Each of the designs is shown to the patient for five seconds. After the five second exposure, the design is withdrawn, and the patient is asked to draw one exactly like it. The patient's performance is evaluated according to the direction of the drawing and its accuracy. The score of 7 and up was found to represent

Variable		CAT scan	MFD
Quantity	Less than 13 ozs.	2.71(1.26)	10.88(6.76)
	13 ozs. and more	3.58(1.02) t = 8.01*	16.4(11.23) t = 3.71*
Years of drinking	less than 31 years	2.51(1.44)	10.58(8.46)
	32 years and more	3.38(1.18) t = 5.44*	17.62(9.56) t = 12.06*
Years of problem drinking	less than 18 years	2.58(1.49)	10.48(7.92)
	18 years and more	3.49(0.98) t = 5.19*	16.46(9.30) t = 7.85*

* $p < 0.001$.

the critical area of brain damage. The MFD test has been used in a number
of diverse clinical investigations and was found to be reliable and valid
(see Graham and Kendall, 1960; May, Urquhart, and Watts, 1970; Kendall,
1966).

RESULTS

Our sample represents a chronic alcoholic population. The mean years
of drinking was found to be 31.32 years (S.D. = 11.00) and the duration of
perceived problematic drinking exceeded 18 years (x = 18.19 yrs.; S.D. =
11.25). The assumed daily alcohol-consumption during the month before
hospitalization was very high (X = 13.42 ozs., S.D. = 6.39). Our group of
patients was characterized by different levels of brain damage.
Nevertheless, it was a group which demonstrated a severe level of gross
cerebral atrophy. The mean score of the CAT scans was 3.15 (S.D. =
1.16), which stands for at least a mild cerebral atrophy. The mean score of
the MFD Test was 15.10 (S.D. = 12.08), which is well into the "critical
area."

It is interesting to note that, in the present study, the correlation
between these two indices was found to be positive and highly significant
(r = .70; p < .001). This finding suggests that the MFD Test, which is a
relatively simple instrument, should be more widely applied in the
evaluation of alcoholic patients. Such a step is important, expecially in
the view of the high cost of CAT scanning.

We assumed that there is a direct relationship between the "dose/
duration effect" and gross cerebral atrophy. Table 1 presents the
comparison of the level of cerebral atrophy for paitents with low and high
dose/duration scores. All comparisons pointed to the same direction -- the
more and the longer you drink the more likely it is that your brain
deteriorates.

DISCUSSION

Atrophy of the brain is one of the most devastating complications and destructive outcomes of alcohol abuse. While many of the alcohol related health and social complications may be reversible, damage to the brain in considered to be irreversible and permanent. Thus, the issue of "dose/duration-effect" relationship between alcohol consumption and gross cerebral atrophy is considered to possess a major clinical importance. Recently, Kish, Hagen, Woody, and Harvey (1980) published a cautious but optimistic report about some components of cerebral atrophy and cognitive impairment that may be reversible. Grant et. al. (1979) found no differences (in neuropsychological performance) between alcoholics and a group of men who drank occasionally. On the other hand, Miller and Orr (1980) summarized their findings by stating that alcoholics resembled patients with organic brain damage. It seems to us that the contradiction between these findings can be primarily attributed to the "dose/duration" variable of alcoholic consumption. Grant et. al. (1979) reported that their subjects used alcohol for an average of 6 years; Miller and Orr's (1980) subjects drank <u>problematically</u> for an average of 6 years, and our subjects for 18 years. These findings indicate that an accumulative effect of alcohol consumption over time on gross cerebral atrophy can be identified, and that there is a "point-of-no-return" beyond which the damage is irreversible.

The results of this study suggest that: (1) damage to the brain in patients with clear alcohol dependency might be more extensive than it has been assumed; and (2) there does seem to be a direct relationship between the number of years of alcohol consumption and the degree of brain damage.

These findings have important implications for the understanding and treatment of alcohol dependence.

Although the effect on the brain of alcohol has long been a subject of interest and concern, for the most part, research has emphasized the more obvious and dramatic changes associated with the end-stage(s) of alcohol dependence.

The impact of alcohol's effect on the brain in earlier stages of alcoholism has not received a great deal of attention. It has been assumed that during earlier stages any effect on the brain would be reversible. What effect various degrees of brain atrophy might have on the future course of the dependency is unknown. The impairment of intellectual functioning due to alcohol intoxification, the development of compulsive and rigid response-sets which may be associated with minimal brain damage, and the reinforcing as well as the anxiety reduction properties of alcohol may be the cornerstones of the development of alcohol dependency.

We have shown that the greater alcohol consumption is, the greater the danger of brain damage one can expect. This association suggests that it is even more important to stress either primary or secondary prevention. The risks of brain damage can be minimized the earlier the diagnosis of alcohol dependency is made and the sooner effective intervention is established. Since most interventions strategies for alcohol dependency are psychotherapeutic in nature, the need for an "intact" well-integrated central nervous system is obvious. Interference with brain function and impairment of cognitive processes might influence the ability of the patient to accurately view and accept his impairment and adversely affect the recovery process. Furthermore, such an impairment coupled with the establishment of compulsive and rigid patterns of response-sets, resulting from the atrophy of the brain, might contribute to the continuation of alcohol consumption. Conversely, we suggest that the alcoholic is engaged in a circular deterioration process. The presence of alcohol in one's central nervous system results in cognitive impairment as well as in effecting the brain tissue, which in turn further increases the cognitive and intellectual impairment. All of these might effect the ability of the alcoholic patient to favorably respond to treatment and maintain sobriety, which are all-to-common phenomena in alcoholics.

Our assumptions raise several crucial issues that our data as well as our clinical observations did not provide sufficient answers for is the alcoholic, with atrophy of the brain, an untreatable person? Should we apply, with such alcoholics, therapeutic techniques which have been proven to be effective with brain-damaged patients and consider the present techniques inadequate and inappropriate? Should we reassess our therapeutic goals and outcome criteria? All of these questions and further research which might prove to be of great importance to the planning and treatment of alcoholism.

REFERENCES

Albert, M. S., Butters, N., and Levin, J. "Temporal gradients in the retrograde amnesia of patients with alcoholic Korsakoff's disease." Arch.Neurol., 1979, 36:211-215.

Brewer, C., and Perrett, L. "Brain damage due to alcohol consumption: An air-encephalographic psychometric and electroencephalographic study." Brit.J.Addict., 1971, 66:70-1832.

Courville, C. B. Effects of Alcohol on the Nervous System of Man. Los Angeles: San Lucas Press, 1955.

Draper, R. J. "Evidence for an alcohol brain damage syndrome." J. of the Irish Med. Assoc., 1978, 71:350-352.

Eaglestone, J., and Mothershead, A. Alcoholism Program Monitoring System Procedures Manual, (Vol. 1), Menlo Park, Calif.: Stanford Research Institute, 1974.

Fox, J. H., Ramsey, R. G., Huckman, M. S., and Proske, A. E. "Cerebral Ventricular Enlargement: Chronic Alcoholics examined by computerized tomography." J.A.M.A., 1976, 236:365-369.

Freund, G. "Chronic central nervous toxicity of alcohol." Ann.Rev.Pharmocol., 1973, 13: 217-227.

Graham, F. K. and Kendal, B. S. "Memory-for-Designs Test: Revised general manual." Perceptual & Motor Skills, 1960, 11:147-188.

Grant, I., Adams, K., and Reed, R. "Normal neuropsychological abilities of alcoholic men in their late thirties." Am.J.Psychiatry, 1979, 136:1263-1269.

Greitz, T. "One years's experience with computer tomography of brain lesions." Acta Radiol., 1975, 346 (suppl.).

Kendall, B. S. "Orientation errors in the Memory-for-Designs Test: Tentative findings and recommendations." Perceptual & Motor Skills, 1966, 22:335-345.

Kish, G. B., Hagen, J. M., Woody, M. M., and Harvey, H. L. "Alcoholics' recovery from cerebral impairment as a function of duration of abstinence." J.Clin.Psych., 1980, 36:584-589.

Lee, K., Moller, L., Hardt, F., Haubek, A., and Jensen, E., "Alcohol-induced brain damage and liver damage in your males." The Lancet, 1979, Oct. 13, 759-761.

May, A. E., Urquhart, A., and Watts, R. E. "Memory-for-Design Test: A follow up study." Perceptual & Motor Skills, 1970, 30:753-754.

Miller, W. R., and Orr, J. "Nature and sequence of neuropsychological deficits in alcoholics." Stud.Alcoholism, 1980, 41: 325-337.

Peterson, G. C. "Psychiatric aspects of chronic organic brain syndrome." Post-graduate Medicine, 1976, 60:162-168.

Ron, M. A. "Brain damage in chronic alcoholism: A neuropathological, neuroradiological and psychological review." <u>Psych.Med.</u>, 1977, 7:103-112.

GASOLINE SNIFFING IN NORTHERN CANADA

Gordon E. Barnes

Department of Family Studies
University of Manitoba
Winnipeg, Manitoba R3T 2N2
Canada

GASOLINE SNIFFING

Gasoline Sniffing in Northern Canada

Gasoline sniffing represents one of the most serious drug use problems in terms of the risk to the user. Numerous deaths have been attributed to solvent abuse (Bass, 1970) and gasoline may be one of the most dangerous substances in use by sniffers. Gasoline is a mixture of hydrocarbons including paraffins, olefins, napthenes, and aromatics. Certain common forms of gasoline also contain tetraethyl lead. Gasoline is particularly dangerous because of the presence of tetraethyl lead and the aromatic hydrocarbon, benzene. Tetraethyl lead has been shown to cause lead poisoning (Angle and Eade, 1975; Boeckx, Postl, and Coodin, 1977; Lynn, 1975). While benzene can have potentially destructive effects on bone marrow cells (Nurcombe, Bianchi, Money, and Cawte, 1970). Several deaths have been attributed to gasoline sniffing (Angle and Eade, 1975; Boeckx et al., 1977, Ferguson, 1975; Nurcombe et al., 1970; Sokol, 1981) with the mechanism being respiratory failure resulting from central nervous system depression and respiratory irritation and bronchiolar obstruction (Nurcombe et al., 1970). The constituents and clinical effects of gasoline are summarized in Table 1.

Prevalence

Not much is known about the actual extent of gasoline sniffing. The earliest reports of gasoline sniffing were primarily case histories. In the first of these papers (Clinger and Johnson, 1951) two cases of intentional inhalation of gasoline vapors were described. While these and other case histories are useful in describing the characteristics and background of

Funding for this project was provided by the Non-Medical Use of Drugs Directorate (now Health Promotion Branch) of Health and Welfare Canada (RODA Grant #1212-6-55). During the design of the project funding was also provided by the Alcoholism Foundation of Manitoba and Medical Services Branch of Health and Welfare Canada. The author was supported by a National Health Research Scholar Award (6607-1155-48) while conducting this research.

Table 1. Gasoline Sniffing Effects[**]

Constituents	Clinical effects	References
Hydrocarbons (mixture C4-C8 saturated hydrocarbons)	Hepatic congestion	Carrol & Abel (1973)
	Anemia	Law & Nelson (1968)
Additives Tricresyl phosphate Trimethyl phosphate Tetraethyl lead[*]	Lead encephalopathy (ataxia, tremor, psychotic behavior)	Boeckx et al. (1977) Law & Nelson (1968)
	Diffuse encephalopathy (delirium, choreiform movements, abnormal EEG)	Carrol & Abel (1973) Easson (1962)
	Fetal gasoline syndrome (profound retardation, hypotonia progressing to hypertonia, scaph-ocephaly, prominent occiput, poor post-natal head growth)	Hunter et al. (1979)
	Deaths (respiratory failure)	Angle & Eade (1975) Boeckx et al. (1977) Ferguson (1975) Nurcombe et al. (1970) Sokol (1981)

[*] Commercially available leaded gasoline contains approximately .53 g of tetraethyl lead per litre (Boeckx et al., 1977) while aviation gas contains 1.4 g of tetraethyl lead per litre (Beattie et al., 1972).
[**] This table is an abridged and updated version of a similar table prepared by Hayden, Comstock, and Comstock (1976).

sniffers, they do not provide much information on the extent of this practice. Clinger and Johnson (1951) for instance, noted merely that their sniffers reported having friends who are engaged in this practice.

Surveys on the actual extent of sniffing seem to focus on particular high risk populations (surveys are summarized in Table 2). Angle and Eade (1975) conducted surveys in two communities in Quebec where sniffing had been reported, Great Whale River and Manouane. In Great Whale River, Angle and Eade (1975) found that 62% of the population under 20 and 53% over 20 admitted sniffing gasoline.

A somewhat different story emerged from the survey conducted in Manouane. Manouane is primarily a native community and is also located in northern Quebec. In Manouane, no gasoline sniffing was detected, although this practice had been quite common at one time. Apparently, a scare resulting from a 1972 fatality due to sniffing in the community had produced a dramatic drop in sniffing. Angle and Eade (1975) noted, however, that residents believed that Manouane was atypical and that the gasoline sniffing problem was quite serious in several other Indian communities.

Similar results to those reported by Angle and Eade, in their Great Whale River study were reported by Lynn (1975) in a study of sniffing in Poplar Hill, Ontario. Lynn found that of 50 Ojibwa Indians surveyed, 23 out of 50 were occasional users and 11 out of 50 were chronic users of gasoline. High lead levels were also observed in chronic and occasional users of gasoline.

Table 2. Summary of Research on Prevalence of Gasoline Sniffing

Author & year	Community	Sample	Sniffing assessed by	Results
Angle & Eade (1974)	Great Whale River, northern Quebec, Canada	338 subjects of Cree & Inuit origin (all ages)	Interview asked if ever sniffed and how often, and blood lead levels tested	62% under age 20 and 53% over age 20 sniffed; high blood lead levels associated with sniffing
	Manouane, southern Quebec, Canada	182 Indian subjects (all ages)	Interview and blood lead level testing	No sniffing detected in interviews or blood lead tests
Boeckx, et al. (1977)	Shamattawa, northern Manitoba, Canada	340 Cree children, 4-18 years of age	Observed by author residing in community for 2 months. 43 suspected sniffers tested by using ereythrocytic ALAD (deltaaminolevulinic dehydrase) as a screening tool	Estimated 53-59% of children sniffed; sniffing confirmed by low ALAD levels among 43 sniffers tested
	Little Grand Rapids, eastern Manitoba, Canada	84 Indian children, 9-17 years of age (suspected sniffers)	ALAD screening	59% showed evidence of sniffing gasoline (i.e., low ALAD activity)
Kaufman (1973)	Pueblo Indian village, New Mexico, U.S.A.	72 Indian children 6-12 years old	Questionnaire asked if ever sniffed	62% had sniffed at least once
Lynn (1975)	Poplar Hill, Ontario, Canada	50 Ojibwa Indians	Survey asked how often used; blood levels tested	46% occasional users; 22% chronic users; high lead levels found in both occasional and chronic users

Gasoline sniffing was also found to be a fairly common practice in a New Mexico, Pueblo Indian village (Kaufman, 1973). Kaufman administered questionaires to 72 children enrolled in elementary school and found that 62% reported that they had sniffed gasoline for its effect at least once.

Gasoline sniffing is not a problem that occurs only in North America. Nurcombe, Bianchi, Money, and Cawte (1970) reported that sniffing gasoline is a common practice among the aboriginal people of the Ecuto Islands in Australia. Although Nurcombe et al., did not report the actual prevalence of this practice, their paper indicated that the sniffing must be fairly common. Groups of as many as 20 youngsters at one time were observed gathering at the petrol pump.

More recently, research conducted by Boeckx, Postl, and Coodin (1977) showed that the problem of gasoline sniffing was fairly wide-spread in the community of Shamattawa, Manitoba, Canada with observer estimates suggesting over 50% of the children involved in sniffing. Clinical signs of lead poisoning associated with gasoline sniffing included decreased deltaaminolevulinic dehydrase (ALAD) activity supported these observations.

Blood tests conducted by Boeckx et al. (1977) in the community of Little Grand Rapids suggested that this problem was not restricted to Shamattawa alone, but might be fairly widespread throughout the province of Manitoba. Because of the serious health hazard presented by gasoline sniffing the present study was originated in an attempt to determine the extensiveness of gasoline sniffing in the province of Manitoba and the possible causes of this behavior.

CHARACTERISTICS OF USERS

Case Histories: Most of the information available on gasoline sniffing is in the form of case histories. An analysis of 30 case histories revealed the following picture of the gasoline sniffer. Of the 30 case histories, 27 were male and 3 were female with an average age of 15.5 years. Most of the cases fell in the age range 11 to 19 although there is one notable exception, a woman 41 years of age reported on by Law and Nelson (1968). The average length of use for these cases when this information was reported was 5.2 years. Most of the cases seemed to come from rural areas suggesting that the availability of gasoline and/or the lack of availability of alternatives such as marijuana could be an important factor. Sniffers generally tend to come from fairly large families (average of 4 children per family). The home situation of the sniffers tended to be disorganized or unhappy with eight case histories reporting broken homes (Clinger and Johnson, 1951; Durden and Chipman, 1967; Faucett and Jensen, 1952; Gold, 1963; Lawton and Malmquist, 1961; Nitsche and Robinson, 1959). In several case histories parental alcoholism seems to have played a role (Boeckx, Postl and Coodin, 1977; Brown, 1968; Easson, 1962; Lawton and Malmquist, 1961). The incidence of delinquent acts seems to be fairly high among sniffers with thefts being the most frequently reported type of delinquent act (Bartlett and Tapia, 1966; Black, 1967; Durden and Chipman, 1967; Gold, 1963; Lawton and Malmquist, 1961; Nurcombe, Bianchi, Money and Cawte, 1970; Oldham, 1961; Tolan and Lingl, 1964)

In the case histories that have reported on the IQ of sniffers the range that has been reported has been from 44 to 120. The average IQ reported has been 94. This average is likely not below normal for people coming from the typical sniffer background. Academic performance by sniffers, however, tends to be poor (Bethell, 1965; Black, 1967; Brown, 1968; kCarrol and Abel, 1973; Durden and Chipman, 1967; Edwards, 1960; Faucett and Jensen, 1952; Gold, 1963; Lawton and Malmquist, 1961; Nitsche and Robinson, 1959).

In most case histories where personality characteristics of sniffers are mentioned, the most frequently observed characteristics have been depression (Brown, 1968; Faucett and Jensen, 1952; Lawton and Malmquist, 1961) and anxiety (Brown, 1968; Easson, 1962; Edwards, 1960; Faucett and Jensen, 1952; Gold, 1963; Law and Nelson, 1968; Lawton and Malmquist, 1961; Neal and Thomas, 1974; Tolan and Lingl, 1964). Introversion or shyness on the part of sniffers has also been a frequently observed characteristic (Black, 1967; Brown, 1968; Clinger and Johnson, 1951; Easson, 1962; Neal and Thomas, 1974).

Surveys: Surveys that have been conducted to date on gasoline sniffing have for the most part confirmed the findings in the case histories. Evidence of extensive gasoline sniffing has been found in populations where the social assets are extremely low. These population include Pueblo Indians (Kaufman, 1973), Australian aboriginal tribes (Nurcombe, Bianchi, Money, and Cawte, 1970), and Canadian Indians (Angle and Eade, 1975; Boeckx, Postl, and Coodin, 1977; Lynn, 1975).

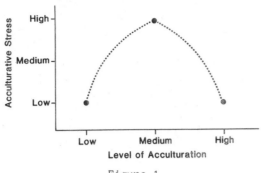

Figure 1

Acculturative stress would seem to be an important factor in all of the communities where extensive gasoline sniffing has been reported. Acculturative stress is greatest when some contact with the white culture has been made but acculturation has only partially occurred. Angle and Eade (1975) have suggested that gasoline sniffing is not a problem in communities where there is little or no contact with white society while heavy abuse occurs where cultural changes are occurring. Nurcombe et al. (1970) have also suggested that cultural changes are important sources of anxiety among the aboriginal tribes of Australia. The hypothesized curvilinear relationship between level of acculturation and acculturative stress is displayed graphically in Figure 1.

Peer and sibling influence seems to be another important factor in native gasoline sniffing. Kaufman (1973) reported that 93% of his sniffers reported sniffing with others. Nurcombe et al. (1970) also reported group gatherings around the petrol pump. Social sniffing has also been found in Canadian Indian groups (Angle and Eade, 1975) and peer influence has been suggested as a possible important causal variable (Scott, 1976) in gasoline sniffing.

Parental use of alcohol seems to be a common practice in the communities where gasoline sniffing occurs but Kaufman (1973) did not find the incidence of alcoholism to be any greater in the families of sniffers than in the families of nonsniffers.

Several comments concerning the possible psychological correlates of gasoline sniffing in native communities have been made but very little hard evidence is available. Nurcombe et al. (1970) found that anxiety was high among gasoline sniffers. Angle and Eade (1975), Nurcombe et al. (1970) and Scott (1976) have suggested that boredom could be a principal cause of sniffing. Very little opportunity for recreation is available in the poor native communities and traditional sources of stimulation are generally not available. Nurcombe et al. have suggested that this situation creates a stimulus hunger which in turn could be an important causal factor in gasoline sniffing. Alienation from parents has also been suggested as a factor in gasoline sniffing (Scott, 1976).

Causal Model of Solvent Abuse

Different forms of drug abuse are invariably related. Drugs which are most similar in their composition and effects tend to be particularly highly

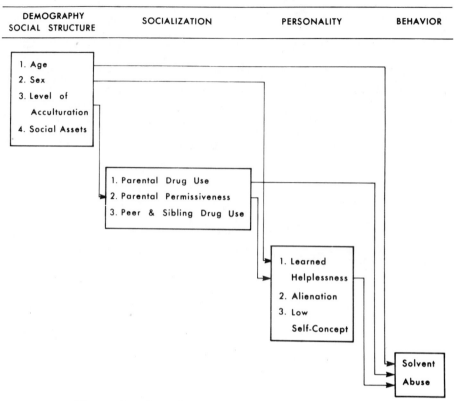

| DEMOGRAPHY SOCIAL STRUCTURE | SOCIALIZATION | PERSONALITY | BEHAVIOR |

Figure 2. Causal Model of Teen-age Solvent Abuse.

associated, and solvent abuse should be strongly correlated forms of drug abuse.* Factors which predict solvent abuse should then be important also in the prediction of gasoline sniffing.

A recent review of the solvent abuse literature (Barnes, 1979) suggested that adolescent solvent abuse is a function of environmental conditions, including background and socialization factors, and personality characteristics that make the person vulnerable to solvent abuse. The model of solvent abuse to be tested in this study (provided in Figure 2) was derived on the basis of this literature review. In this model gasoline sniffing is viewed as a function of (1) a person's place in the demographic-sociocultural system with respect to certain important factors such as age, sex, level of acculturation, and social assets; socialization factors, including parental drug use, parental permissiveness vs. control, and peer and sibling drug use; and personality variables including a learned helplessness syndrome comprised of the characteristics of hopelessness, passivity, anxiety, and depression, alienation, and low self-concept.

---

*This does not preclude patterns of multiple drug use in which drugs that are very dissimilar in their effects (e.g., stimulants and barbiturates) may be used. In that type of multiple drug use a general sensation seeking tendency for all types of new experience seems to be important (see for instance, Kohn, Barnes, and Hoffman, 1979).

METHOD

Research Design

It became apparent early in the planning of the Northern Sniff project that it would not be possible to obtain a random sample of communities. Since a random sample was not possible the next best alternative seemed to select several samples that would be as diverse as possible. Samples were required that would include native students that were both high and low in their level of acculturation. Although most of the reports on gasoline sniffing seemed to be coming from the poorer, less acculturated, native communities, there were fears at the time that this practice might be wide-spread. A sample of nonnative students was also required to determine whether the predictors of solvent and drug use which applied in native samples also predicted drug use in a non-native sample.

Because of the nature of the sample included in this study which was young and sometimes fairly unacculturated, steps had to be taken to ensure that the variables included in this survey were measured in a reliable and valid fashion. Since the author had no way of knowing in advance how sophisticated the sample would be in their English usage, pretests had to be conducted. Communities that were chosen were relatively low in their level of acculturation. If a questionnaire could be used in these communities then it would be possible to use this technique in the other more acculturated communites.

Pretests showed that a questionnaire format could be utilized if the questions were kept simple. Students also seemed to answer the questions honestly. In Shamattawa where blood lead testing had shown depressed ALAD levels (indicating gasoline sniffing, Boeckz et al., 1977) a pretest was conducted. In this pretest 20 children aged 10 to 15 were questioned concerning their use of gasoline, other solvents, marijuana, tobacco, and alcohol. In this sample 50% of the children admitted sniffing gasoline while none of the children said they used marijuana, a drug that was unavailable in the community.

After the pretests had been conducted the final questionnaire was prepared for administration and six communities were selected for inclusion in the study. Techniques of data collection varied somewhat from place to place to ensure maximum participation in the study. In the community of Easterville where absenteeism was known to be high (note: attendance rates had dropped from around 80% to approximately 60% the year before the study) a local person was hired to conduct the survey on a door to door basis. In the other communities where school attendance rates were higher (generally above 80%) surveys were administered during class time with additional questionaires left with the classroom teacher for students who were absent on the day of testing to complete and leave in a sealed envelope for mailing to Winnipeg. While this procedure allowed for some students to complete the survey in a somewhat different context, it was felt that the increased representativeness of the sample obtained in this fashion warranted the risk. Subjects were told that the project was being conducted for the purpose of determining the prevalence of drug use and factors associated or not associated with use. Anonymity of responses was guaranteed and maintained.

Subjects

Questionnaires for 623 students were collected from six northern Canadian communities (shown in Figure 3) including:

1. Berens River. N = 66, includes 88% of students enrolled in school in
this age group. A Saulteaux community located 180 miles north of Winnipeg
(approximate population = 988).

2. Easterville. N = 122, includes approximately 74% of community
population in this age group. A Cree community located 216 miles northeast
of Winnipeg (approximate population = 565).

3. Gypsumville. N = 50, includes 89% of students enrolled in school in
this age group. An unincorporated community located approximately 167 miles
north of Winnipeg. The school draws students from a 15 mile radius.

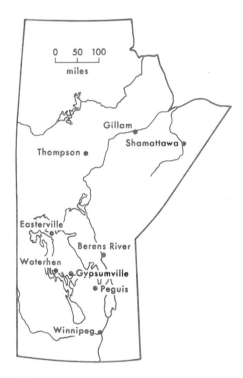

Figure 3. Community Locations.

4. Peguis. N = 162, includes 77% of students enrolled in school in
this age group. A mixed Saulteaux and Cree community located 120 miles
north of Winnipeg (approximate population = 1,208).

5. Gillam. N = 168, includes 80% of students enrolled in school in
this age group. A northern white community located 453 miles north of
Winnipeg (approximate population = 2,500).

6. Waterhen. N = 55, includes 93% of students enrolled in school in
this age group) - A small community located 170 miles northeast of Winnipeg.
The school services the white community of Waterhen (approximate population -
152), the Saulteaux community at Skownan (approximate population = 293),
and Metis community at Mallard (approximate population = 196).

Data were collected during the fall and winter of 1977-78 school term. The sample included 334 Indians, 92 Metis, and 197 non-Indian students.* Students ranged in age from 10 - 19 years and were in grades from 4 through 10. There were 136 males and 307 females for a total of 623 respondents.

Follow-up data were collected in the spring of 1978 in Berens River. The spring 1978 survey was completed by 21 males and 26 females.** The ages in this sample ranged from 12 to 17 with a mean of 14 years. The follow-up data will only be used in this report to provide additional validity data on the school self-acceptance measure.

Measurement

Social Assets: This scale was based on a test developed by Luborsky, Todd, and Katcher (1973) and included the following items:

1. Mother's health

2. Father's health

3. Own health

4. Whether they had enough money for clothes and food

5. Number of friends

6. Family size

7. House ownership

8. Radio ownership

9. Availability of running water

10. Mother working outside of home or not

Items that would overlap with other scales were excluded. A question of school performance was excluded for this reason as were questions on parental education levels. The weighting system advocated by Luborsky, Todd, and Katcher (1973) was also utilized.

Actual scores on the test were based on the person's total obtained score divided by the total possible score for the scale based on the number of items answered. Scores were not computed for anyone answering less than one-half of the questions. Scores on the social assets scale ranged from .29 to 1.00 with a high score indicating greater social assets.

Acculturation: An acculturation scale was constructed based on the following questionnaire items:

---

* The students were allowed to indicate their own perceived status as either (1) an Indian person; (2) a Metis person; (3) a white person; or (4) none of the above. It is these self-classifications that are used in this report. The term "native person" is used to describe the combined Indian and Metis groups.
**Original plans called for follow-up data on the entire sample but these plans had to be abandoned when difficulties were encountered gaining access to communities. There also seemed to be a distrust of the idea of number coded questionaires and the ability to protect anonymity in some communities.

1. Language spoken at home

2. Father's education

3. Mother's education

4. Ownership of a TV

5. Father's wage employment history

Lower scores were assigned to students whose parents had less education, did not speak English at home, did not own a TV, and where the fathers did not work regularly for wages. Scores on this scale were computed in a similar fashion to the way scores were computed on the social assets scale. Actual scores ranged from .20 to 1.00 with a high score indicating more acculturation.

Parental Control: Parental control scales were developed by taking the items from the Children's Reports of Parental Behavior Inventory (Schaefer, 1965) that weighted the highest on the firm control-lax control factor. Seven items for father control and seven items for mother control were utilized in our questionnaire. The format of the scale was also simplified to a yes-no format. Because of the high correlation between the mother and father control scales (r = .66), these scales were combined into a parental control scale. This was done in order to avoid possible problems with multicollinearity in the planned regression analyses. Scores were computed in the same fashion utilized with the previous scales (i.e., total obtained score divided by total possible score) with actual scores ranging from .50 to 1.00 with a high score indicating greater parental control. In one sample the Cronbach's alpha for this combined scale was .79.

School Self-Acceptance: The 20-item school self-acceptance scale was taken directly from the Instructional Objectives Exchange (1972) Self-Appraisal Inventory (intermediate level). Sample items include:

1. My teachers usually like me.   True   Untrue

2. I often get upset in school.   True   Untrue

Scale scores were once again computed out of 1 with the range in actual scores being from .50 to 1.00 with a high score reflecting higher school self-acceptance. In our sample the Cronbach's alpha for this scale was .72.

Drug Use: The drug use questions used in this survey included 5 questions involving the six month reported frequency of use for the 5 subtances: tobacco, alcohol, marijuana, gasoline, and other solvents. Other substances such as heroin etc. were not included because these substances are generally not available in remote communities.

As a measure of total involvement with drugs a Gutman scale was constructed as follows:

5 = use of solvents other than gasoline

4 = use of gasoline

3 = use of marijuana

2 = use of alcohol

```
1 = use of tobacco

0 = use of no drugs
```

The coefficient of scalability for this scale was .60 and the coefficient of reproducibility was .93. Both of these figures are generally considered to be satisfactory indicators of a unidimensional scale. A score of 5 on this scale indicated the student used solvents plus all of the other substances; a score of 4 indicated the student used solvents plus all of the other substances; and so on down to a score of 0 which indicated the students used no drugs at all.

Data Analysis

The philosophy for analyzing the data in this study has been to:

1. Check for reliability and validity of measures wherever possible,

2. Investigate the prevalence of gasoline sniffing and other solvent use,

3. Examine the univariate relationships between individual variables, scales, and the use of gasoline and other solvents,

4. Determine the multivariate relationships between predictors (demographic-cultural, socialization, and personality) and gasoline sniffing and drug use utilizing the techniques of multiple regression and path analysis.

For the last stage of the data analysis the sample was divided into three subsets: (1) low social assets-native; (2) high social assets-native; and (3) nonnative.

The low social assets sample was comprised of native people living in Easterville, Berens River, and Waterhen. The high social assets-native sample was comprised of native people living in Gypsumville, and Peguis and the non-native sample was comprised of non-native students attending school in Gillam. In this paper the results presented will focus on the prediction of drug use in the low social assets-native sample where most of the sniffing occurred.

RESULTS

Validity of Scales

Because of the nature of the sample in this study, young and sometimes fairly unacculturated, a considerable amount of time and effort was put into assessing the validity of our scales. Several analyses were conducted with this purpose in mind. First, Indian, white and Metis comparisons shown in Table 3 provide evidence for the predictive validity of our scales. Predictions were that Indians would score lower than whites on social assets, acculturation, and parental control. All of these predictions were confirmed by the data. It was also predicted that Metis people would score similar to Indians on some variables and similar to whites on others. In fact, the Metis students resemble the Indian students in their acculturation level and social assets, but score closer to the white students in their parental control scale scores.

It was particularly important to validate the school self-acceptance

Table 3. Indian, Metis, and White Comparisons
on Social Assets, Acculturation, Parental Control-
Permissiveness, and School Self-Acceptance

| | Means | | | Analysis of variance | |
Variable	Indian N = 334	Metis N = 92	White N = 197	Df	F
Social assets	.73	.72	.81	2/613	91.30*
Acculturation	.73	.75	.88	2/613	54.05*
Mother control	.83	.86	.89	2/575	13.58*
Father control	.83	.86	.87	2/537	7.98*
School self-acceptance	.78	.77	.84	2/594	17.22*

Lines under means show which groups do not differ significantly
(p < .05) using Duncan's Muliple Range Test.
***p <.001.

Table 4. Correlation Coefficients for School Self-Report Measures
and School Records: Berens River Sample

| | Self reports | | | School records | | | | | |
	1	2	3	4	5	6	7	8	9
1. School performance	–								
2. How far going in school	.45**								
3. School self-acceptance	.49***	.42**							
4. Language arts	.46***	.45**	.37*						
5. Math	.47***	.49***	.38*	.71***					
6. Science	.53***	.32*	.47**	.69***	.70***				
7. Social studies	.50***	.49***	.36*	.69***	.72***	.51***			
8. School attendance	.40**	.44**	.50***	.48***	.38**	.42**	.47**		
9. Grade point average	.58***	.50***	.46**	.91***	.82***	.84***	.86***	.53***	–

N = 40

*p ≤ .05. **p ≤ .01. ***p ≤ .001.

measure because researchers in the past have been critical of measures used in this area (e.g., Cockerham and Blevins, 1976; Fuchs and Havighurst, 1973). Measures such as the Coopersmith Self-Esteem Inventory have been shown not to be valid for use with Indian samples (Cress and O'Donnel, 1975). In fact Cockerham and Blevins (1976) have argued that the literature concerning Indian self-esteem is seriously flawed and in need of reevaluation because of problems in the validity of measures utilized.

In order to assess the validity of the Instructional Objectives Exchange school self-acceptance measure for use in our sample the following two techniques were employed. First, each questionnaire in both the fall 1977 and the spring 1978 surveys, included items asking the children how far they wanted to go in school and how they were doing in school. Correlations between the scores on the school self-acceptance scale and these two items were computed for both the fall 1977 and the spring 1978 surveys. Analysis of the fall 1977 data revealed that correlations between students' school

Table 5. Number of Sniffers Found in Communities Studied

Community	Date	Total	Gas or gas & other solvents	Other solvents only	Total	%
Pretests						
Shamattawa	Fall, 1976	20	10		10	50
Berens River	Spring, 1977	45	2	8	10	22
Survey						
Berens River	Fall, 1977	66	3	4	7	11
Gillam	Winter, 1978	175	1	5	6	3
Easterville	Winter, 1978	122	12	2	14	11
Peguis	Winter, 1978	158	1	1	2	1
Gypsumville	Winter, 1978	46	0	0	0	0
Waterhen	Winter, 1978					
Waterhen		16	0	0	0	0
Skownan		19	1	1	2	11
Mallard		18	1	1	2	11

Table 6. Shamattawa: Gasoline and Other Solvent Use
(6-Month Prevalence)

Drug frequency	Gasoline N	Gasoline %	Other solvent N	Other solvent %
Not at all	10	50	14	73.7
About 3 times or less	1	5	1	5.3
About 4-6 times	4	20	1	5.3
More than once/month	1	5	0	0.0
More than once/week	1	5	2	10.5
Almost every day	3	15	1	5.3
Total	20	100	19	100.1

Table 7. Berens River: Gasoline and Other Solvent Use
(6-Month Prevalence)

Drug frequency	Gasoline N	Gasoline %	Other solvent N	Other solvent %
Not at all	43	96	34	79
Less than once/month	0	0	3	7
Several times/month	2	4	2	5
Several times/week	0	0	4	9
Almost every day	0	0	0	0
Total	45	100	43	100

self-acceptance scores and ratings of how far they wanted to go in school were significant for both the native ($r = .39$; Df = 408; $p < .0001$) and nonnative ($r = .39$, Df = 190; $p < .0001$) samples. Similar results occurred for the correlations between school self-acceptance and students' perceptions of how they were doing in school (native sample: r = .40; DF = 409; p < .001; nonnative sample: r = .54; Df = 189; p < .0001). Similar significant correlations were found in the spring 1978 survey (see Table 4).

In a second attempt to validate the school self-acceptance measure, school grades (average for the year) and absenteeism records (days absent/school year) were obtained for the Berens River spring 1978 sample, and correlations between these measures and school self-acceptance scores computed. These correlations were provided in Table 4. Results show that the student self-reports correlate quite highly with (1) each other; and (2) the school records. Students who have higher school self-acceptance report that they are doing better in school (r = .49) and expect to go farther in school (r = .42) than students with lower school self-acceptance. The students' self-reports on school self-acceptance, school performance, and how far they expect to go in school are all highly correlated with their school records, including attendance and grade point average.

In order to validate the solvent-abuse reports provided by students, a separate investigator visited several of the communities and conducted interviews with key informants. The results of this investigation are reported elsewhere (Stryde, 1977) and will not be discussed at length here except to say that estimates of the number of sniffers arrived at by the two different techniques (key informant survey and self-report questionnaires) were remarkably similar.

Frequency of Gasoline and Other Solvent Use

The number of sniffers found in each of the surveys conducted to date, including pretests, is summarized in Table 5. The highest percentages of sniffers occurred in the Shamattawa and Berens River pretests. The frequencies of gasoline sniffing and other solvent use in these two samples are summarized in Tables 6 and 7. Gasoline sniffing was more common in Shamattawa, while the use of other solvents was reported more frequently in Berens River.

In the six communities survey conducted in the fall and winter of 1977-78, the heaviest prevalence of gasoline sniffing was reported in the community of Easterville. This community, like Shamattawa and Berens River, is also fairly low on both acculturation and social assets. The community has an additional problem in that they were forced to relocate several years ago on account of a hydro project. Community residents are still having problems adjusting to the new location of their community, which is viewed as being less satisfactory than the "old post." Low rates of solvent use were reported in several communities, including the high social assets and high acculturation Indian community of Peguis.

The overall frequency of gasoline sniffing and other solvent use in the six communities survey is summarized in Table 8. These results show that the frequency of sniffing in the overall sample is quite low, with most of the people involved in this practice sniffing relatively infrequently (see Table 8).

Univariate Prediction of Sniffing

Demographic-Cultural Predictors. Examination of the univariate relationships between demographic-cultural factors and sniffing (gas and other solvents) in the overall sample, using X^2 and F tests as appropriate, revealed the following findings. Gasoline sniffers were predominantly young and male. Sniffers tended to come from broken homes that had no running water or radio, and were more likely to report that they did not have enough money for clothes and food. They were more likely to report that their

Table 8. Gasoline and Other Solvent Use in Overall Sample
(6-Month Prevalence)

Drug frequency	Gasoline		Other solvent	
	N	%	N	%
Not used at all	591	96.9	590	96.9
Less than once/month	14	2.3	15	2.5
Several times/month	3	0.5	1	0.2
Several times/week	1	0.2	2	0.3
Almost every day	1	0.2	1	0.2
Total	610	100.0	609	100.0

Table 9. Standardized Multiple Regression Coefficients and
R^2 Values for the Gasoline Sniffing, Solvent Abuse,
and Drug Use (6-Month Prevalence) Dependent Variables in the
Low Social Assets-Native Sample

Predictor variables	Gasoline	Other solvents	Drug use
Age	-.05	-.03	.30**
Sex	-.09	-.06	-.06
Social assets	.13	.15	.01
Acculturation	-.14	.01	-
Parental drug use	.21**	.21**	.27**
Sniff influence	.27**	.32**	.31**
Parental control	-.01	.01	-.08
School self-acceptance	-.09	.02	-.14*
R^2	.18	.15	.32
F-Value	5.29**	4.25**	13.11**

* = p < .05. ** = p < .01.

mother's health was poor and less likely to report that their mothers worked.
Sniffers, and particularly those who used gasoline, tended to be low on
acculturation (i.e., more often reporting that they used a native language
at home, that their mothers had low education, and that they did not possess
a TV).

Socialization Predictors. Socialization variables that were included
in this study were: parental drug use, parental control, and sniff
influence items and scales. Within the parental drug use items the only
variable that was significantly associated with sniffing was the mothers'
use of tobacco, with more gasoline sniffing reported by children whose
mothers used tobacco. Most of the sniff influence variables were
significantly associated with solvent abuse. Students who reported that
they had more friends who sniffed, and who encouraged them to sniff, and
more family members who sniffed, were more likely to sniff gas and other
solvents themselves. Of the two parental control scales (mother control and
father control), the mother control scale was significantly associated with
sniffing while the father control scale was not.

Personality. The school self-acceptance measure was significantly
associated with solvent use with gasoline sniffers reporting lower school
self-acceptance.

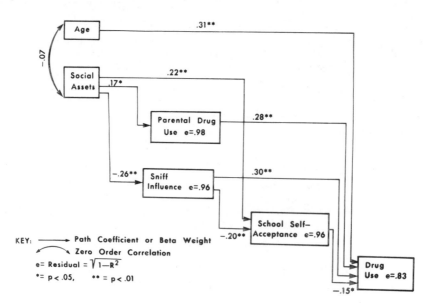

| DEMOGRAPHY SOCIAL STRUCTURE | SOCIALIZATION | PERSONALITY | BEHAVIOR |

Figure 4. Path Analytic Model of Teen-Age Drug Use
in Low Social Assets-Native Sample.

Table 10. Zero-Order Correlations between Age, Social Assets,
Parental Drug Use, Sniff Influence, School Self-Acceptance, and
Drug Use in the Low Social Assets-Native Sample

Variable	1	2	3	4	5	6
1. Age	–					
2. Social assets	-.07	–				
3. Parental drug use	.06	.17*	–			
4. Sniff influence	-.01	-.26	-.05	–		
5. School self-acceptance	.01	.27**	-.06	-.26	–	
6. Drug use	.32**	-.09	.29**	.32**	-24	–

* = p < .03. ** = p < .01.

Multivariate Prediction of Sniffing and Drug Use in General

For the multivariate prediction of drug use the sample was divided into three groups (low social assets-native, high social assets-native, and high social assets-white) with regression analyses conducted within each of the three groups separately. In this paper, the results presented are based on the low social assets-native sample, where most of the sniffing occurred. Results of the regression analyses predicting gasoline sniffing, other solvent use, and overall drug use in this sample are provided in Table 9. These results show that there were two significant predictors of both gasoline sniffing and other solvent use. These predictors were parental drug use and sniff influence. Students who came from an environment where the parents used more drugs and where more solvent users were present were most likely to report sniffing themselves.

The predictors taken as a whole were able to account for 18% of the variance in gasoline sniffing and 15% of the variance in other solvent use.

Results of the regression and analyses predicting overall drug use in the low social assets-native sample showed that four variables made a significant contribution in the prediction of overall drug use. These variables, in order of magnitude, were: sniff influence, age, parental drug use, and school self-acceptance. Drug use was greater among students who had reported more sniff influences in their environment, who were older, who reported more parental drug use, and lower school self-acceptance.

Path Analytic Interpretation of Results

To further test the model of solvent abuse provided in Fig. 1, the technique of path analysis (Asher, 1976) was utilized. Although time ordering of the variables in a path analytic model is necessary to prove a causal sequence, path analysis can still be used for interpretative purposes when time ordering of the variables is not present.

Since solvent abuse occurred primarily in the low social assets-native sample, the data from this sample were used in the path analytic model. Since the drug use variables formed a Gutman scale, this drug use scale was used in the path analytic model. The higher the person's score on the drug use scale the more serious and/or deviant is the drug use pattern, with the highest scores being solvent abusers.

To test the model, the demography-social structure variables (age, sex, social assets, level of acculturation) were first entered into regression equations to predict the socialization variables (parental drug use, parental control, and sniff influence). Next, the demography-social structure and socialization variables were entered into a regression equation together to predict the personality variable and school self-acceptance. Finally, all of the variables were entered into a regression equation to predict the drug use variable.

After the initial regression analyses were performed, the model was trimmed by eliminating the nonsignificant paths and dropping any variables that did not fit into the model (i.e., did not predict or were not significantly predicted by other variables). The final reduced model is provided in Fig. 4. The zero-order correlations for the variables included in this model are provided in Table 10.

The path analysis results in Fig. 4 indicate that four variables contribute significantly in the prediction of drug use. Drug use is greater for students who are older, have parents who use drugs, have more social influence to sniff, and lower school self-acceptance. School self-acceptance, in turn, is higher among the students with higher social assets and less influence to sniff. Parental drug use and sniff influence are not explained to a very great extent by the variables in the model, although students with higher social assets reported more parental drug use and less social influence to sniff. Variables which dropped out of the model included sex, level of acculturation, and parental control.

Although the model in Fig. 4 does not prove the causal sequence, it is suggestive. The demography-social structure variables are time ordered and must occur at the first level in the model. Other variables, although not time ordered, seem to fit into the sequence as hypothesized. Variables assigned intermediate positions are predicted by variables occurring at an earlier stage in the model then in turn predict variables ahead in the model.

The model may be useful in explaining such things as the role of social assets in predicting drug use. Although social assets does not predict drug use directly, it has both positive and negative indirect effects that seem to cancel each other out. Social assets has a positive relationship to drug use by producing more parental drug use, which in turn predicts drug use. Social assets also has a negative effect on drug use, however, by predicting lower social influence to sniff and higher school self-acceptance.

SUMMARY AND CONCLUSIONS

Validity of Data

In this study, several strategies were employed for determining the validity of students' responses. Results generally seemed to indicate that the students were responding honestly and accurately. In communities such as Shamattawa, where gasoline sniffing was known to exist, the prevalence of sniffing found in the survey was much higher than in other areas. Students were not likely overestimating their use, since no students indicated using marijuana in Shamattawa, where this substance was not available. It is also unlikely that any large number of sniffers were missed by this survey. Sniffing is a fairly visible type of drug use. Community interviews conducted by Stryde (1977) obtained estimates of the number of sniffers in four communities that closely paralleled the survey results.

The strongest evidence for the validity of the data collected in this study is provided by the high correlations found in the Berens River sample between school self-acceptance scores and behavioral measures of school absenteeism and grades, obtained from the school.

Prevalence of Drug Use

Gasoline Sniffing. Apart from the community of Shamattawa, where gasoline sniffing was prevalent, the prevalence of gasoline sniffing found in this study was quite low. Only 3.1% of the overall sample reported any use of gasoline and most of the sniffers (i.e., 14 out of 19) reported sniffing less than once a month.

Other Solvent Use. The use of other solvents followed a similar pattern to the gasoline sniffing with the overall prevalence being quite low (3.1% users) and the frequency of use by sniffers also reported to be quite low (i.e., 15 out of 19 did so less than once per month).

Predictors of Use

Gasoline Sniffing. In the gasoline-sniffing literature sniffers are generally characterized as being predominantly male, from low social assets environments, and communities undergoing acculturative stress. Their socialization is characterized as being low in controls against deviance and high in deviant models (i.e., parent, peer, and sibling models for drug use). School performance is also poorer among sniffers.

The univariate analyses performed in this study generally tended to support this description of the gasoline sniffer. Gasoline sniffers tended to be predominantly male in this study. They also tended to come from homes that were lower on a variety of social assets, including (1) more broken homes; (2) no running water; (3) not enough money for clothes and food; and (4) poorer mother's health.

Acculturative stress also seemed to be an important factor. In communities such as Pequis, which were already highly acculturated, gasoline sniffing did not occur. In the communities undergoing rapid acculturation, gasoline sniffing was more prevalent. Four out of the five variables used in this study to measure acculturation level were significantly associated with gasoline sniffing. Gasoline sniffers tended to come from homes where the language spoken at home was a native language, where the mothers had low education, where there was no television, and the father did not work for wages regularly.

All of the socialization variables were also significantly associated with the gasoline sniffing variable in the univariate analyses. Gasoline sniffers tended to come from homes where more parental drug use occurred, particularly mothers smoking. Gasoline sniffers also tended to come from homes where less parental control existed. The social influence to sniff also tended to be higher for gasoline sniffers than for nonsniffers. Gasoline sniffers reported that they had more friends who sniffed, more friends that encouraged them to sniff, and more often said they had other family members who sniffed.

In the low social assets-native sample, where the majority of sniffing occurred, personality also played a role in predicting gasoline sniffing. Sniffers tended to score lower in their level of school-acceptance.

Multivariate analyses had not been conducted before in any studies of gasoline sniffing. Results in this study are therefore unique in providing some indication of the relative importance of the various predictors in predicting gasoline sniffing. Results showed that the sniff influence and parental drug use predictors are the most important factors in predicting gasoline sniffing.

Other Solvent Use. The predictors which were expected to predict gasoline sniffing were also expected to predict other solvent use. In general the results were found to be in the same direction in the other solvent use analyses, but were somewhat weaker. There were no sex differences in other solvent use. Users were slightly lower in their social assets with the variables significantly associated with use being somewhat different than for gasoline sniffing. Sniffers were predominantly from the low social assets communities. Acculturation factors were less important in predicting other solvent use than in predicting gasoline sniffing.

Of the socialization variables, sniff influence was strongly associated with other solvent use, while parental drug use and parental control factors played a less important role.

Personality did not play as strong a role in the prediction of other solvent use. The results, therefore, do not support the findings by Annis et al. (1971) that sniffers are lower on self-acceptance than nonsniffers.

Multiple regression analyses revealed that the same two predictors that carried the most weight in predicting gasoline sniffing were also the most important factors in predicting other solvent use. Sniffing of other solvents was predicted best by the sniff influence and parental drug use variables.

Drug Use. The five drug-use variables were incorporated into a Gutman scale with a high score on this scale indicative of illicit drug use. In the low social assets sample predictors accounted for 32% of the variance in drug use. The amount of variance explained was fairly high considering the number of predictors used.

The technique of path analysis was used to test the viability of the proposed causal model. Results of the path analysis suggested that the proposed causal model was indeed viable. Socialization is predicted by the person's place in the demography-social structure. Personality is predicted by the person's place in the demography-social structure and socialization. Each level of analysis, demography-social structure, socialization, and personality contributes in the prediction of drug use.

IMPLICATIONS OF FINDINGS AND SUGGESTIONS FOR FUTURE RESEARCH

Implications

1.   The problem not as bad as feared.

At the completion of this research funders were somewhat relieved to find that the prevalence of gasoline sniffing was not as serious as some had feared. Apart from Shamattawa, where serious problems existed, the use of gasoline was fairly uncommon, with most users tending to be irregular rather than regular users.

2.   Gasoline sniffing is not a separate phenomenon divorced from other types of drug use and the laws (variables, processes, etc.).

The finding that gasoline sniffing fitted a Gutman scale with the other drugs tested in this study, and was predicted by factors such as parental drug use and peer and sibling influence, supports the conclusion that the general laws governing drug use also apply in the case of gasoline sniffing. There is a tendency among whites to view Indian ways as being "profane" (Braroe, 1975). This is especially true in the case of gasoline sniffing, which is viewed with distaste and horror by middle-class whites, while our own patterns of drug use are more readily tolerated. Our results suggest that most Indians use gasoline in much the same way whites use marijuana and other drugs. Most people use gasoline occasionally while those who have problems may become heavier users. The selection of the type of drug to be used seems to depend on availability and peer influence. In Shamattawa, where marijuana is unavailable and the alcohol supply is jealously guarded by adults, the greatest amount of gasoline sniffing occurred.

3.   Solutions to the problem(s) of gasoline sniffing must be complex.

The findings in this study suggest that drug use is a function of complex interrelationships between the sociocultural, socialization, and personality systems. Intervention strategies must take these three different levels into account. To a certain extent the interventions attempted to date in Manitoba have already done this. It was realized early on in the Shamattawa experience that attempts to deal with adolescent gasoline sniffing would be futile without also considering the problem of parental drug use. In Berens River attempts were made to impact on the solvent-abuse problem by the incorporation of a program directed at improving school self-acceptance. In this program students were given an opportunity to work outside the school in different work settings (stores, airports, logging camps) as part of the regular school curriculum. The hope and underlying logic to this program was that they would experience positive reinforcement in these roles that they normally were not experiencing in the school system. Through these experiences it was hoped that self-concepts would be raised and the need for using drugs reduced. Although this program showed promise in its initial stages, it had to be dropped when the program originator left the community and her replacement could not keep the program going.

Future Research

1.   The need for longitudinal research.

In this study the author had hoped originally to test the causal relationships between variables by employing a longitudinal panel survey approach followed up with cross-lagged correlational analyses to test the causal sequence of the variables in the model. Since this objective was not accomplished in this study, the need for longitudinal research on solvent use is still great. Questions that need to be asked include: Do solvent users mature out of this type of drug use? If they do mature out of solvent use, do solvent users then become heavy users of other drugs? Although there is some suggestive evidence that most sniffers mature out of this problem, cases have been described to the author where heavy sniffers of gasoline have retained a preference for this substance over others into adulthood.

2. The need for program evaluation studies on intervention programs and policies designed for sniffing.

There are very few studies that have ever been conducted investigating the effectiveness of intervention strategies for reducing sniffing. One of the problems seems to be that sniffing has a fairly low prevalence in many areas, which tends to work against the introduction of programs. Since this does not seem to be the case in parts of the Canadian North, these areas provide a possible proving ground for intervention strategies. In Shamattawa, experiences to date suggest that the prevalence of sniffing can be reduced by introducing community patrols, community development workers, drop-in centers, and other programs designed to help improve the community as a whole. Strategies such as evacuating sniffers from the community do not seem to be successful either from the point of view of the sniffers (they are unhappy away from home and still sniff), the home community (they resent the "kidnapping" of children), or the host community for evacuees (sniffers cause problems such as risks of starting fires in foster homes and enticing others into trying sniffing).

REFERENCES

Angle, M. R., & Eade, N. R. Gasoline sniffing and tetraethyl lead poisoning in a northern native community. Epidemiological and Social Research Division, Research Bureau, Non-Medical Use of Drugs Directorate, Health and Welfare Canada, Report No. ERD-74-19, March, 1975.

Annis, H. M., Klug, R., & Blackwell, D. Drug use among high school students in Timmins. Unpublished manuscript. Toronto: Addiction Research Foundation, 1971.

Asher, H. B. Causal Modelling. Sage University paper series on quantitative applications in the social sciences, 07-003. Beverly Hills: Sage Publications, 1976.

Barnes, G. E. "Solvent abuse: A review." International Journal of the Addictions, 1979, 14: 1-26.

Bartlett, S., & Tapia, F. "Glue and gasoline `sniffing,' the addiction of youth." Modern Medicine, 1966, 63(4): 270-272.

Bass, M. "Sudden sniffing death." Journal of the American Medical Association, 1970, 212: 2075-2079.

Beattie, A. D., Moore, M. R., & Goldberg, A. "Tetraethyl-lead poisoning." The Lancet, July 1, 1972, 12-15.

Bethell, M. R. "Toxic psychosis caused by inhalation of petrol fumes." British Medical Journal, 1965, 2: 276-277.

Black, P. D. "Mental illness due to the voluntary inhalation of petrol vapour." Medical Journal of Australia, 1967, 2: 70-71.

Boeckx, R. L., Postl, B., & Coodin, F.S. "Gasoline sniffing and tetraethyl lead poisoning in children." Pediatrics, 1977, 60: 140-145.

Braroe, N. E. Indian and White. Stanford, California: Stanford University Press, 1975.

Brown, N. W. "Gasoline inhalation." Journal of the Medical Association of Georgia, 1968, 57: 217-221.

Carrol, H. G., & Abel, G. G. "Chronic gasoline inhalation." Southern Medical Journal, 1973, 66: 1429-1430.

Clinger, O. W., & Johnson, N.A. "Purposeful inhalation of gasoline vapors." Psychiatric Quarterly, 1951, 25: 557-567.

Cockerham, W. C., & Blevins, A. L. "Open school vs traditional school: self-identification among native American and white adolescents." Sociology of Education, 1976, 49: 164-169.

Cress, J. N., & O'Donnell, J. P. "The self-esteem inventory and the Oglala Sioux: A validation study." Journal of Social Psychology, 1975, 97: 135-136.

Durden, W. D., Jr., & Chipman, D. W. "Gasoline sniffing complicated by acute carbon tetrachloride poisoning." Archives of Internal Medicine, 1967, 371-374.

Easson, W. M. "Gasoline addiction in children." Pediatrics, 1962, 29: 250-254.

Edwards, R. V. "A case report of gasoline sniffing." American Journal of Psychiatry, 1960, 117: 555-557.

Faucett, R.L., & Jensen, R. A. "Addiction to the inhalation of gasoline fumes in a child." Journal of Pediatrics, 1952, 41: 364-368.

Ferguson, C. A. "Chemical abuse in the north." University of Manitoba Medical Journal, 1975, 45: 129-132.

Fuchs, E., & Havighurst, R. J. To Live on This Earth. Garden City, New York: Doubleday and Company, Inc., 1973.

Gold, N. "Self-intoxication by petrol vapour inhalation." Medical Journal of Australia, 1963, 2: 582-583.

Hayden, J. W., & Comstock, E. G. "The clinical toxicology of solvent abuse." Clinical Toxicology, 1976, 9: 169-184.

Hunter, A. G. W., Thompson, D., & Evans, J. A. "Is there a fetal gasoline syndrome?" Teratology, 1979, 20: 75-80.

Instructional Objectives Exchange. Measures of Self-Concept-Grades K-12 (Revised edition). Los Angeles, California, 1972.

Kaufman, A. "Gasoline sniffing among children in a Pueblo Indian village." Pediatrics, 1973, 51: 1060-1064.

Kohn, P. M., Barnes, G. E., & Hoffman, F. M. "Drug-use history and experience seeking among adult male correctional inmates." Journal of Consulting and Clinical Psychology, 1979, 47: 708-715.

Law, W. R., & Nelson, E. R. "Gasoline sniffing by an adult." Journal of the American Medical Association, 1968, 204: 1002-1004.

Lawton, J. J., Jr., & Malmquist, C. P. "Gasoline addiction in children." Psychiatric Quarterly, 1961, 35: 555-561.

Luborsky, L., Todd, T. C., & Katcher, A. H. "A self-administered social assets scale for predicting physical and psychological illness and health." Journal of Psychosomatic Research, 1973, 17: 109-120.

Lynn, H. Gasoline sniffing among Indian children. Paper presented at the inaugural Symposium of Community Health, University of Toronto, October, 1975.

Neal, C. D., & Thomas, M. I. "Petrol sniffing, a case study." British Journal of Addiction, 1974, 69: 357-360.

Nitsche, C.J., & Robinson, J. F. "A case of gasoline addiction." American Journal of Orthopsychiatry, 1959, 29: 417-419.

Nurcombe, B., Bianchi, G. N., Money, J., & Cawte, J. E. "A hunger for stimuli: The psychosocial background of petrol inhalation." British Journal of Medical Psychology, 1970, 43: 367-374.

Oldham, W. "Deliberate self-intoxication with petrol vapour." British Medical Journal, 1961, 2: 1687-1688.

Schaefer, E. S. "Children's reports of parental behaviour: An inventory." Child Development, 1965, 36: 413-424.

Scott, M. A. Gasoline sniffing in Shamattawa. Medical Services, Unpublished manuscript, 1976.

Sokol, J. Solvent abuse among juveniles. Paper presented at the
    International Congress on Drugs and Alcohol, Jerusalem, Israel,
    September 13-18, 1981.
Stryde, W. A report on gasoline sniffing in Manitoba. Unpublished report.
    Winnipeg: Health Promotion Branch Regional Office, 1977.
Tolan, E. J., & Lingl, F. A. "`Model psychosis' produced by inhalation of
    gasoline fumes." American Journal of Psychiatry, 1964, 120: 757-761.